Springer Series on Polymer and Composite Materials

Series editor

Susheel Kalia, Dehradun, India

More information about this series at http://www.springer.com/series/13173

Gordana Marković · Visakh P.M.
Editors

Rubber Nano Blends

Preparation, Characterization and Applications

Springer

Editors
Gordana Marković
Technical Rubber Good
Pirot
Serbia

Visakh P.M.
Department of Ecology and Basic Safety
Tomsk Polytechnic University
Tomsk
Russia

ISSN 2364-1878 ISSN 2364-1886 (electronic)
Springer Series on Polymer and Composite Materials
ISBN 978-3-319-48718-2 ISBN 978-3-319-48720-5 (eBook)
DOI 10.1007/978-3-319-48720-5

Library of Congress Control Number: 2016955688

Printed on acid-free paper

This Springer imprint is published by Springer Nature
The registered company is Springer International Publishing AG
The registered company address is: Gewerbestrasse 11, 6330 Cham, Switzerland

Contents

Chapter 1
Rubber Nanoblends: State of the Art, New Challenges, and Opportunities

Visakh P.M.

Abstract The present chapter deals with a brief account on various topics in rubber nano blends preparation, characterization and applications. This chapter discussed with different topics such natural rubber nano blends, nitrile rubber latex blends, polyurethane rubber based nano blends, chlorosulphonated rubber based nano blends, polybutadiene rubber based nano blends, styrene butadiene rubber based nano blends, polychloroprene rubber based nano blends and ethylene propylene diene rubber nano blends.

Keywords Nano blends · Rubber nano blends · Rubber blends · Polyurethane rubber blends · Chlorosulphonated rubber blends · Styrene butadiene rubber blends · Ethylene propylene diene rubber nano blends

1.1 Natural Rubber Nanoblends

From the literature we can see that the natural rubber nanocomposites (NRN) have been using for many applications in the field of industry and technology. Many types of nanomaterials have been using as fillers and antioxidation agents to enhance the mechanical, thermal, electrical, and magnetic properties of the NR in literature. Nanomaterials have very special properties such as high surface area, antioxidation properties, mechanical strength, wear, electronic confinement, and tear and aberration resistance. Many scientists are working for developing high-performance natural rubber composites for many industrial applications. Many research works in the field of nanotechnology give a boost up in the development of natural rubber nanocomposites.

The work of Suil et al. [1] used the treated CNT as fillers to improve the thermal stability, dynamic compression properties, rebound resilience, and storage modulus

Visakh P.M. (✉)
Department of Ecology and Basic Safety, Tomsk Polytechnic University, Lenin av. 30, 634050 Tomsk, Russia
e-mail: visagam143@gmail.com

G. Marković and Visakh P.M. (eds.), *Rubber Nano Blends*, Springer Series on Polymer and Composite Materials,
DOI 10.1007/978-3-319-48720-5_1

of the natural rubber-based nanocomposites. Nanostructured fillers are attracting enormous attention due to the environmental concerns and health issues related to the conventional chemicals. Nanofillers have unique properties of high surface-to-volume ratio because of their shape and size. Many nanofibers are used for making nanocomposites with natural rubber. From biological resources, plants are having important bio-nanomaterials called cellulose, and this provides alternate resources for the manufacturing of green natural rubber nanocomposites. This kind of nanomaterials and nanocomposites provides an economically viable method to enhance the mechanical- and load-bearing properties of natural rubber nanocomposites.

An interfacial layer surrounding the filler restricts the mobility of fillers; when the addition of cellulose whiskers in NR matrix, this interfacial layer makes better mechanical properties compared to its bulk counterparts [2]. The synthesized NRN showed ferromagnetic behavior which can be attributed to the characteristic contribution from the magnetic property of nickel nanoparticles incorporated into the NR matrix [3]. Nanocomposite with gold nanoparticles and natural rubber makes new photoluminescence functionalities within the sensible range suitable for the development of nanosensors [4]. Natural rubber nanocomposites have been prepared using dynamic vulcanization techniques to make super-toughened natural rubber nanocomposites, and it has great industrial importance. Chen et al. reported [5] the development of bio-based polyactide/NRN with continuous cross-linked phase through a facile dynamic vulcanization method. The developed natural rubber nanocomposites can be used as environmental friendly and alternative for non-degradable polyolefins that facilitate the remediation of pollution caused by the deposition of synthetic polymers derived from hydrocarbons [6]. Natural rubber nanocomposite properties such as physicochemical and mechanical can be modified by using organoclay fillers. Many organic fillers can be modified by alkyl functional groups with varying chain length. Alkylammonium groups containing organoclay fillers can be used as improving cross-linking density of NR. In skim latex mixing process, the natural latex rubber is centrifuged to obtain the protein-rich byproduct called as skim natural rubber latex. This latex rubber is treated with surfactant and alkali to remove the aqueous phase. The multi-component rubber nanocomposite blends were prepared by the latex/melt intercalation method. For this, nanofillers are needed to be dispersed in distilled water to form an aqueous dispersion. Another nanocomposite-preparing method is freeze drying technique; it is used at lower temperature. For the preparation on natural rubber composites, rubber latex mixes with the common vulcanization ingredients with continuous stirring by using a homogenizer for 30 min at 70 °C, and then mixture was mixed with 1 w% nanofiller solution and stirred well for 2 h followed by 30-min homogenizing process.

The obtained nanocompounds were kept in a cylindrical glass shell and subjected to liquid nitrogen flow for freezing the compound quickly. In the in situ sol–gel method, the natural rubber is made into the sheet form using conventional two-roll milling process, and the required amount of natural rubber was swollen for 24 h in the required amount of organic solvent, and the alkoxide precursor material was added into this swollen NR and then immersed in the solvent and mixed well in order to facilitate the homogenous mixing of alkoxide precursor with the swollen rubber. The addition of fillers or additives into the NR matrix reinforces and enhances the physicochemical and mechanical properties of the natural rubber nanocomposites. Nanofillers have unique properties with high surface-to-volume ratio, tunable surface energy, and physicochemical properties which provide large surface interface area for the NR matrix to cross-link and interact with the filler. The natural rubber–nickel nanocomposites can be used as flexible magnetic shielding materials for safety applications. The nanoengineered elastomers of nickel metal nanoparticles can be used for application in fabrication of low-frequency electronic peripherals.

1.2 Nitrile Rubber Latex Blends

Specific properties of nitrile rubbers are resistant to non-polar solvents, fats, oils motor fuel, and chemicals. NBR is a type of statistical copolymer containing butadiene and nitrile units, which provides room to control the proportions of the constitutional units in balance with the properties for various end-use applications. The main drawback of NBR is poor ozone and weather resistance, which makes it unsuitable for applications where these qualities would be required. NBR has an operating temperature range of around −30 to +100 °C, within which NBR can withstand most of the severe applications. Blending NR and NBR would aim to obtain the oil resistance of nitrile rubber while retaining the durability or NR. The ratio of NR to NBR would determine the presence of each of the rubber's prominent qualities within the final mixture of latex.

The effects of different constitutional ratios between NR and NBR were analyzed with blends of NR and NBR, and oil swelling properties were studied. Latex blends of NR and NBR latex were dried and mixed with vulcanizing agents, and subsequently vulcanized. Mechanical and chemical properties of the vulcanizates were thus examined. The result was oil swelling and hardness properties of these blends increased with an increase of NBR content. Finally, clay or layered silicates were added to this blend and analyzed the properties. This type of blend can be prepared through different intercalation methods, either individually or combined, such as latex shear blending methods [7]. When the addition of layered silicates, tensile mechanical properties and tear strengths of NBR/silicate nanocomposites were increased compared to neat NBR. Specifically, the tensile and tear mechanical properties of the NBR/layered silicates increase by 200 and 60 %, respectively.

Using DLS techniques, Alex and Nah [8] carried out a study focusing on NR/NBR/organoclay nanocomposites and the particle size of rubber latex was determined using a particle size analyzer. This method allows for deeper analysis of the properties of natural rubber and nitrile rubber nanocomposites. Weight percentage of layered silicate as 7.5 wt% in NBR/silicate nanocomposites was achieved exfoliated nanocomposite with the use of high shear force generated in the ball-milling process. In food industry, NBR/NR blends have been using because of their unique oil and fat resistance [9]. In contrast, NBR does not crystallize when stretched and so requires reinforcing fillers for optimum tensile strength, abrasion resistance, and tear resistance. SBR/NBR blends reduce the overall oil resistance but increase the resistance to polar liquids in proportion to the SBR content.

1.3 Polyurethane Rubber-Based Nanoblends

Elastomer polyurethane (TPU) (Thermoplastic) is a linear segmented block composed of hard and soft segments. The soft segment polyurethane is aliphatic polyester polyols or aliphatic polyether polyols are coupled with hard segment formed from aromatic di-isocyantes and hard diols urethane linkages [10]. The field synthesis nanopartical polyurethane elastomer modification was also studied by some of the research groups [11–13]. TPU is an outstanding contributor of polymer blends. Addition of TPU to polyvinyl chloride (PVC) improves compression set, abrasion resistance, and low-temperature flexibility.

Typical rubber particle sizes for effective toughening should be in nanoscale range. Rubber particles with microsize have been found to be relatively inefficient, although they might be active in crack bridging. Hence, many plastics alloys blends are commercially available. The blends of nylon-6 with TPUR in different proportions have been reported [14]. At higher organoclay content of bentonite, the layered filler is arranged in intercalated layers. The effect of TPU molecules on the organically modified inter-lamellar spacing could not be occurred exfoliation of the silicate layers in TPU [15]. Thermoplastic elastomers have been used to replace the traditional elastomers due to their high mechanical efficiency, easy processability, and low cost. TPU has very much interesting industrial applications such as automobile industry and military, medical component, electronics, and civil construction. TPU-based nanofiller blends and composites show improved thermal resistance, mechanical properties, and physical properties. So, the feasibility of using TPU- nanofillers blending is clearly demonstrated for rocket motor insulation, rocket nozzle ablatives, carbon/carbon composites, and damage tolerant high-performance epoxy fiber-reinforced composite systems. TPU has been used in many applications including caster wheels, automotive instrument panels, sporting goods, power tools, medical devices, inflatable rafts, drive belts, footwear, and a variety of extruded film, sheet, and profile applications.

1.4 Chlorosulphonated Rubber-Based Nanoblends

Chlorosulfonated polyethylene (CSM) has been used in many applications such as sheeting cable and geomembrane because of the resistance to deterioration by heat, oils, ozone, and oxidation [16]. Clorobutyl rubber (CIIR) and bromobutyl rubber (BIIR) are the most important derivatives of butyl rubber. Their predominant application is in tire inner liners with the performance of the modern radial tire being impacted by the amount of halobutyl rubber used in the inner liner compound. The thermal stability of pure CR/CSM rubber blend and CR/CSM rubber blends filled with nano- and micro-silica particles has been discussed. The incorporation of CSM in the blend would enhance cross-linking as M100 is a measure of degree of cross-linking. CSM/CIIR blend gives a higher M100 value than the CSM/IIR blends. This observation may be attributed to greater enhancement effect of CSM on CIIR. The thermal stability of pure CR/CSM rubber blend and CR/CSM rubber blends filled with nano- and micro-silica particles has been discussed.

The improvement of tensile properties for nano-silica-filled composites due to the improved dispersion of the nano-silica into the rubber matrix leads to a higher bound rubber formation [17]. Miscible blends of rubbers that consist of a single rubber phase with microscopically uniform cross-linking and distribution of reinforcing agents reflect a compositionally weighted average of the microscopic properties of the constituents. Many reviews [18–20] are focused on immiscible blends of rubbers. Isobutylene-based rubbers include butyl rubber and halogenated butyl rubber; due to their impermeability and resistance to heat and oxidation, these polymers find application in tire inner liners, inner tubes, curing bladders and envelopes, and other specialty applications where air retention and resistance to heat and oxidation are desired [21].

Chlorobutyl (CIIR) and bromobutyl (BIIR) rubbers are manufactured similarly to butyl rubber, but with an added subsequent halogenation step. Due to the nearly equivalent reactivity ratios of the monomers and low concentration of isoprene, the isoprenyl units in these polymers are randomly distributed along the polymer chain. Commercially, chlorobutyl rubber (CIIR) and bromobutyl rubber (BIIR) are the most important derivatives of butyl rubber. Their predominant application is in tire inner liners with the performance of the modern radial tire being impacted by the amount of halobutyl rubber used in the inner liner compound. Typical performance requirements of the tire inner liner are air retention, adhesion to tire carcass compound, resistance to fatigue and cracking, heat resistance, high tensile, and tear strength [22]. In this study, fourteen different SBR/CSM formulations were produced by mechanical mixing and vulcanized. The mechanical properties such as the hardness, the modulus, the tensile strength, and the abrasion loss were investigated.

From the results, the tensile strength increases with increasing nano- and micro-silica loading in CR/CSM rubber compound. The thermal stability of pure CR/CSM rubber blend and CR/CSM rubber blends filled with nano- and micro-silica particles has been discussed. Two stages of thermal decomposition occurred in CR/CSM rubber blend and nano- and micro-silica particle size-filled CR/CSM rubber blends. In the FTIR spectra of NR/CSM and NBR/CSM, the researchers in the field elastomers based on polyisoprene, acrylonitrile, and chlorosulphonated polyethylene rubbers [23] did not report the appearance of new bands related to these processes in the conjugated double-bond region. The blend of natural rubber (NR) and chlorosulphonated polyethylene rubber (CSM) is intended to produce a vulcanizate with the best properties from each component.

1.5 Polybutadiene Rubber-Based Nanoblends

Polybutadiene rubber/PS block copolymers can be epoxidized by using meta-chloroperoxy benzoic acid [24]. Then, in the presence of methylene dianiline as a hardener, the blends were prepared with epoxy resin, boehmite clay, and layered silicate nanofiller. Corish and Powell [25] observed various advantages for polybutadiene rubber/NBR blend such as better heat stability and abrasion resistance. Tire components are comprised of a polybutadiene rubber/SBR blend containing exfoliated clay nanoplatelets, which has been already patented by the Goodyear Tire and Rubber Company in USA [26]. A crystallizable blend of polybutadiene rubber can also be compatibilized by the addition of 5–10 % amorphous diblock copolymers of (1,2 PBR/1,4 PBR). By the solution intercalation method, PBR/Organomontmorillonite (OMMT) nanocomposites were prepared. Go and Ha [27] prepared 98 % cis content polybutadiene rubber blend with maleic anhydride (MA)-treated EPDM and ethylidene norbornene (ENB) as a termonomer. MA-g-EPDM was prepared by reacting EPDM with 2.0 phr of MA in the presence of 0.2 phr dibenzothiazyl disulfide in a banbury mixer with speed of 40 rpm at 230 °C for the duration of 3 min. Zhang et al. [28] prepared two elastomers PBR and brominated poly(isobutylene-co-p-methylstyrene) (BIMS), p-methylstyrene (PMS), and p-bromomethylstyrene (BrPMS). Consequently, in standard rubber tires, hard sidewall and the soft inner liner can be made of carbon-reinforced BIMS blended with butyl and NR. Vulcanization of the polybutadiene rubber/NR/nanoclay blend was taking place at the highest temperature, and the other rubber nanoblends showed a lower vulcanization temperature than that of the pure PBR/NR blend. Effects of expanded graphite (EG) and isocyanate modified graphite (i-MG) are on the dynamic mechanical properties of PBR, SBR, and PBR/SBR blends by DMA cum thermal analysis.

The storage modulus of PBR/SBR blends was found to be increased at very-low-temperature region and then decreased from a low-to-high temperature region which was because of very low transition temperature of PBR. Bottazzo et al. [29] reported that the thermal degradation of pure PBR/NR blend has two steps. He reported that the first decomposition of the NR matrix occurs between 280 and 410 °C, and the second step decomposition is attributed to the PBR phase degradation and it is observed in the range 410–520 °C. Saeed et al. [30] assessed dispersion of silica in the PBR blend to select a suitable mixing time for incorporating the silica in the rubber blend. The resulted rubber nanoblends were examined in SEM to study surface morphology and nanofiller dispersion. Small pieces of the uncured PBR blends were placed in liquid nitrogen for 3 min and then fractured to create two fresh surfaces.

Debapriya et al. [31] studied the SEM of the tensile-fractured samples of PBR/NR/RR nanoblends filled with nanosilica to get an idea about the dispersion of silica nanoparticles. It is also reported that RR content plays a vital role in nanosilica dispersion in the ternary rubber nanoblends, which is corroborated from the evaluation of mechanical properties. PBR–elastomer blends are being used in most rubber products for a long time. This is evident in the competition of patents on rubber blends by largest tire manufacturers. A blend of PBR/NR gives high curing temperature while PBR/NR/SBR-based blends give improved heat stability. Blend of liquid PBR gives ease processing and physical properties. Addition of 10–30 % of SBR, NR, and IR to cis-PBR along with alkaline salt of fatty acid improves cutting resistance and vulcanizates. The addition of butyl and halobutyl rubber to PBR/NR blends can lower resilience and improve skid resistance. When BIMS is added to carbon black-filled PBR/NR blends, the loss tangent increases at low temperatures at −30 to 10 °C, but decreases above 30 °C and the addition of BIMS improves resistance to tread wear [32]. Pneumatic tire is a highly engineered product made of several components, including chafer, apex, black sidewall, belts, radial ply, bead, and liner, and these parts are generally prepared from PBR/NR (50/50) and (30/70) blends for passenger tire-tread applications and PBR/SBR and PBR/NR (20/80) for truck-tire applications.

1.6 Styrene Butadiene Rubber-Based Nanoblends

A styrene butadiene rubber (SBR) is an important type of elastomeric material. SBR nanoblends are classified as miscible and immiscible, depending on the constituents present in the matrix. The SBR nanoblends have attractive properties such as high thermal resistance, mechanical properties and damping at high-frequency, and mechanical and viscoelastic properties than single rubber [33–37]. Solution of styrene butadiene rubber (S-SBR) (styrene content of 21 % and vinyl content of 63 %), nitrile butadiene rubber (NBR) (nitrile content of 34 %), and natural rubber (NR) were used for the preparation of blend. The 50/50 (wt%) blend of NR and SBR latexes was prepared by slowly adding the given amount of SBR latex into the NR latex under mixing for 30 min. Then, a given amount of (50/50) NR/SBR blend latex was added into the aqueous clay suspension and stirred for 30 min.

In recent years for lighter, stiffer and stronger materials are in demand which can be used in different sectors such as aeronautics and civil engineering and in various structural applications. SBR nanoblends with different compositions can be prepared by two-roll mill [38]. Elongation at break of NR/SBR blend is also improved when the addition of clay into the rubber matrix. In addition, the improvement can be increased by increasing the amount of clay up to 6 phr and then decreased slightly in the nanoblends containing 10 phr clay. In the case of SBR nanoblend gels, the TS gives maximum value at 6 phr of Na-MMT loading for both NR and SBR nanoblend gels. The EB values generate different trends for NR and SBR nanoblend gels. The effect of different type of nanomaterials on thermal properties of SBR base nanoblends was discussed [39, 40].

The effect of different wt% loadings of CNTs on thermal behavior of SBR nanoblends was studied by Hosseini et al. [41] using DSC. Heat capacity ratios at transition steps of NR and SBR in the blends were also correlated with the blend composition. Stephen and co-workers [42] prepared and studied SBR-based nanoblends with different layered silicate nanomaterials and characterized their morphological properties by transmission electron microscope (TEM). The configuration and the entropy are increased during the process of expulsion of SBR, and the reason for the decrease in gallery height should be ascribed to the displacement of SBR molecules from the interlayer during the process of curing. The signal distribution profile shows clearly a large and flexible domain for the pure SBR and SBR/TiO_2 nanoblend, but a rigid domain for the SBR/TiO_2 photodegraded nanoblend. The viscoelastic properties of SBR depend on structure, crystallinity, and extent of cross-linking. Dynamic mechanical analyzer (DMA) is used in tension mode to analyze the viscoelastic properties of SBR nanoblends. Tancharernrat and co-workers [43] prepared and studied the styrene butadiene copolymer (SBR)–SiO_2 nanoparticles with core–shell morphology via differential microemulsion polymerization. And finally they have studied the effect of encapsulated SiO_2 on the mechanical properties of the NR/(SBR–SiO_2) nanoblend. Mohamed and his co-workers [44] prepared composites and studies the effect of organomontmorillonite clay with modified by hexadecyltrimethyl ammonium bromide on the mechanical properties of acrylonitrile butadiene rubber (NBR)/SBR blend nanocomposites.

1.7 Polychloroprene Rubber-Based Nanoblends

The degree of crystallinity in polychloroprene rubber is largely dependent upon the amount of trans-configurations in the polymer. CR has a better resistance to swelling in mineral, animal, and vegetable oils and fats. CR with blends of neoprene and natural rubber (NR) is often used to provide good hot tear strength and building tack, and blends with polychloroprene rubber to improve nonstaining static ozone resistance. Because of the incompatible nature of the polymers, the split masterbatch mixing technique is considered essential. This could be explained in light of inefficient co-cure due to curative migration and filler transfer from non-polar NR to more polar CR phase. Chlorine atom in CR deactivates the double bond in the oxidative aging process. So it is expected that under accelerated thermal oxidative aging better resistive performance could be achieved for the vulcanizates containing greater amount of CR. Since CR is a polar rubber, its presence always enhances the dielectric constant values of a blend containing non-polar rubber. It is interesting to note that the values are higher for the vulcanizates obtained from the one-stage process as compared to that of the two-stage process especially in the case of the blend vulcanizates containing CR and NR. The ethylene–propylene rubber (EPR) or EPDM has better resistance to heat, ozone, and cut growth [45]. However, these CR/EPR or EPDM blends are incompatible [46].

A typical mixing procedure was followed to obtain the CR/EPDM blend compounds. Chloroprene rubber (CR) concentration on the cure characteristics, microstructure, and mechanical and rheological properties of isobutylene–isoprene rubber (IIR)/CR blend was investigated. A blend of chloroprene rubber in less than

40 % is preferred with due consideration of good ozone, weathering, aging, and oil resistance of the vulcanizates. It ought also to increase the degree of fire resistance [47]. Polychloroprene rubber-based nanoblends will play a key role in medical technology of the future. Polychloroprene rubber can be implanted in the body to provide special prosthetic functions or in diagnostic, surgical, and therapeutic applications without causing adverse effect on blood and other tissues. Polychloroprene rubber-based nanoblend has several valuable properties and works well in applications requiring resistance to gasoline, sunlight, ozone, oxidation, and oil. In the automotive world, neoprene rubber-based nanoblend is used for many under-hood and under-body parts and mid-performance polymer with a good all-around balance of performance properties. Polychloroprene rubber-based nanoblend sheets with stand a temperature range of 34 to 202 °C. They can be compounded for an excellent resistance to most of the acids and alkalies as well as to open flame. These sheets are used for many applications such as in refineries, food industry, chemical, and process units. For some of the mail applications, polychloroprene rubbers are blended with synthetic resins to promote mechanical strength and heat stability phenolic polychloroprene. Phenolic polychloroprene is the one of the most common uses for military applications. It also is used to make protective gloves and similar articles because it is resistant to many chemicals and oils. Neoprene's resistance to abrasion is why it is commonly used to make protective covers and its lightweight makes advantageous in most of its uses.

The most effective screening pigments for coloring of polychloroprene-based rubber nanoblends are rutile titanium dioxide and red or yellow iron oxides but they are far less efficient than carbon black. But its need to be avoided for long-term direct sunlight exposure of colored polychloroprene-based rubber nanoblends. Polychloroprene-based rubber nanoblends have been used for many applications such as cable sheathing, structural gaskets and bridge bearing pads. Antiozonants should be nonvolatile and non-extractable by water and have minimum effect upon compound bin storage stability. Chloroprene rubber (CR) has a polar chlorine part and a nonpolar hydrocarbon part. Chloroprene rubber (CR) has many advantageous properties, such as oil resistance, toughness, a dynamic flex life, and adhesion capacity. In the study of CR as a compatibilizer in NR–NBR blend preparation, efforts were made to prepare 20:80 NR–NBR blends with CR with the aim of studying the effect of poly(ethylene oxide) (PEO)-coated nanocalcium silicate along with nano-N-benzylimine aminothioformamide and stearic acid-coated nano-zinc oxide in the sulfur vulcanization of the blends. Chloroprene rubber (CR) has a backbone structure similar to that of NR but is more polar because of a chlorine substituent. The solubility parameter value of CR is intermediate to that of NR and NBR. In the presence of a dipole within the repeat unit allows the possibility of the interaction of the acrylonitrile NBR rubber. Polychloroprene rubber-based nanoblends were the first commercially successful synthetic elastomer blends. Polychloroprene rubber-based nanoblends are used in coating, military, spacing, biomedical, and aerospace applications. Polychloroprene rubber latex-based nanoblends are used in waterborne adhesives and dipped goods.

1.8 Ethylene Propylene Diene Rubber-Based Nanoblends

Ethylene–propylene–diene monomer (EPDM) is a saturated, nonpolar rubber. The selection of EPDM depends upon on the properties of the end product. Different ratios of monomers grant different characteristics to EPDM [48]. The EPDM polymers are produced by varying of the molecular weight, molecular weight distribution, ethylene/propylene ratio and level, and the type of diene termonomer. EPDM rubber is used for thermal insulator in space applications with high ethylene content [49]. The Tg of EPDM varied with the content of the ethylene and generally increased with the increase of propylene content.

EPDM is a rapid growing synthetic rubber used in many industrial applications and it is more stable than other conventional elastomers. For EPDM, there are some disadvantages, its poor solvent resistance and adhesion properties. The blending of NBR and EPDM rubber was performed to achieve the best properties from each component [50]. The cross-linking process of rubber is happened through the sulfur bond formation and EPDM and NBR co-cross-linking during the C–C bonds is formed on the border of their phases. When the thermal aging, the NBR/EPDM nanocomposites show the small changes in mechanical properties, but in the case of silica-reinforced nanocomposites exhibited higher changes in all mechanical properties than carbon black-reinforced nanocomposites. The degradation of EPDM and its blends has been studied under a variety of conditions including non-isothermal and isothermal degradation under a nitrogen atmosphere, photooxidation, radiochemical oxidation, thermo-oxidation, and ultraviolet irradiation. Nevertheless, since the small amount of EPDM in the mixture of NBR/EPDM is presented, it should be noted that in accordance with Kurian et al. [51], the EPDM structure comprises rigid domains and flexible regions. The rigid domains, related to polyethylene blocks and cross-linked dienes, probably inhibit the decomposition, increasing the E_a value. In the case of the NBR/EPDM rubber blend reinforced with 70 phr of silica degradation process, one cannot expect to get appropriate model if the calculation is only based on the theoretical equation [52].

For the NBR/EPDM rubber blend reinforced with CB/silica (35/35) and NBR/EPDM rubber blend reinforced with CB/silica (50/20) degradation processes, the equations of Turmanova et. al. [53] referring to the diffusion-controlled reactions were established as the most appropriate kinetic models. This means that NBR/EPDM blends are completely immiscible and large EPDM domains are dispersed in the NBR matrix. The interphase adhesion between NBR and EPDM is very weak. According to poor compatibility between NBR and EPDM, they have poor adhesion with carbon black. Valentini et al. [54] investigated grapheme/EPDM nanocomposites for military components subjected to high-frequency, high-magnitude shock pulses. In aerospace missions, pyroshocks occur due to controlled explosions of ordnance devices enabling the functionality of space modules. Ethylene–propylene–diene terpolymer rubber and graphene nanoplatelets nanocomposites were prepared via a vulcanization process. It was observed how such hybrid nanocomposite has a higher number of cross-links than carbon black nanocomposites.

The use of EPDM as an insulator for case-bonded solid rocket motors was evaluated more than a decade ago and found feasible [55]. EPDM with low specific gravity and it has emerged as a novel material for manifold applications including insulation for solid rocket motors. Different composites with EPDM with filler-based rocket motor insulation compound have been widely reported in patents and many research articles [56–60]. Light polymer nanostructured materials are necessary for improved performance for use in rocket insulation applications.

References

1. Suil G et al (2008) Preparation and properties of natural rubber composites reinforced with pretreated carbon nanotubes. Polym Adv Technol 19:1543–1549
2. Bendahou A, Kaddami H, Dufresne A (2010) Investigation on the effect of cellulosic nanoparticles morphology on the properties of natural rubber based nanocomposites. Eur Polymer J 46:609–620
3. Jamal EMA et al (2009) Synthesis of nickel–rubber nanocomposites and evaluation of their dielectric properties. Mater Sci Eng B 156:24–31
4. Flavio C et al (2014) Organic acids and protein compounds causing the photoluminescence properties of natural rubber membranes and the quenching phenomena from Au nanoparticle incorporation. Luminescence 29:1047–1052
5. Chen Y, Yuan D, Xu C (2014) Dynamically vulcanized biobased polylactide/natural rubber blend material with continuous cross-linked rubber phase. ACS Appl Mater Interfaces 6:3811–3816
6. Parulekar Y, Mohanty AK (2006) Biodegradable toughened polymers from renewable resources: blends of polyhydroxybutyrate with epoxidized natural rubber and maleated polybutadiene. Green Chem 8:206–213
7. Paul DR, Robeson LM (2008) Polymer nanotechnology: nanocomposites. Polymer 49:3187–3204
8. Alex R, Nah C (2006) Studies on natural rubber/acrylonitrile butadiene rubber/organoclay nanocomposites. Plast Rubber Compos 35:219–225
9. Lewan MV (1998) NR/NBR blends—compounding for food contact applications. In: Tinker AJ, Jones KP (eds) Blends of natural rubber. Chapman & Hall, London, pp 94–105
10. Tatjana R, Saulute B, Krzysztof P, Jan P (2006) Application of polyurethane-based materials for immobilization of enzymes and cells: a review. Chemija 17:74–89
11. Sonnenschein MF, Guillaudeu SJ, Landes BG, Wendt BL (2010) Comparison of adipate and succinate polyesters in thermoplastic polyurethanes. Polymer 51:3685–3692
12. Robert CK (2008) In: Pascal F (ed) Handbook of specialty elastomers (Chapter 4), CRC Press, Raton, pp 134–154
13. John S (1997) In: Arcella V, Ferro R (eds) Modern fluoropolymers (Chapter 2), Wiley, West Sussex, pp 71–90
14. Palanivelu K, Sivaraman P, Sharma SK, Verma SK (2003) Int J Plast Technol 7:133
15. Abdeen Z (2014) Enhanced recovery of $Pb2^{+}$ ions from aquatic media 4 by using polyurethane composite as adsorbent, environmental process, pp 1–15. doi:10.1007/s40710-014-0048-0
16. Flory PJ (1946) Ind Eng Chem 38(4):417–436

17. Fu SY, Xi-Qiao F, Bernd L, Yiu-Wing M (2008) Effects of particle size, particle/matrix interface adhesion and particle loading on mechanical properties of particulate–polymer composites. Compos B 39:933–961
18. Marković G, Samaržija-Jovanović S, Jovanović V, Marinović-Cincović M (2010) J Therm Anal Calor 100:881–888
19. Essawy H, El-Nashar D (2004) Polym Test 23(7):803–807
20. Marković G, Radovanović B, Marinović-Cincović M, Budinski-Simendić J (2006) Kautch Gummi Kunst 5:251–255
21. Voet A (1980) J Polym Sci 15:327–373
22. Zhang L, Wang Y, Wang Y, Sui Y, Yu D (2000) J Appl Polym Sci 78(11):1873–1878
23. Coran AY, Hamed P, Goettler LA (1976) Rubber Chem Technol 49:1167–1181
24. Pandit R, Berkessel A, Lach R, Grellmann W, Adhikari R (2012) Synthesis and characterization of nanostructured blends of epoxy resin and block copolymers. Nepal J Sci Technol 13(1):81–88
25. Corish PJ, Powell BDW (1974) Elastomer blends, rubber, chemistry and technology, vol 47 (03), pp 481–510
26. Zanzig DJ, Yang X, Cohen MP (2010) Tyre with component comprised of a blend of polybutadiene rubber and composite of styrene/butadiene elastomer which contains exfoliated clay platelets, US 7,714,055 B2 to Goodyear
27. Go JH, Ha CS (1996) Rheology and properties of EPDM/BR blends with or without a homogenizing agent or a coupling agent. J Appl Polym Sci 62(03):509–521
28. Zhang Y, Ge S, Tang B, Koga T, Rafailovich MH, Sokolov JC, Peiffer DG, Li Z, Dias AJ, McElrath KO, Lin MY, Satija SK, Urquhart SG, Ade H, Nguyen D (2001) Effect of carbon black and silica fillers in elastomer blends. Macromolecules 34:7056–7065
29. Bottazzo J, Guglielmi M, Polizzi S, Brusatin G (2013) Natural rubber/cis-1,4-polybutadiene nanocomposites: vulcanization behavior, mechanical properties, and thermal stability. In: Polymer engineering and science, pp 671–678
30. Saeed F, Ansarifar A, Ellis R, Meskel Y, Irfan MS (2012) Two advanced styrene-butadiene/polybutadiene rubber blends filled with a silanized silica nanofiller for potential use in passenger car tire tread compound. J Appl Polym Sci 123:1518–1529
31. Debapriya D, Panda PK, Bhunia S, Roy M (2014) Effect of sol-gel-derived nano-silica on the properties of natural rubber-poly butadiene rubber-reclaim rubber ternary blends/silica nanocomposites. Polym Plast Technol Eng 53(11):1131–1141
32. Mangaraj D (2002) Elastomer blends. Rubber Chem Technol 75(03):365–427
33. Shanks RA, Kong I (2013) General purpose elastomers: structure, chemistry, physics and performance. In: Advances in elastomers I, advanced structured materials, vol 11. Springer, Berlin. doi:10.1007/978-3-642-20925-3_2
34. Deepalekshmi P, Visakh PM, Mathew AP, Chandra AK, Thomas S (2013) Advances in elastomers I, advanced structured materials 11, Springer, Heidelberg. doi:10.1007/978-3-642-20925-3_1
35. Friedrich K, Fakirov S, Zhang Z (2005) Polymer composites from nano- to macro-scale
36. Apostolo M, Triulzi F (2004) Properties of fluoroelastomer/semicrystalline perfluoropolymer nano-blends. J Fluor Chem 125:303–314
37. Hu GH, Cartier H (1999) Reactive extrusion: toward nano blends. Macromolecules 32:4713–4718
38. Boonmahitthisud A, Chuayjuljit S (2012) Effects of nanosized polystyrene and polystyrene-encapsulated nanosilica on physical properties of natural rubber/styrene butadiene rubber nanocomposites. Poly Plast Technol Eng 51(3):311–316
39. Prasertsri S, Lagarde F, Rattanasom N, Sirisinha C, Daniel P (2013) Raman spectroscopy and thermal analysis of gum and silica-filled NR/SBR blends prepared from latex system. Polym Test 32:852–861
40. Pal K, Rajasekar R, Kang DJ, Zhang ZX, Pal SK, Das CK, Kim JK (2010) Effect of fillers on natural rubber/high styrene rubber blends with nano silica: morphology and wear, Mater Des 31:677–686

41. Hosseini SM, Madaeni SS, Khodabakhshi AR (2010) Preparation and characterization of PC/SBR heterogeneous cation exchange membrane filled with carbon nano-tubes. J Membr Sci 362:550–559
42. Stephen R, Ranganathaiah C, Varghese S, Joseph K (2006) Sabu Thomas, Gas transport through nano and micro composites of natural rubber (NR) and their blends with carboxylated styrene butadiene rubber (XSBR) latex membranes. Polymer 47:858–870
43. Tancharernrat T, Rempel GL, Prasassarakich P (2014) Preparation of styrene butadiene copolymer–silica nanocomposites via differential microemulsion polymerization and NR/SBR–SiO_2 membranes for pervaporation of water–ethanol mixtures. Chem Eng J 258:290–300
44. Mohamed RM (2013) Radiation induced modification of NBR and SBR montmorillonite nanocomposites, J Ind Eng Chem 19:80–86
45. Deuri AS, Bhowmick AK (1987) J Appl Polym Sci 34:2205–2222
46. Ramesan MT, Mathew G, Kuriakose B, Alex R (2001) Eur Polym J 37:719–728
47. Lewitzke C, Lee P (2001) Application of elastomeric components for noise and vibration isolation in the automotive industry. SAE technical paper 2001-01-1447, doi:10.4271/2001-01-1447
48. Brydson JA (1978) Rubber chemistry. Applied Science Publishers, London
49. Deuri SA, De PP, Bhowmick AK, De SK (1988) Studies on the ageing of EPDM based rocket insulator compound by stress relaxation and the effect of propellant binder. Polym Degrad Stab 20(2):135–148
50. Jovanović V, Samaržija-Jovanović S, Budinski-Simendić J, Marković G, Marinović-Cincović M, Composites based on carbon black reinforced NBR/EPDM rubber blends. Compos Part B 45:333–340
51. Kurian T, De PP, Khastgir D, Tripathy DK, De SK, Peiffer DG (1995) Reinforcement of EPDM-based ionic thermoplastic elastomer by carbon black. Polymer 36:3875–3884
52. Montserrat S, Málek J, Colomer P (1998) Thermal degradation kinetics of epoxy-anhydride resins: I. Influence of a silica filler. Thermochim Acta 313:83–95
53. Turmanova SCh, Genieva SD, Dimitrova AS, Vlaev LT (2008) Non-isothermal degradation kinetics of filled with rise husk ash polypropene composites. Express Polym Lett 2:133–146
54. Valentini L, Bolognini A, Alvino A, Bon SB, Martin-Gallego M, Lopez-Manchado MA (2014) Pyroshock testing on graphene based EPDM nanocomposites. Compos B Eng 60:479–484
55. Bhuvaneswari CM, Sureshkumar SM, Kakade SD, Gupta M (2006) Ethylene-propylene diene rubber as a futuristic elastomer for insulation of solid rocket motors. Defense Sci J 56:309–320
56. Fan JL, Tsai S-H, Tu F-H, Tu Y-T (2007) Low density rocket motor insulation, United States patent application, Publication No. US2007/0112091 A1, 17 May 2007
57. Uyarel AY, Pektas I (1996) A thermal analysis investigation of new insulator compositions based on EPDM and phenolic resin. J Therm Anal 46:163–176
58. Deuri AS, Bhowmick AK (1986) Ageing of rocket insulator compound based on EPDM. Polym Degrad Stabil 16:221–239
59. Deuri AS, Bhowmick AK (1987) Degradation of rocket insulator at high temperature. J Therm Anal 32:755–770
60. Harvey AR, Ellertson JW (2003) Rocket motor insulation containing hydrophobic particles, United States patent no. US 6,606,852 B1, Aug 2003

Chapter 2
Natural Rubber Nanoblends: Preparation, Characterization and Applications

Oriparambil Sivaraman Nirmal Ghosh, S. Gayathri, P. Sudhakara, S.K. Misra and J. Jayaramudu

Abstract Natural rubber nanocomposites are emerging as one of the key players in the industry and research due to its novel material properties and nanofunctionalities. The development of nanotechnology provided a wide range of methods for the synthesis of novel multifunctional nanofillers and blending them with natural rubber to make high performance natural rubber nanocomposites. In this chapter we summarizes the advanced methods for synthesis, characterization and applications of natural rubber nanoblends using various techniques including skim latex mixing, latex/melt intercalation, freeze drying and in-situ non-aqueous sol-gel method. A detailed description of advanced nanomaterial characterization methods like XRD, XPS, XRF, UV–Vis spectroscopy, Photoluminescence spectroscopy, Raman spectroscopy, TGA-DTA, SEM-EDX, EPMA, TEM-SAED, VSM and FTIR are also elaborated for understanding the physicochemical characteristics of nanofillers and natural rubber nanocomposites. A brief outline of current trends in the development of nanomaterials for filler applications was summarized. The fundamental aspects of design and utilization of various nanostructured materials including carbon nanotubes, clay, iron oxide, zinc oxide, nickel, silica, gold, silver and graphene based filler materials are discussed in detail.

Keywords Natural rubber · Nanoblend · Nanocomposite · Nanofiller · Advanced characterization technique

O.S. Nirmal Ghosh (✉) · S. Gayathri
Nanophotonics and Nanoelectronics Research Laboratory, Centre for Nanoscience and Technology, Madanjeet School of Green Energy Technologies, Pondicherry University, Kalapet 605014, India
e-mail: nirmalgvr@protonmail.ch

P. Sudhakara · S.K. Misra
Regional Centre for Extension and Development, CSIR—Central Leather Research Institute, Leather Complex, Jalandhar 144021, Punjab, India

J. Jayaramudu
Indian Rubber Manufacturers Research Association, Thane 400604, Maharashtra, India

G. Marković and Visakh P.M. (eds.), *Rubber Nano Blends*,
Springer Series on Polymer and Composite Materials,
DOI 10.1007/978-3-319-48720-5_2

2.1 Introduction

Natural rubber (NR) is an elastomer which is derived from a milky colloidal suspension from the sap of typical plants. The chemical structure of the NR is given in Fig. 2.1. NR is generally known as *cis*-1,4-poly (isoprene). Hevea brasiliensis plant is originated from Brazil. Among the rubber plants, the Hevea brasiliensis is the only sap tree having commercial and industrial significance. Hevea trees are the natural source with high molecular weight approximately 97 % *cis*-1,4-poly (isoprene), and due to this high molecular weight, the NR polymer chain continues to be on the same side of the double bond. Due to its regularity of the structure, the NR undergoes crystallization when it is stored below 20 °C or subject to stretching. The specific gravity of NR is 0.93 at 20 °C.

Hevea trees produce high-yield latex for a tapping period of 30 years. The NR latex contains 60 % water, 36 % of total solid content, 33 % dry rubber content, 1–2.5 % of proteinaceous substances, 1–2.5 % of resinous substance, 1 % sugar, less than 1 % ash, and 0.5 % of salts of phosphorus, potassium, and magnesium. Due to the sticky and unsaturated bonding nature of NR, it does not have any industrial significance up to 1839 (Fig. 2.2).

In 1839, Thomas Hancock and Charles Goodyear have independently invented vulcanization process to chemically modify the NR and establish cross-links into the NR chains. The vulcanization process is a systematic process of converting the NR into a non-sticky and hard polymeric material by adding sulfur, accelerators, antioxidants, and fillers. It is an irreversible chemical process. The mechanical

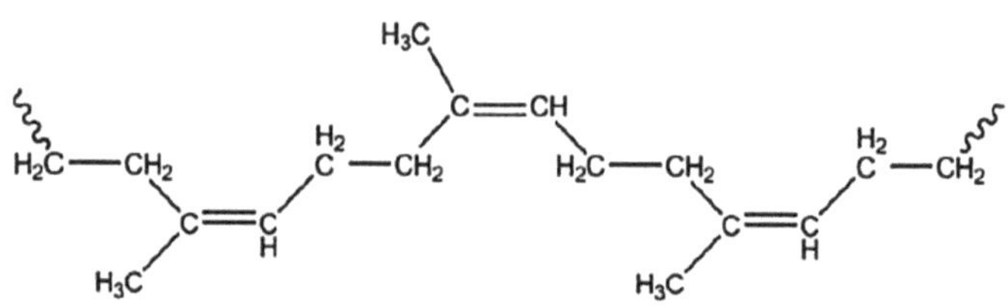

Fig. 2.1 Chemical structure of natural rubber (*cis*-1,4-poly(isoprene))

Fig. 2.2 Schematic of sulfur vulcanization of natural rubber

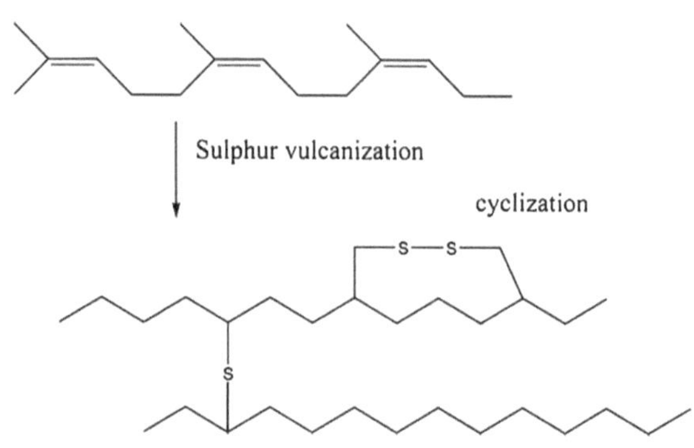

properties of the NR increase during the vulcanization process due to the cross-linking of NR chains. The vulcanized rubbers have very high resistance to organic solvents with improved elastic and flexible nature. During the advanced vulcanization process, NR is compounded with different concentrations of fillers, low concentration of accelerators, aromatic amines, and phenols for antioxidation purposes and 5–8 % of sulfur followed by the temperature treatment. Natural rubber nanocomposites (NRNs) are having wide range of applications in the field of industry and technology. NRNs containing required amount of Montmorillonite structures composed of nanostructured alumino-silicate layers which acts as filler to reinforce the NR matrix. As a result of this clay-based filler reinforcement, the water permeability of NR decreases considerably with simultaneous enhancement in superior thermal and mechanical properties. Dynamic vulcanization techniques are used to manufacture NR-based polymer blends. Therefore, through this method, it is possible to develop high-performance thermoplastics with improved mechanical and thermal properties with low hardness [1].

Nanoscience and technology deals with the development of nanomaterials and studying its characteristics for various applications. The rise of the nanoera in science and technology also has great impact in the development of rubber technology. Generally, nanomaterials have low-dimensional aspects and size varying from 1 to 100 nm. All the electrons present in the nanomaterials are available on the surface due to its high surface to volume ratio. A wide range of multifunctional nanomaterials are used as fillers and antioxidation agents to enhance the mechanical, thermal, electrical, and magnetic properties of the NR. The nanofillers act as the major reinforcement agent in NR matrix. It is found that the fillers help to enhance the tensile strength, tear resistance, aberration resistance, and tensile modulus of the NR. The physicochemical and mechanical properties of NR are closely related with that of filler materials used. Utilization of nanostructured materials as fillers leads to the development of NRNs and nanotechnology revolution in rubber technology. The performance of filler in the NR matrix is governed by the surface interface characteristics which include concentration, particle size, shape, structure, and degree of interactions with rubber matrix and surface of the particle agglomerates. Compared to the bulk filler materials, nanostructured filler materials exhibit special properties such as high surface area, mechanical strength, wear, tear and aberration resistance, antioxidation properties, control over dimensions, and electronic confinement. Due to the large surface area of nanomaterials, higher level of surface interface interaction between rubber matrix and nanofillers is possible in NRNs. The formation of surface interface between NR and nanofillers is governed by the volume fraction and size of the constituent nanoparticles. The degree of bonding between NR matrix and nanofiller particles is the major influencing factor in determining the degree of reinforcement in elastomer properties of NRN.

A NRN is a multiphase solid material where one of the phases has dimension below the range of 100 nm. During the synthesis of NRN, the nanofillers are added into the natural rubber at rates from 1 % to 10 wt%. The nanofillers are categorized into various types according to the nanodimension and behavior of the fillers. Generally, one-dimensional nanofillers are in the form of laminas, shells, and plates.

Nanotubes and fibers with diameter lesser than 100 nm are considered as two-dimensional nanofillers. The template like isodimensional nanostructured materials is considered as three-dimensional nanofillers. Due to the importance and advantages of nanostructured materials in NRN, scientific communities across the globe are attracted very much toward the emerging field of research in NRN and various techniques for the production and characterization of NRN. In this regard, we tried to summarize the current literature available in the field of NRN research in this chapter. The fundamental aspects of physicochemical characterization of NRN and nanostructured materials are also included and discussed in this chapter for the benefit of scientists and beginners in NRN research.

2.2 Recent Works on Natural Rubber Blends

Due to the wide attention given to the natural rubber blends, a lot of literatures have been evolved in past few years. Scientists from various fields are working together on developing high-performance natural rubber composites for different industrial applications. Recent development in the field of nanotechnology research provided a boost up in the development of NRNs. Development and utilization of nanofillers and functional additives as reinforcement materials helps to provide nanofunctionalities to the NRN that includes high mechanical stability, aberration resistance, resistance against organic liquids and solvents, and enhanced thermal stability. Thus, the novel properties of the nanomaterials can be extended to its natural rubber composites also. J. Wu et al. recently demonstrated the enhancement in vulcanization kinetics of NRN through the addition of graphene during the vulcanization process. In a typical synthesis process, they have mixed the graphene efficiently with the NR to reduce the induction period of vulcanization process. It is identified that upon addition of graphene the induction period was suddenly suppressed. They recognized that the graphene loading has influence in vulcanization kinetics in such a way that the low graphene loading enhances the vulcanization kinetics, and upon increasing the graphene loading, there is a slight increase in induction period of vulcanization process. They found that the graphene reduces the melting point of sulfur and provides enhancement in cross-linking of NR. This enhanced cross-linking results an increase in cross-linking density of graphene NRN. It is identified that the addition of graphene leads to the accelerated reaction between accelerator, sulfur, and activator [2]. In the similar fashion, G. Sui et al. explored the ability of carbon nanotubes as a reinforcement agent [3]. In this extensive study, Sui group used the pretreated CNT as fillers to improve the dynamic compression properties, thermal stability, rebound resilience, and storage modulus of the NRN. The CNTs were prepared using chemical vapor deposition (CVD) followed by purification in hydrofluoric acid (HF) for 24 h and washed continuously by using deionized water until the pH become neutral. Further, they have adopted a typical treatment method to improve the reinforcement through better cross-linking of fillers and NR matrix. CNT-NRN was prepared by solvent mixing and two-roll milling. Interesting observations have been made in this study proves that the utilization of pretreated CNT as a filler in NR helps to improve the mechanical properties. This is due to the high surface to volume ratio of CNT and formation of better coordination bonding between NR matrix and CNT [3].

Fine-tuning the electrical and mechanical properties of NRN is taking great importance due to wide range of applications in energy storage devices, sensors, etc. International scientist community have paid lot of attention to this area to develop novel nanoblend-based NRN. Recently, Thomas et al. demonstrated the effect of functionalized CNTs in enhancement of electrical properties of NRN. They have found that the dielectric constant of the NRN increases with increase in CNT loading. 1-octadecanol was used to modify the CNT for better dispersibility. It is found that the 1-octadecanol-functionalized CNT fillers show better stress transfer from NR matrix to the filler. Due to this reason, the fillers tensile strength and elongation at break of the NRN with functionalized CNT was better than that of NRN with normal CNT as fillers [4]. Efforts have been made to manipulate the thermal properties of NR using graft copolymerization techniques. Recently, Nakason et al. was successfully achieved the copolymerization of maleic anhydride and NR to improve the glass transition temperature of the graft copolymers of NR and anhydride [5]. Dispersion of fillers in the NR matrix is a very important factor which influences the mechanical, thermal, and electrical properties of NRN. Multiwall carbon nanotubes (MWCNTs) are one of the key players as fillers which not only reinforce the NR but also it facilitate improvement in stiffness of the NR matrix. Bokobza et al. found that the good dispersion of MWCNTs leads to the formation of NRN with high electrical and mechanical properties. According to Halpin–Tsai model, the aspect ratio of the MWCNTs should be 90 in order to achieve an optimal dispersion of MWCNTs in NR matrix to form a NRN with superior material properties such as electrical conduction and mechanical reinforcement [6].

Nanostructured fillers and curing agents are attracting enormous attention due to the environmental concerns and health issues related to the conventional chemicals. Nanofillers, antioxidants, and curing agents have unique properties of high surface to volume ratio due to their nanosize and shape. This will help the NR industry to meet out the large requirement of chemicals for NR production and manufacturing. Using the nanoadditives which help us to reduce the consumption of chemicals, since they provide the same performance at low weight percentage compared to the bulk counter parts. In this regard, Kim et al. conducted an exclusive investigation to understand the effect of nanostructured ZnO on the mechanical properties and cure characteristics of silica-filled NRNs. In this study, they have used nano-ZnO with particle size varies in the range of 30–40 nm and specific surface area of 25–50 m^2/gm. In contrary to the conventional volume of bulk ZnO, they have added only 20 wt% nano-ZnO into the unfilled system and achieved the same level of cure characteristics and mechanical properties achieved by adding higher volume of bulk ZnO. This is due to the large specific surface area of nano-ZnO, which led to the increase in the degree of cross-linking in NR matrix [7]. Nanofibers are also used as reinforcing agents in NR. Nanofibers of cellulose generated from biological resources like plants are having important role in providing alternate resources for the manufacturing of green NRN. This approach provides an economically viable method to enhance the mechanical and loadbearing properties of NRN. Recently, Abraham et al. was reported the fabrication of NRN using cellulose nanofiber and natural rubber latex. In this work, the researchers have isolated cellulose nanofibers (CNF) from raw banana fibers by steam explosion process. The CNF-based NRN exhibits improved Young's modulus and tensile strength. Dynamic mechanical analysis showed that with the addition of CNF in NR there is a considerable change in the storage modulus of the NR matrix. This is due to the reinforcement effect of CNF in the NR latex [8].

In a similar study carried out by Bendahou et al., cellulose whiskers and microfibrillated cellulose were extracted from the rachis of the palm from Phoenix dactylifera L as a reinforcement agent in natural rubber matrix, and they have found that the swelling behavior of NRN by toluene was decreasing considerably with the addition of only 1 wt% of cellulose in NR matrix. From their results, it is obtained that the higher filler–matrix interaction and adhesion modify the physicochemical properties of the NRN. Compared to NR, the cellulose-NR composites exhibit enhanced material properties including high stiffness, low shrinkage and reduced water intake. The addition of cellulose whiskers created an interfacial layer surrounding the filler which restricts the mobility of fillers in NR matrix resulting better mechanical properties compared to its bulk counterparts [1]. Making of blends NR with polymers like acrylonitrile butadiene rubber (NBR) is an easiest way to achieve better thermal stability. Kumari et al. reported that blending of NBR with NR helps to increase the decomposition temperature of NR–NBR composite, which is due to the enhancement in interfacial adhesion between the NR and NBR blend components. Also they have observed that the addition of compatibilizer like neoprene does not affect the thermal stability of the blends, and NR–NBR blends are thermodynamically immiscible even after addition of compatibilizer [9]. Fillers are widely used to reinforce the natural rubber matrix and make the NR suitable for real time applications by uplifting its performance. The nanostructural properties of fillers have a great influence over the linear viscoelastic properties. Recently, Zachariah et al. reported the rheological properties of NR–clay nanocomposite (NRCN) in which the organoclay nanomer 1.44P was treated with surfactant containing hydrogenated tallow, quaternary ammonium. It revealed that the viscosity of the NRCN is directly proportional to the nanofiller component in the NRCN [10].

Due to the development in design and product development, there is a huge demand for elastomer-based novel nanocomposites for flexible electronics and electrical industry. Metallic nanoparticle-dispersed flexible elastomers are of particular interest due to their semiconducting and nanofunctional properties. The insulating properties of natural elastomer were modified by tuning the dielectric properties by impregnating the nickel nanoparticles in NR matrix. This method provides a novel approach in changing the magnetic as well as electrical properties of NR to incorporate the desired physicochemical properties matching for technological applications. There is an appreciable change in elastic modulus of the NRN with the increase in nickel metallic nanoparticle-based filler concentration.

The synthesized NRN showed ferromagnetic behavior which can be attributed to the characteristic contribution from the magnetic property of nickel nanoparticles incorporated in the NR matrix [11]. Current advancements in nanomaterials opened a bright corridor in ultrasensitive membrane-based sensors, actuators, and transducers. The size tunable photoluminescent properties of nanomaterials make them amicable for developing sensors. The surface plasmon resonance properties and band gap confinement of metal nanoparticles such as gold and silver have potential applications in development of metal nanoparticle–NR nanocomposites for ultrathin membranes for nanosensors. Recently, Cabrera et al. reported the fabrication of NR-Au nanoparticle-based membranes for optical sensing. The gold nanoparticles show a tendency to agglomerate together during interaction with organic acids and proteins, which leads to the optical quenching in photoluminescence due to the increase in band gap resulted from the increase in nanocluster size. They had fabricated a NR membrane by casting method, and gold nanoparticle was incorporated through a facile direct reduction of Au(III) in the NR membrane itself. The casted NR membrane was dipped into a solution containing gold precursor, and the solvated gold ions are reduced by the constituent organic acids and proteins present in the NR membrane. The reduced gold nanoparticles formed on the NR membrane surface attributes novel photoluminescence functionalities with in the sensible range suitable for the development of nanosensors [12]. In a similar fashion, silver–NR nanocomposites were prepared by reduction of silver salt followed by UV light exposure. This method provides an economical way of preparing Ag-NR nanocomposites for antibacterial, electrical, thermal, and optical applications [13]. Zakaria et al. observed that the NR matrix reinforced with alumina boron carbide nanofillers shows excellent mechanical properties due to the high surface energy of the filler particles. While keeping the electrical insulation properties in its lower limits, the thermal conductivities of aluminum oxide and boron carbide provide superior heat transfer ability to the NRN and make them highly desirable for high-temperature electrical insulation purposes [14]. High-performance-grade fluoroelastomer rubber (FR) is usually used for applications in automotive, aerospace, and petrochemical industries. The FR usually consists of vinylidene fluoride (VDF), which facilitates the increase in bond dissociation energy of the C–F bond present in the FR. Due to this, the fluoroelastomers have long-lasting specific resistance to oil and chemicals at high temperature (200 °C). Due to this, high thermal stability and resistance against chemical degradation, FR is widely used in O-rings, hoses and seals. FR-blended NRNs along with nanofillers have particular interest due to their good interfacial bonding, elongation at break values, enhanced thermal and mechanical stability at vigorous performing conditions [15]. Latest development in the NRN-based soft electromechanical generators for sustainable energy development provides an economically viable way for power production. This novel approach can help us to partially meet out the huge energy demand in the current scenario through the utilization of sustainable energy resources like ocean energy [16].

Quitmann et al. observed that constrained shape memory natural rubber (SMNR) acts as stress generator upon the exposure of selected solvent vapor and regains their shape after removing the solvent exposure. The stress generated during the exposure of solvent vapor is proportional to the concentration of the solvent vapor and also this process is reversible. These peculiarities made them an interesting candidate for wide range of applications in elastomers for shape memory applications [17]. Development of novel NRN for applications in petrochemical industry has particular interest due to the key role of these materials in hydrocarbon transport, storage, and supply systems. Raghavachari et al. observed that the NRN can be used as good alternative for oil spill recovery. The environmental pollution caused by the oil spill from crude carriers and oil wells in sea is projecting great threat to the marine ecosystem and sustainable habitat of rare marine species. In this regard, they have used nanocomposites of epoxidized natural rubber with magnetite for adsorbing petrol and recovering it [18]. New approaches in dynamic vulcanization techniques resulted in the formation of supertoughened NRN have great industrial importance. Chen et al. reported the development of biobased polyactide/NRN with continuous cross-linked phase through a facile dynamic vulcanization method. Their research group demonstrated the fabrication of ultra-toughened PLA/NRN with continuously cross-linked network of NR. The enhancement in interfacial compatibilization resulted from the PLA grafting onto NR was the major reason for supertoughness of the developed NRN [19].

Great efforts are there to replace the synthetic polymers with their biobased counterparts. Mohanty et al. demonstrated the fabrication of NRN of polyhydroxybutyrate (PHB) blended with epoxidized NR and maleated polybutadiene. This revolutionary approach will be very helpful in the development of biodegradable toughened bacterial polyester of epoxidized NR. The developed NRN can be used as environmental friendly and alternative for non-degradable polyolefins that facilitate the remediation of pollution caused by the deposition of synthetic polymers derived from hydrocarbons [20]. Nowadays, lightweight, flexible, and environment-friendly electromagnetic wave absorbing elastomers are attracting enormous attention due to the spiking demand of such materials in applications in aerospace engineering, telecommunication devices, electrical components, electronic gadgets, and machineries for biomedical and military applications. Kong et al. observed that the nanocomposites of iron oxide with NR can be used as microwave absorbing material [21]. Physicochemical and mechanical properties of the NRN can be modified by using organoclay fillers. Intercalation functionalities of these organoclay fillers can be modified by organic functionalization using alkyl functional groups with varying chain length. It is observed that the cross-linking density of NR was drastically improved by using organoclay fillers containing alkylammonium groups. The dispersion of clay fillers in NR matrix and vulcanization reaction of NR was accountably increased due to the organic functional modification of clay fillers [22]. NRNs of graphene is one of the hottest areas of NRN due to the attractive material properties of graphene. The exfoliated layers of graphite provide great interaction between the filler and NR matrix, which improves the mechanical properties to a great extent. The special electrical properties of graphene also provide electrical and conducting properties to the graphene-based NRNs [23]. Reduced graphene oxide (RGO) particles are also widely used as filler material to enhance the thermal, mechanical, and electrical properties of NR composites. Ruoff et al. observed that cocoagulation of RGO and NR generates web-like surface morphology due to the interaction of RGO flakes with latex particles. As a result of incorporation of RGO in NR matrix, the conductivity and stiffness of the RGO NRN were greatly increased. The conducting elastomers have potential upthrust in flexible electronics industry upsurges the utilization of graphene like nanomaterials in the development of NRN [24]. In the near future, the rapid advancements in nanotechnology will lay down the stepping stones for the development of smart and intelligent NRNs for functional applications in unconventional areas in science and technology.

2.3 Preparation of Natural Rubber Nanoblends

2.3.1 Skim Latex Mixing

In the skim latex mixing process, the natural latex rubber is centrifuged to obtain the protein rich by product called as skim natural rubber latex. This skim natural latex rubber is treated with surfactant and alkali to remove the aqueous phase by forming the creamy skim latex. The nanofillers are added into this creamy skim latex and mixed well followed by the addition of formic acid to initiate the coagulation process and obtain a consolidated mass, followed by drying and vulcanization process to obtain the NRN. Rosamma Alex et al. reported the preparation of NRNs using organoclay nanofillers using skim latex mixing. They reported that in a typical procedure the skim latex, surfactant, and alkali are mixed together at a fixed concentration and definite time to initiate the protein hydrolysis, which produces the polypeptides and amino acids. Also during the skim latex mixing process, the surfactant is adsorbed over the rubber particles formed which lead to the creaming of natural rubber latex. The aqueous serum layer formed was removed by filtration. The nano-organoclay-based fillers are dispersed in water and mixed with the creamed skim latex followed by mechanical stirring. The serum formed during this process was removed, and reaction temperature was maintained at 35 °C constantly throughout the process. After thorough mixing of nanostructured organoclay fillers, required amount of formic acid is added into the reaction mixture to coagulate the skim latex and obtain the organoclay-filled coagulum. Further, the obtained coagulate was washed with distilled water to remove the remaining acid content completely. The coagulum was then processed by drying in an air oven at 70 °C for 24 h and the dried rubber had a pale yellow color. This dried rubber was mixed with the required compounding ingredients including zinc oxide, stearic acid, polymerized 2,2,4-trimethyl-1,2-dihydroquinoline, *N*-cyclohexy-2-benzothiazole sulfenamide, and sulfur. After proper mixing, the obtained compound was vulcanized at 150 °C to their optimum cure time using a hydraulic press having electrical heating setup to heat the operating metal plates for providing the fixed heat of 150 °C to vulcanize the compounded rubber nanocomposite [25].

2.3.2 Latex/Melt Intercalation Method

The latex/melt intercalation method is used to prepare the multicomponent rubber nanocomposite blends. In this process, first, the required nanofillers are dispersed in distilled water to form an aqueous dispersion of the nanofillers. Further, the natural rubber latex was mixed with this nanofiller aqueous solution and stirred well. Then, the required amount of acid is added into the above obtained mixture to coagulate the consolidated solid mass of rubber particles mixed with nanofillers. The obtained compound was washed with distilled water to remove the acid content thoroughly followed by heating at 80 °C for 24 h under reduced pressure to obtain the NRN. The polymer which is to be blended with the natural rubber was mixed thoroughly with pure natural rubber and this compound was mixed with the already prepared natural rubber mixed with nanofillers. Thus, the two compounds are mixed together and compounded with other required compounding ingredients at 60–70 °C followed by the addition of accelerator, sulfur, and antioxidant. The obtained rubber compound was vulcanized at 150 °C for the optimum cure time (t90) determined by an oscillating disk rheometer. The vulcanization and compounding processes were carried out using a hydraulic press with electrical heating plates kept at a constant temperature of 150 °C.

Recently, Mahdi Abdollahi et al. reported the preparation of natural rubber/butadiene rubber (NR/BR) blend/sodium-montmorillonite nanocomposites by combined latex/melt intercalation method. In a typical process, they have prepared the clay nanocomposites of NR/BR blend. 75 wt% of natural rubber was mixed with the 2 wt% aqueous solution of montmorillonite clay by using latex mixing method. The obtained natural rubber latex/nanofiller solution was coagulated by adding required amount of sulfuric acid and washed several time to remove the remaining acid content. The coagulum was further heated at 80 °C for 24 h under reduced pressure. Simultaneously, the BR was mixed with NR in a 6-in. two-roll mill. In a typical preparation method to make NR/BR/clay nanocomposites, firstly, NR and NR/clay mixtures containing various amount of Na MMT were mixed with BR and compounding ingredients according to the designed composition including zinc oxide 5 phr, stearic acid 2 phr, and dibenzothiazole disulfide 0.5 phr and heated at 60–70 °C; then, the diphenyl guanidine 0.5 phr, tetramethylthiuram disulfide 0.2 phr, sulfur 2 phr, and *N*-isopropyl-*N*-phenyl-*p*-phenylenediamine 1 phr were added together with the compound and mixed well followed by vulcanization process carried out at 150 °C for optimum cure time (*t*90), which is determined by an oscillating disk rheometer. The standard mold (2 mm × 110 mm) at 20 ton pressure was used for the compounding process of NR/BR/clay nanocomposites [26].

2.3.3 Freeze-Drying

Freeze-drying is the processing technique used to prepare NRNs at lower temperature. Generally, the natural rubber latex was mixed with the common vulcanization ingredients with continuous stirring by using a homogenizer for 30 min at 70 °C; after this, the obtained mixture was mixed with 1 wt% nanofiller solution and stirred well for 2 h followed by 30 min homogenizing process. The coarse particles were filtered out to obtain the mixture solution of natural rubber latex, vulcanization ingredients, and nanofillers. The viscosity of this mixture solution was constantly monitored using suitable viscometer. Further, the obtained mixture was immediately frozen inside a cylindrical glass shell using liquid nitrogen flow. The frozen shells were loaded into a lyophilizer maintained at −54 °C with vacuum of less than 400 mtorr. The liquid phase of the compound in the ice form was sublimed to dry the compound followed by heating at 100 °C for 30 min to complete the vulcanization process and obtain the NRNs.

In a recent study, Rathanawan Magaraphan et al. demonstrated the synthesis of natural rubber latex/clay aerogel nanocomposites using freeze-drying process. In a typical synthesis process, they have mixed the natural rubber latex with vulcanization ingredients. They have used a formulation consists of 60 % natural rubber, 10 % potassium hydroxide solution 0.5 phr, 50 % zinc oxide dispersion 1 phr, 50 % calcium carbonate dispersion 10 phr, 50 % sulfur dispersion 1 phr, 50 % zinc diethyl dithiocarbamate 0.75 phr, 50 % Wingstay-L dispersion 1 phr, and 1 % clay dispersion varying from 1 to 3 phr. All the ingredients listed above were thoroughly mixed together by continuous stirring by homogenizer at 70 °C for 30 min. The obtained solution was kept in tight plastic bottle. To prepare the nanocomposites with various concentrations of clay nanofillers, different weight percentage solutions of pristine clay fillers were added into the prevulcanized NR latex solution with vigorous stirring for 2 h, followed by stirring for 30 min at room temperature using homogenizer. After mixing, the prevulcanized NR latex solution was filtered to remove the coarse particles to obtain the nanocompounds. The obtained nanocompounds were kept in a cylindrical glass shell and subjected to liquid nitrogen flow for freezing the compound quickly. The frozen compound was attached to a lyophilizer maintained at −54 °C in a vacuum less than 400 mtorr for 36 h to sublime the ice in the compound. After subliming the ice completely out from the compound, the freeze-dried nanocompound was vulcanized at 100 °C for 30 min to obtain the natural rubber/clay nanocomposite [27].

2.3.4 *In Situ Non-aqueous Sol–Gel Method*

Due to the emerging trends in synthesis of rubber nanocomposites, unconventional methods like in situ non-aqueous sol-gel methods are inviting research interest due to its flexibility to make rubber nanocomposites incorporated with alkoxide-based fillers. Generally, in the in situ sol-gel method the natural rubber is made into the sheet form using conventional two-roll milling process, and the required amount of natural rubber was swollen for 24 h in required amount of organic solvent and the alkoxide precursor material was added into this swollen NR immersed in the solvent and mixed well in order to facilitate the homogenous mixing of alkoxide precursor with the swollen rubber. Sequentially, the required amount of alcohol was added to the solution and after 30 min an appropriate amount of dilute acid solution was introduced into the above solution and stirred vigorously for a particular time. The reaction of alcohol with acid produces water, and this process will initiate the hydrolysis and condensation of alkoxide precursor. Thus, the in situ generation of alkoxides was propagated throughout the NR polymer matrix which leads to the in situ formation of nanostructured alkoxide fillers in the NR matrix. The final suspension was cast in a petri dish to remove the volatile content using aspiration hood. The obtained product was heated at 70 °C in a vacuum to prepare the NRNs followed by two-roll milling process to prepare them in sheet from. The in situ generated nanofillers are very much attractive due to their better surface interfacing properties which lead to the overall enhancement in mechanochemical properties of the NRNs.

Laura Wahba et al. recently reported the preparation of in situ sol-gel synthesis of NRNs loaded with silica nanofillers. They have reported the in situ non-aqueous sol-gel procedure in which the natural rubber sheets were swollen in required amount of toluene followed by the addition of TEOS precursor with vigorous stirring. As a result of thorough mixing, the silica precursor was intercalated into the NR polymer matrix, after the mixing process, required amount of ethanol is added into the toluene solution in which the NR was immersed and stirred for 30 min at 80 °C. Further, the required amount of carboxylic acid was added into the reaction mixture to initiate the in situ generation of water molecules, which activates the hydrolysis and condensation of TEOS precursor loaded in the NR matrix leads to the in situ formation of silica nanofillers. Then, the suspension was casted in a petri dish and kept inside an aspiration hood overnight to remove the volatiles. In conclusion, the obtained product was heated at 70 °C for 1 h under vacuum to remove the by-products to obtain the final NR nanocomposite product. The final nanocomposite product was reprocessed into the form of sheets using a conventional two-roll mill [28].

2.4 Characterization of Nanofillers and Natural Rubber Nanoblends

Synthesis, fabrication, and characterization of natural rubber nanoblends are highly interdisciplinary research area. In this section, we would like to discuss the basic and advanced characterization techniques to understand the material properties for the benefit of the readers from different perspectives of science and technology. The physicochemical characterization of nanofillers can be done using various analytical techniques including X-ray diffractometry, UV–Visible spectrometry, photoluminescence spectrometry, vibrating sample magnetometry, Raman spectroscopy, scanning electron microscopy, energy-dispersive X-ray spectroscopy, transmission electron microscopy, selective area electron diffraction, thermo gravimetric and differential thermal analysis (DTA), differential scanning calorimetry, nanoindentation,

X-ray photon emission spectroscopy, electron probe microanalysis, X-ray fluorescence spectrometry, atomic force microscopy, and Fourier-transformed infrared spectroscopy.

2.4.1 X-ray Diffractometry (XRD)

The X-ray diffraction technique is the most widely used characterization technique to investigate the crystal structure of the materials. In an ideal crystalline material, all the constituents are perfectly arranged in lattice and the lattice can extend in three-dimensional space. But in reality, the arrangement of constituents in a crystal are not perfectly ordered due to the lattice imperfections and stress which arises as a result of impurities, intrinsic, extrinsic defects, and lattice mismatch. Thus, the physicochemical properties of the materials are greatly affected by the crystal structure of the materials. The lattice planes in a crystal lattice of solid are of the order of 1 Å, and due to this, the X-ray wavelengths are compatible to interact with these lattice planes in the crystal. In XRD technique, the monochromatic X-ray beams generated from a controlled X-ray source is used to make a constructive interference pattern. According to Bragg's law, $n\lambda = 2d\sin\theta$ where λ is the wavelength of the X-ray radiation, d is the interplanar distance between two successive atomic planes, θ is the Bragg's angle (angle in which X-ray falls on the sample), and n is an integer number denotes the order of diffraction. The Miller indices (h, k, l) of a hexagonal crystal structure can be related to Bragg's law using the following equation

$$1/d = 4/3\left[\left(h^2 + hk + k^2\right)/a^2\right] + \left(l^2/c^2\right)$$

where a and c are the lattice constants.

The X-ray diffraction setup generally consists of four basic components: X-ray tube, Goniometer, sample holder and an X-ray detector. Generally, the Cu metal is used as the anode material for producing X-ray. The filament in the X-ray tube is heated to produce electrons and these electrons are accelerated by high voltage around 30–60 kV and guided to hit on target material fixed as anode. During the collision, the electron in the innermost K shell of the Cu metal will leave their space which will be occupied by the electrons in the outer shells such as L or M. Thus, the energy differences of the electrons are emitted as X-ray radiation generally named as Kα (energy difference of electrons from L shell) and Kβ (energy difference of electrons from M shell). We will deploy a filter to select Kα or Kβ radiation as per our convenience and requirement. Generally, Cu Kα (1.5405 Å) is used as a source due to the experimental convenience. The generated X-ray pattern can be compared with the standard diffraction pattern provided by the Joint Committee for Powder Diffraction Standards (JCPDS) or International Center for Diffraction Data (ICDD). Each crystalline material has its characteristic X-ray diffraction pattern, and by comparing the obtained pattern with the standard pattern, we can identify the crystal

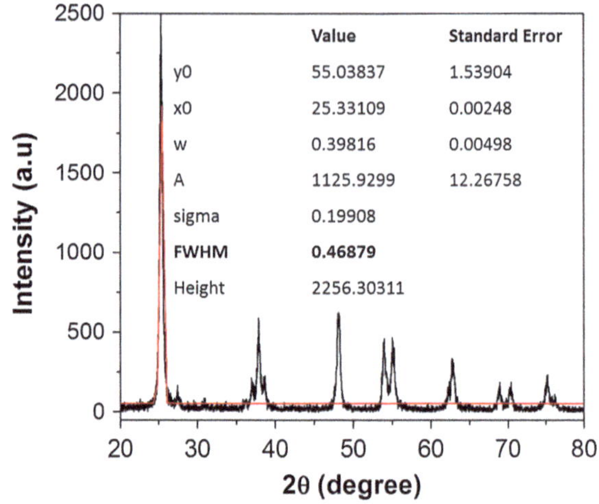

Fig. 2.3 X-ray diffractogram of TiO_2 nanoparticles

structure and phase of the crystalline material. In this way, XRD can be considered as the finger print technique to identify the material, structure, and phase. Considering the nanotechnological aspects, the crystallite size of the nanocrystalline material can be computed from the XRD patterns.

We can use different software packages to analyze the X-ray diffractogram. Generally, Origin software can be used to plot the XRD pattern and by selecting the signal processing option in the tool bar we can fit a Gaussian curve on the obtained XRD pattern which will give the full width half maximum (FWHM) for the dominant peak as shown in Fig. 2.3. Similarly, we can use various fittings for multiple peaks also to obtain their corresponding FWHM values. From the FWHM of the peaks in XRD pattern, the crystallite size of the nanomaterials can be calculated using Scherrer formula $D = K\lambda/\beta \cos \theta B$ where D is the average crystallite size, β is the FWHM in radians of 2θ, λ is the wavelength of the X-ray used, θB is the Bragg angle of the peak, and K is a constant equal to $2(\ln 2/\pi)^{1/2} \approx 9$. The NRNs can be characterized using XRD to understand the crystalline nature of the composite materials. The crystalline structure of nanofillers used to reinforce the natural rubber matrix can be identified from the XRD pattern, and we can get an idea about the influence of material processing and polymerization in growth and agglomeration of the nanofillers.

Generally, inorganic nanofillers used as the reinforcing agent have particular crystal structure and crystallite plane orientation which favors the catalytic nature of the fillers and these properties have direct influence on the physicochemical characteristics of the NRNs. The changes occurred in the crystal structure of the nanofillers are due to the exposure to heat, mechanical stress and strain. The crystal structure and average crystallite size of the nanofillers can be identified using XRD analysis. X-ray diffraction analysis helps us to confirm the incorporation of filler nanoparticles in the natural rubber matrix. In particular, the nanodimensional aspects of the fillers can be speculated from the geometrical behavior of the obtained XRD pattern. The more broad the peaks in the XRD pattern, we can consider that the dimension of the fillers is projecting more toward the nanoregime and vice versa. The common mistakes take place during the XRD analysis of NRNs can be divided into two different categories. One is due to the fast scanning rate of XRD with a fixed sample stage. Since the amount of filler added into the natural rubber matrix is very small compared to the total volume of the composite, we have to provide enough exposure time of X-ray beams for effective diffraction in atomic planes that acquire more data points. In this case, fast scan is not preferred for analyzing NRN samples. To achieve XRD profiles with large data sets, a very small step size in the slow scan mode is advisable for the NRNs. If you take an XRD in fast scan mode, sometimes you will not be able to detect the corresponding diffraction peaks for the nanofillers or nanostructured reinforcing agents.

XRD technique can be used for understanding the structural changes happening to the filler during the interaction with NR, composite additives such as acids and activators. Figure 2.4a depicts the XRD patterns of MMT and Fig. 2.4b–d corresponds to the acid-treated MMT. The interaction of MMT with HCl leads to an increase in the crystallinity of the MMT filler, and this can be identified from the intensity of (001) plane. If the aim of the XRD analysis is to identify the effect of blending and processing techniques in the size and crystalline nature of the nanofillers, a systematic stepwise approach is required. Initially, we have to record the slow scan XRD pattern of natural rubber matrix and fillers separately, after that the XRD of the composite is recorded. We can utilize various types of computational packages such as X'Pert Pro, X'Pert High Score Plus, and Jana to compare graphical parameters obtained through the computational analysis of XRD spectra and derive the results through processing of simple algorithms consists of differential equations and polynomial functions. We can adopt different types of refinement methods to obtain a clear picture regarding the crystalline nature of the NRNs and their constituent components.

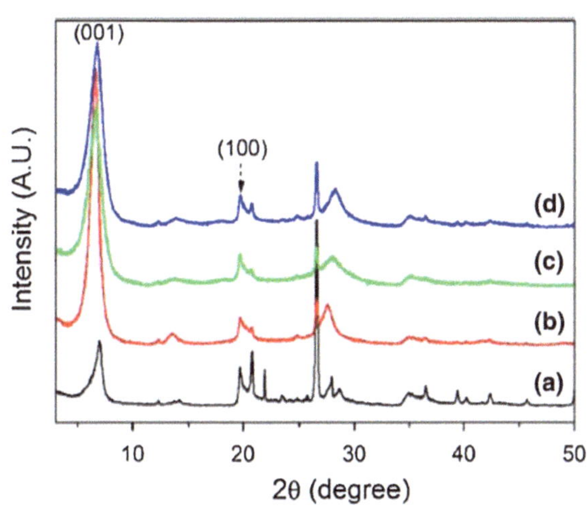

Fig. 2.4 X-ray diffraction patterns of **a** pristine MMT, **b** 0.02 M HCl-treated MMT, **c** 0.1 M HCl-treated MMT, and **d** 0.5 M HCl-treated MMT "Adapted with permission from (doi:10.1021/am504426j | ACS Appl. Mater. Interfaces 2014, 6, 18769–18779). Copyright (2014) American Chemical Society"

2.4.2 X-ray Photoelectron Spectroscopy (XPS)

XPS analysis technique is used for the detection of dopants and impurities incorporated in the NRNs. Due to the advancements in nanotechnology, the development of NRNs for surface sensitive flexible membranes is fascinating enormous research interest among scientific communities across the globe. To analyze the physico-chemical properties of surface interface engineered NRNs, XPS can be used as a strong analytical tool. XPS is a versatile, nondestructive, chemical surface sensitive technique due to the limited ability of the low-energy X-ray radiation used in XPS to probe the uppermost atomic layers of the sample's surface. Generally, the probing depth of X-rays used in XPS is limited to 10 nm and the concentration of the elements of interest should exist in a minimum order of 0.1 atomic % or higher (Fig. 2.5).

As a result of inelastic scattering, the photoelectrons ejected during the XPS scan can travel only a very short distance to escape from the material and arrive at the detector without losing their kinetic energy. Due to this, we can get information regarding the materials present only in the few upper layers of atoms on the surface of the sample. In the XPS measurements, we measure the characteristic binding energies of the ejected core electrons using the Einstein's equation for photoelectric effect $E_k = hv - E_b - W$, where E_k and E_b are the kinetic and binding energy of the ejected electrons, respectively, hv is the energy of the incident X-ray photon and W is the combined work function of the sample and the spectrometer (Fig. 2.6).

The photoejected electrons are collected using electrostatic or electromagnetic lenses and then sent to an energy analyzer where the electrons are separated according to their corresponding kinetic energies. The received electrons are detected using a multichannel detector. The binding energy of the core-level electrons expressed in x-axis belong to the different atomic orbits while in the y-axis

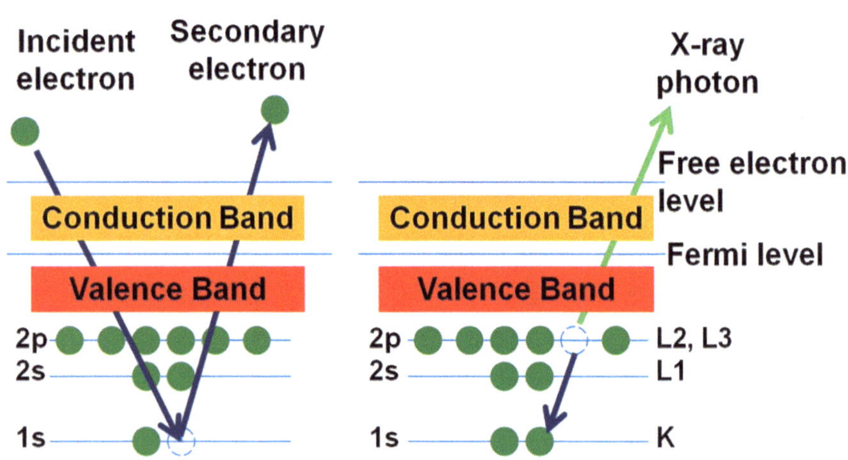

Fig. 2.5 Schematic of X-ray photoelectron emission process

Fig. 2.6 X-ray photoelectron spectrometer

a series of photoelectron peaks where their relative intensities indicate the number of the outgoing electrons at each particular binding energy. The spectroscopic notation for the XPS spectral lines conventionally takes the form nl_j which is different from the (2S+1) L_J atomic notation which describe the electron energy states in an atom. The letters n, l, j are the principal, the orbital angular momentum and the total angular momentum quantum numbers, respectively. Generally, the orbital angular momentum number is designated by the letter in lower case. The XPS is utilized to obtain both qualitative and quantitative information. Generally, a survey scan is recorded over a broad range of energies using wide steps (usually between 50 and 100 eV). Using this fast scan, we can identify the presence of elements and the amount of elements present which are enough for the detection range. The elemental surface concentration can be consistently assessed by integrating peak areas after excluding the background contribution caused by inelastic scattering.

Since the innermost electrons are shielded by the electrons in the outer shell, XPS measurement can be used to explore the oxidation state of the atom and the local chemical environment around the atom. The binding energies of core electrons are greatly affected by the chemical structure giving a small shift in the peak positions. This small shift is called a chemical shift, and it is related to the chemical state of the material. During the fast scan, we will be able to identify the binding energy range for typical chemical state of the particular atom. To identify the chemical shift, we have to perform a high-resolution scan over a selected narrow energy range, around one of the core-level peak, at small step sizes, typically (10–20) eV. The detailed line shape and small shifts in peak position can be easily determined in the high-resolution XPS spectra. The ability to identify different oxidation states and chemical environments is considered as one of the major strength features of the XPS technique. The measured peak positions are compared to those documented standard database in order to calculate the chemical shift. Since the photoejected electrons are scattered by the gas molecules present in the analytical chamber of XPS instrument high vacuum is required to perform XPS experiments. High vacuum will enhance the mean free path of XPS electrons and make it possible to accurately measure the kinetic energy of the electrons. The charge cloud formation on the surface is a general problem in XPS measurements which causes the electrostatic "charging effect." Because of the charging phenomena, XPS peaks would appear at higher binding energy than they should be. In order to overcome this static charging issue, charge neutralizer system was used to produce low-energy electrons to correct for the built-up charges. XPS technique can be used to analyze thin film, metal/natural rubber membranes, NRNs, etc. The NRN sample should be sectioned using a sample sizer before the XPS analysis. From the XPS spectra, we can identify the oxidation state of inorganic and organic materials and fillers present in the NRNs, and it will help us to understand the chemical bonding between the constituents in NRNs. We can use XPS technique to understand the chemical texture of noble metals implanted on the surface of the flexible natural rubber membranes of nanosensors. The XPS analysis is suggested as a strong tool to understand the chemical interaction between filler/reinforcement agent/curing agent and the natural rubber matrix (Figs. 2.7 and 2.8).

For example, Fig. 2.9a elucidates the wide-scan C1s XPS spectrum of olive residual and Fig. 2.9b shows high-resolution spectrum of the olive residual. From the high-resolution XPS spectra, we can find the different binding states of carbon. We can find out the chemical and oxidation states of elements present in the sample

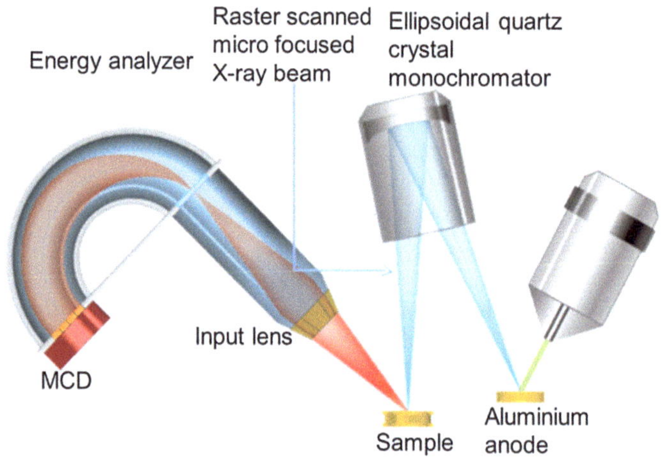

Fig. 2.7 Schematic of energy analyzer system in XPS

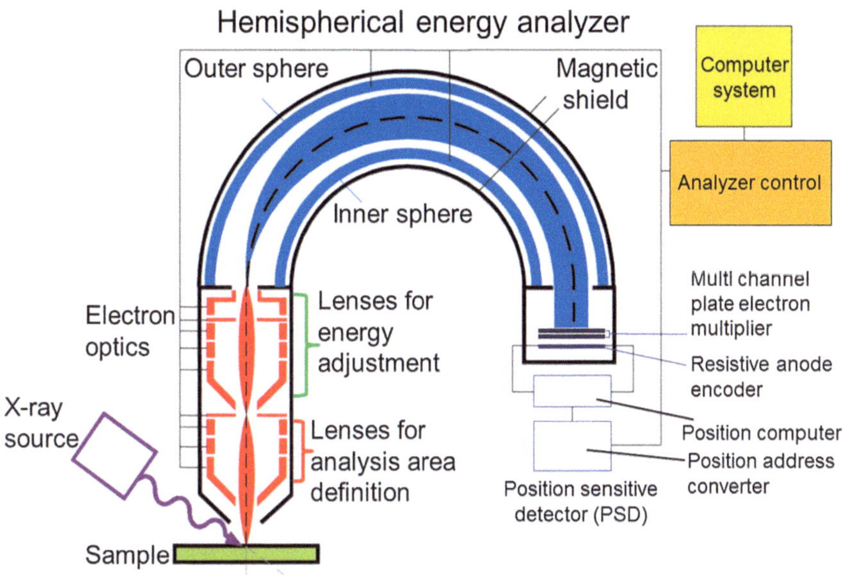

Fig. 2.8 Block diagram of energy analyzer system in XPS

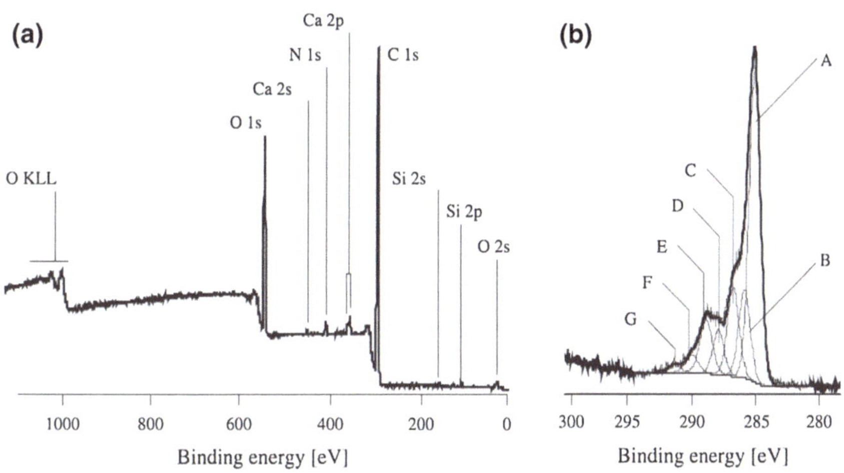

Fig. 2.9 **a** Wide-scan and **b** high-resolution C1s XPS spectra of olive residual sample. "Adapted and licensed under a Creative Commons attribution-type from (Materials Research. 2012; 15(4): 671–678| doi:10.1590/S1516-14392012005000086"

from their characteristic binding energy peaks. In a similar fashion, the physio-chemical characteristics of the nanofillers can be identified from the XPS analysis, and it also helps us to identify the surface polarity of the prepared nanocomposites.

2.4.3 X-Ray Fluorescence Spectroscopy (XRF)

X-ray fluorescence spectroscopy (XRF) is a nondestructive method used for elemental analysis of materials. An X-ray source is used to irradiate the specimen and to cause the elements in the specimen to emit (or fluoresce) their characteristic X-rays. A detector system is used to measure the positions of the fluorescent X-ray peaks for qualitative identification of the elements present and to measure the intensities of the peaks for quantitative determination of the composition. The technique is nondestructive, rapid, precise, and potentially very accurate. X-rays are electromagnetic radiation of very high energy (or short wavelength). When an X-ray photon strikes an atom and knocks out an innershell electron, if the incident photon has energy greater than the binding energy of the innershell electron, a readjustment occurs in the atom by filling the innershell vacancy with one of the outer electrons and simultaneously emitting an X-ray photon. The emitted photon (or fluorescent radiation) has the characteristic energy of the difference between the binding energies of the inner and the outer shells. The XRF wavelengths and relative intensities of a given element are constant to first approximation. Small changes may occur when the distribution of the outer (or valence) electron changes. A major area of research in XRF involves the use of "soft" X-ray emission (or long-wavelength XRF) spectra for chemical state analysis. Soft X-ray peaks often exhibit fine structure, which is a direct indication of the electronic structure (or chemical bonding) around the emitting atom. Thus, the shift in peak position, change in intensity distribution, or appearance of additional peaks can be correlated with a variety of chemical factors, including the oxidation state, coordination number, and nature of covalently bound ligands. The NRNs can be characterized using XRF technique to obtain the elemental and composition of the constituents in the rubber nanocomposite. The exact amount of additives and nanostructured fillers present in the composites can be identified quantitatively using this technique (Fig. 2.10).

2.4.4 UV–Visible Spectroscopy

The spectroscopic technique in which the ultraviolet and visible range of signals in the electromagnetic spectrum is utilized to study the samples is known as UV–Vis spectroscopy. We need a light source which gives the entire visible spectrum plus the near ultraviolet so that you are covering the range from about 200 nm to about 800 nm. (This extends slightly into the near infrared as well.) We cannot get this range of wavelengths from a single lamp, and so a combination of two is used—a deuterium lamp for the UV part of the spectrum, and a tungsten/halogen lamp for the visible part. When radiation interacts with matter, a number of processes can occur, including reflection, scattering, absorbance, fluorescence/phosphorescence (absorption and re-emission), and photochemical reaction (absorbance and bond

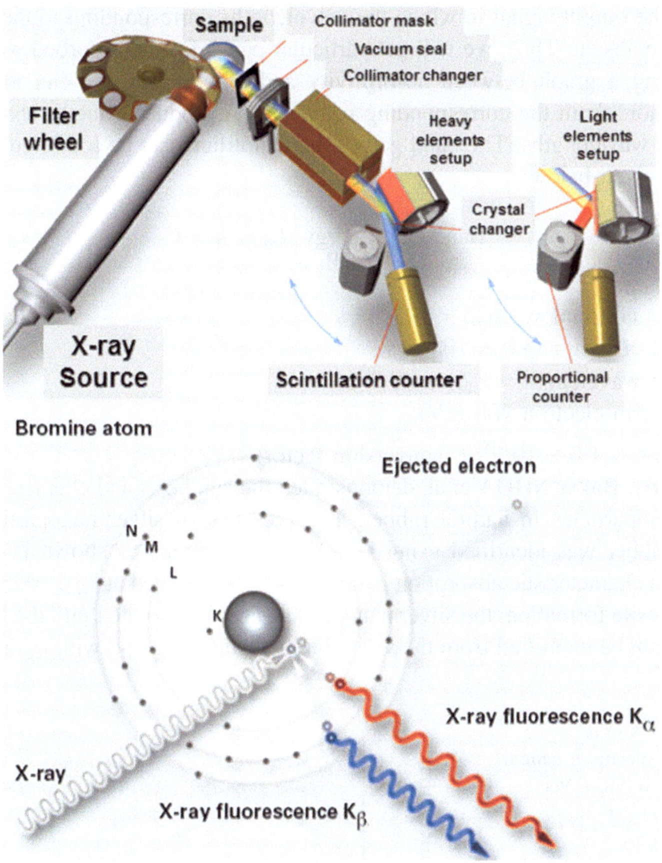

Fig. 2.10 Schematic of X-ray fluorescence spectrometer

breaking). In general, when measuring UV–Vis spectra, we want only absorbance to occur. Because light is a form of energy, absorption of light by matter causes the energy content of the molecules (or atoms) to increase. The total potential energy of a molecule generally is represented as the sum of its electronic, vibrational, and rotational energies: $E_{total} = E_{electronic} + E_{vibrational} + E_{rotational}$. The amount of energy a molecule possesses in each form is not a continuum but a series of discrete levels or states. The differences in energy among the different states are in the order; in some molecules and atoms, photons of UV and visible light have enough energy to cause transitions between the different electronic energy levels. The wavelength of light absorbed is that having the energy required to move an electron from a lower energy level to a higher energy level.

When the input signal passing through the sample, the sample absorbs the energy from input UV–Vis signal and the electrons in the lower molecular orbitals jump into the excited higher molecular orbitals using the absorbed energy. So we will get the output signal in which the wavelengths corresponding to the absorbed energy are absent. Thus, we will get particular peaks for the absorbed wavelength. By plotting a graph between absorptivity and wavelength, we can identify the chromophore from the corresponding peaks (λ_{max}) produced due to absorption of particular wavelengths. The band gap of the nanofillers can be found out using the equation given below:

$$\text{Band Gap Energy}\,(E) = h \times C$$

h Planks constant = 6.626×10^{-34} J s
C Speed of light = 3.0×108 m/s
λ Cutoff wavelength = $\lambda_{max} \times 10^{-9}$ m
E Energy in electron volt (eV)

where $1\,\text{eV} = 1.6 \times 10^{-19}$ J (conversion factor)

Recently, Baker NHHA et al. demonstrated the synthesis and characterization of silver nanoparticles in natural rubber. The presence of silver nanoparticle in the natural rubber was identified using UV–Vis spectroscopy as shown in Fig. 2.11. The broad characteristic absorption band of NR was obtained at λ_{max} = 239. During the composite formation, the silver nanoparticle are incorporated into the NR matrix and this can be identified from the variation in absorption band with in the range of

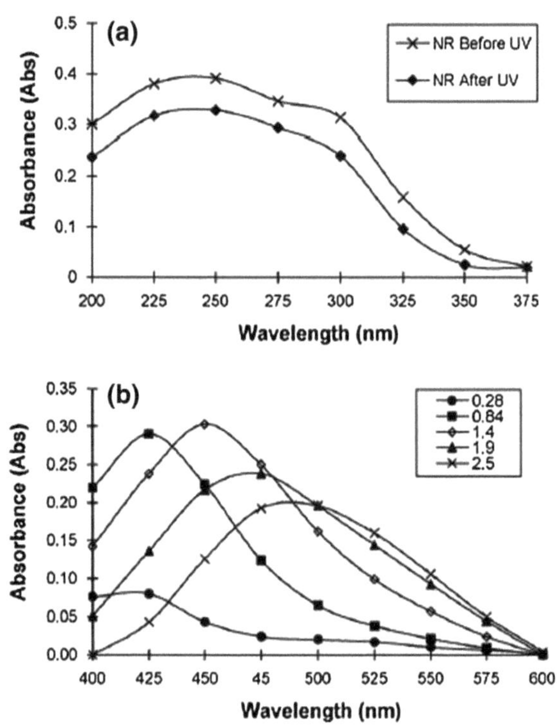

Fig. 2.11 UV–Vis absorption spectra: **a** natural rubber and **b** NR–silver composite at various concentrations, $\times10^{-3}$ M. "Adapted from N.H.H. Abu Bakar et al. Materials Chemistry and Physics 104 (2007) 276–283"

425–484 nm with varying silver nanofillers concentration. UV–Vis-based spectroscopic techniques can be easily adopted for identifying the presence of optically active nanofillers in the prepared NRNs.

2.4.5 Photoluminescence Spectroscopy

Photoluminescence spectroscopy is considered as one of the most important method to identify the optical luminescence of the nanomaterials (Fig. 2.12).

Photoluminescence (PL) is the spontaneous emission of light from a material under optical excitation. PL measurement is a kind of powerful and nondestructive technique to identify the semiconducting and nanometal fillers in NRNs. Various ranges of wavelength are used to carry out the PL measurements. Generally, during PL measurements the NRNs was excited with a selective wavelength around the band gap (E_g) of the composite, which were already determined by using UV–Vis spectroscopy methods.

When the light of sufficient energy is illuminated a composite consisting of semiconducting or metal nanofillers, photons are absorbed and excitations are created. These excited carriers will undergo relaxation by emitting a photon. These emission spectra can be analyzed to identify the optical and surface interfacing properties of the nanofillers. In PL measurements of NRN, only the photons with equal to or higher energy than the band gap of the nanocomposites/nanofiller were absorbed. Therefore, we have to choose different excitation source to do the measurements according to different filler/nanocomposites material with different electronic band structure. The PL peak positions reveal the energy transitions, and the PL intensity indicates the relative rates of radiative and non-radiative recombinations occurring during the excitation process. We can change other external parameters during the PL measurement, such as temperature, excitation power, and applied external perturbation such as magnetic field, electrical field, and pressure.

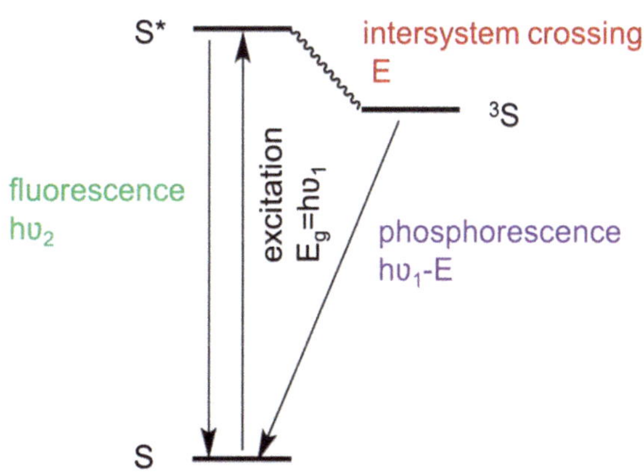

Fig. 2.12 The fluorescence and phosphorescence energy transfer

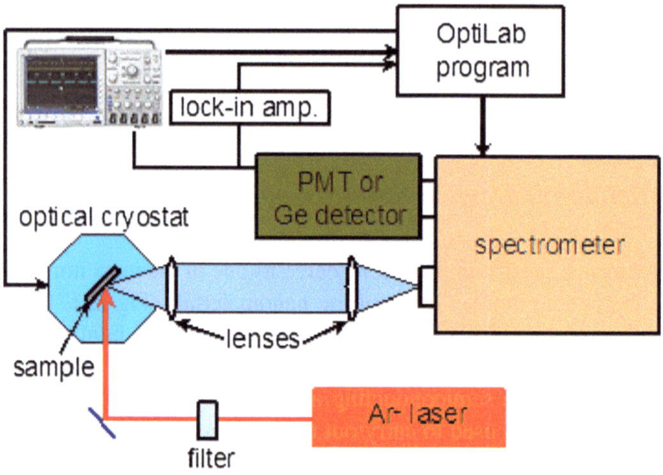

Fig. 2.13 Block diagram of a photoluminescence spectrophotometer

This will help us for understanding the electronic states and bands. Various information including band gap of nanofillers, impurity levels and defects, material quality, recombination mechanism, and surface plasmon resonance properties of nanofillers can be obtained from the PL spectra (Fig. 2.13).

Recently, Cabrera FC et al. demonstrated the Au/NR nanocomposite synthesis in which the authors were successfully synthesized the NR/Au nanocomposite. The research group collected the emission spectra of natural rubber excited at 380 nm and obtained various emission maxima as shown in Fig. 2.14. The bands are at 371 and 382 nm. The shift in this PL bands can provide various information including recombination kinetics and dynamics of natural rubber, nanofillers, and NRNs. The same method can be used for identifying and understanding the change in fluorescence of the NRN due to the incorporation of nanofillers such as metal nanoparticles and metal oxides.

2.4.6 Raman Spectroscopy

Raman spectroscopy is widely used to analyze structural and chemical properties of the nanofillers. In Raman spectroscopy, the sample is illuminated by monochromatic light like laser light. During the illumination process, a very small proportion of the incident radiation interacts with the vibrational energy levels in the sample and exchanges energy as a result of this interactions. The scattered light that has undergone an energy change is called as Raman scattered light. Raman scattered light has energies lower or higher than those of the incident light, and this energy change is equal to the vibrational energies of the sample. This energy results a rise to the peaks in a Raman spectrum and the amount of the peak shift which is referred

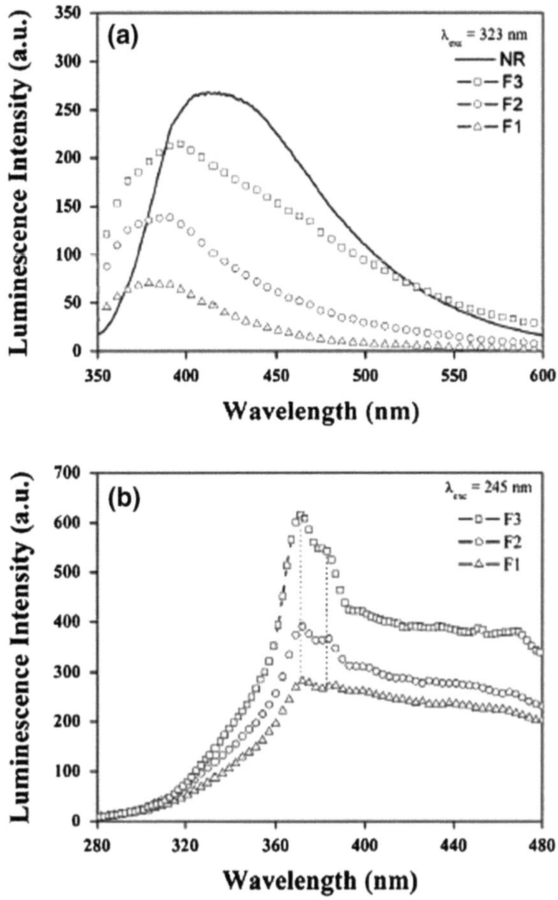

Fig. 2.14 Emission spectrum evaluated using excitation light at wavelengths **a** 323 nm and **b** 245 nm. Measured for both NR membranes and different latex phases thin films NR, natural rubber. "Adapted from Cabrera FC et al. Luminescence 2014; 29: 1047–1052"

as shift in energy from the incident light is called as the Raman shift. In a detailed aspect, we can consider Raman spectroscopy as one of the advanced characterization technique which measures the inelastic scattering of the light over the sample. Raman spectroscopy is used to measure the vibrational modes present in a molecule. A key difference between the Raman and infrared processes is that, in the former process, the photons involved are not absorbed or emitted but rather shifted in frequency by an amount corresponding to the energy of the particular vibrational transition. In the Stokes process, which is the parallel of absorption, the scattered photons are shifted to lower frequencies as the molecules abstract energy from the exciting photons; in the anti-Stokes process, which is parallel to emission, the scattered photons are shifted to higher frequencies as they pick up the energy released by the molecules in the course of transitions to the ground state. In addition, a substantial number of the scattered photons are not shifted in frequency. The process which gives rise to these photons is known as Rayleigh scattering.

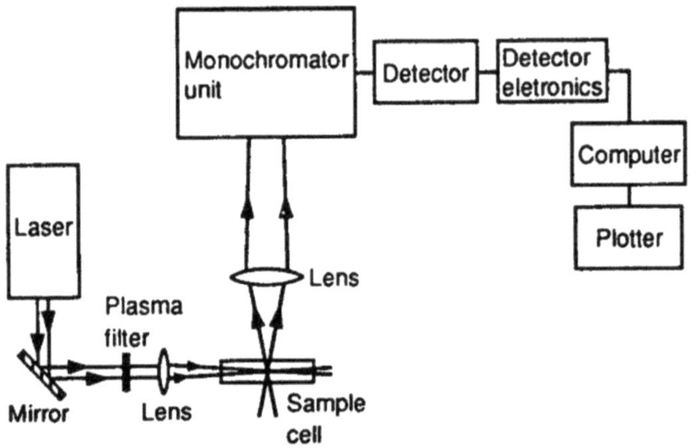

Fig. 2.15 Schematic diagram of a Raman spectrometer

To be active in the infrared spectra, transitions must have a change in the molecular dipole associated with them. For Raman activity, in contrast, the change has to be in the polarizability of the molecule. A typical Raman spectrometer consists of the following basic components: (1) an excitation source, usually a laser; (2) optics for sample illumination; (3) a double or triple monochromator; and (4) a signal processing system consisting of a detector, an amplifier, and an output device. A diagram showing various components of the Raman spectrometer is shown in Fig. 2.15. A number of stages are involved in the acquisition of Raman spectrum. A sample is mounted in the sample chamber and laser light is focused on it with the help of a lens.

Generally, liquids and solids are sampled in a Pyrex capillary tube. The scattered light is collected using another lens and is focused at the entrance slit of the monochromator. Monochromator slit widths are set for desired spectral resolution. The monochromator effectively rejects stray light and serves as a dispersing element for incoming radiation. The light leaving the exit slit of the monochromator is collected and focused on the surface of a detector. This optical signal is converted to an electrical signal within the detector and further manipulated using detector electronics. The Raman spectrum will give a clear picture about the chemical composition, crystallite characteristics such as orientation, lattice strain, and crystal orientations. The spectral parameters concomitant with the Raman spectra can be attributed to various physicochemical aspects of the sample in such a way that the characteristic Raman frequency gives an idea about the composition and identity of the material, polarization of Raman peak provides information about the crystal symmetry and orientation, changes in frequency of Raman peak correspond to the stress or strain in the crystal lattice of the analyzed sample, and width and intensity of the Raman peak deliver the information regarding quality and quantity of the constituent materials in the sample (Fig. 2.16).

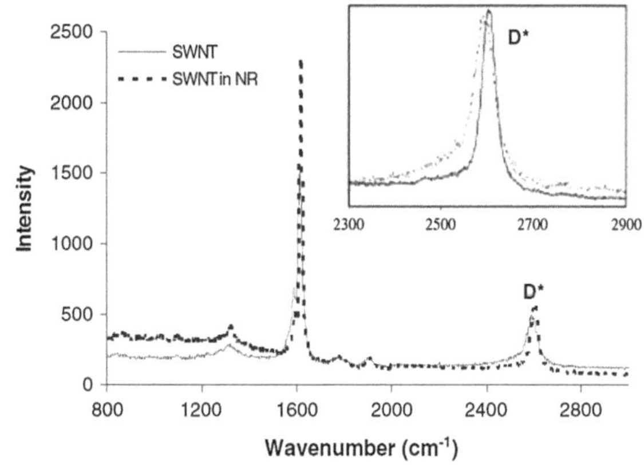

Fig. 2.16 Typical Raman spectra of SWNTs in air (*dotted line*) and SWNT/NR composite (*solid line*). The *insert graph* is the enlarged spectrum around the D* peak. "Adapted from Zhao Q et al. Carbon 44 (2006) 1740–1745"

Recently, Zhao Q et al. used Raman spectroscopic techniques to understand the vulcanization process in single-wall nanotube and natural rubber. Surface interface properties of the nanocomposite determine the degree of polymer molecular slippage over the SWNT. The composite formation of SWNT fillers in NR matrix can be identified from the D* peak shift of the SWNTs to a higher frequency when it is embedded in the NR matrix.

2.4.7 Thermogravimetric Analysis and Differential Thermal Analysis (TGA-DTA)

Thermal analysis is the analysis of a change in a property of a sample, which is related to an imposed change in the temperature. The sample is usually in the solid state, and the changes that occur on heating include melting, phase transition, sublimation, and decomposition can be analyzed using thermal analysis. The analysis of the change in the mass of a sample on heating is known as thermogravimetric analysis (TG). TG measures mass changes in a material as a function of temperature under a controlled atmosphere. Its principal uses include measurement of a material's thermal stability and composition. TG is most useful for dehydration, decomposition, desorption, and oxidation processes. The most widely used thermal method of analysis is DTA. In DTA, the temperature of a sample is compared with that of an inert reference material during a programmed change of temperature. The temperature should be the same until thermal event occurs, such as melting, decomposition, or change in the crystal structure. In an endothermic event takes place within the sample, the temperature of the sample will lag behind that of the reference and a minimum will be observed on the curve. On the contrary, if an exothermal event takes place, then the temperature of the sample will exceed that of the reference and a maximum will be observed on the curve (Fig. 2.17).

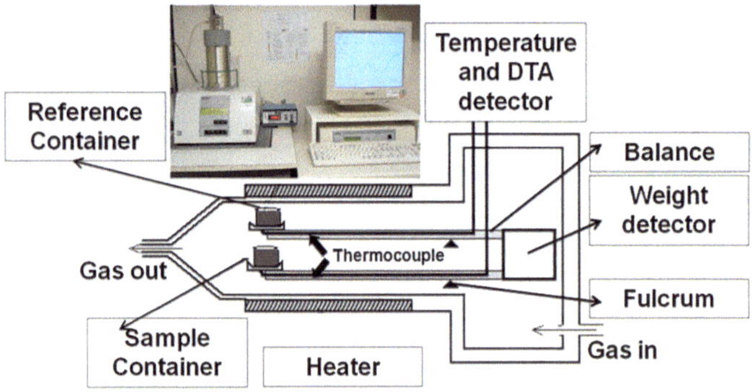

Fig. 2.17 Schematic of TG/DTA

The TG/DTA apparatus consists of 2 cuvettes, one for the sample to be analyzed and the other is the reference. In TGA analysis, the weight loss of the sample is measured as a function of temperature. But, in the case of DTA analysis, the change in temperature of the sample and the reference is measured as a function of the temperature.

The most widely used thermal method of analysis is DTA. In DTA, the temperature of a sample is compared with that of an inert reference material during a programmed change of temperature. The temperature should be the same until thermal event occurs, such as melting, decomposition or change in the crystal structure. In an endothermic event takes place within the sample, the temperature of the sample will lag behind that of the reference and a minimum will be observed on the curve. On the contrary, if an exothermal event takes place, then the temperature of the sample will exceed that of the reference and a maximum will be observed on the curve.

The TG/DTA apparatus consists of 2 cuvettes, one for the sample to be analyzed and the other is the reference. In TGA analysis, the weight loss of the sample is measured as a function of temperature. But, in the case of DTA analysis, the change in temperature of the sample and the reference is measured as a function of the temperature. TG/DTA can be used to find out various material properties such as melting temperatures, transition enthalpies, phase transformations, phase diagrams, crystallization temperatures, degree of crystallinity, glass transition temperatures, decomposition effects, reaction kinetics, and purity determinations.

Carrado et al. determined the percentage of final polymer loading in polymer clay by DTG analysis. Figure 2.18 displays DTG curves for synthetic Li-hectorite and its polymer hectorite derivatives. DTG peak at 67 °C corresponds to the surface water loss, and at 730 °C dehydroxylation of the clay layers occurs. The percentage of weight loss over the temperature range varying from 200 to 600 °C provides the information regarding the polymer loading.

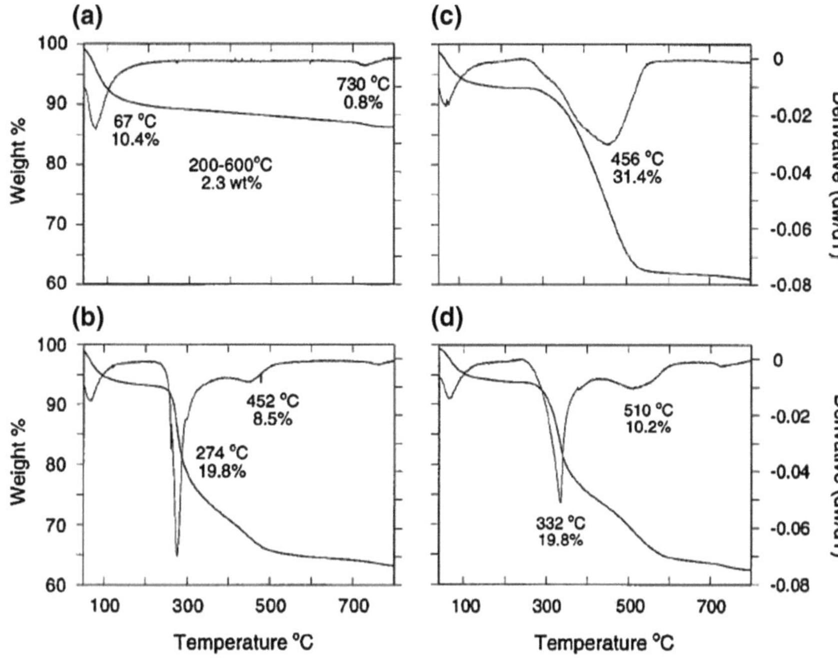

Fig. 2.18 TGA and DTG curves for synthetic hectorite preparations: **a** Li, **b** HPMC-2 (20 %), **c** PANI-2 (20 %), and **d** PVP-3 (20 %). "Adapted from Carrado KA et al. Chem. Mater. 1998, 10, 1440–1445"

2.4.8 Scanning Electron Microscopy-Energy-Dispersive X-Ray Spectroscopy (SEM-EDX)

The scanning electron microscope (SEM) is a type of electron microscope that helps in forming an image of the sample surface by scanning. The electrons in the beam interact with the atoms in the surface to generate signals that provides properties such as composition, topography, and electrical conductivity. The signals produced by an SEM include secondary and back—scattered electrons, characteristic X-rays, specimen current, and light due to cathodoluminescence. All SEMs usually have the ability to detect secondary electrons. SEM helps in obtaining high-resolution images of samples ranging in size from those visible to the naked eye to those which are just a few nanometers in size. In most of the applications, the data collected are over a preselected area of the sample surface, and following this, a 2D image is generated that shows the various spatial variations. Conventional SEMs with a magnification range of 20X–30000X with a spatial resolution of 50–100 nm can scan areas which vary from 1 cm to 5 μm in width. SEMs also have the ability to analyze particular points as can be seen during electron energy X-ray diffraction (EDX) operations which help in determining the chemical composition of the sample concerned. Energy-dispersive X-ray analysis, also known as EDS, EDX or EDAX, is a technique used to identify the elemental composition of a sample. During EDS, a sample is exposed to an electron beam inside a scanning electron microscope (SEM). These electrons collide with the electrons within the sample, causing some of them to be knocked out of their orbits. The vacated positions are filled by higher energy electrons which emit X-rays in the process. By analyzing the emitted X-rays, the elemental composition of the sample can be determined. EDS is a very handy tool for performing the constitutional analysis of any kind of material. Secondary electrons and backscattered electrons are most commonly used for the purpose of imaging samples: secondary electrons depict morphology and topography of the samples while backscattered electrons depict contrasts in composition in multiphase samples. X-rays are produced when incident electrons collide in elastically with the electrons of the atoms in the sample. As the excited electrons get back to their previous positions, they emit X-ray which is the characteristic property of a material. The schematic diagram of a SEM is shown in Fig. 2.19.

Essential components of all SEMs include the following:

- Electron source (“gun”),
- Electron lenses,
- Sample stage,
- Detectors for all signals of interest, and
- Display/data output devices.

Infrastructure requirements:

- Power supply,
- Vacuum system,
- Cooling system,
- Vibration-free floor, and
- Room free of ambient magnetic and electric fields (Fig. 2.20).

Figure 2.21 shows the cryogenic fracture surfaces of the NRN loaded with various concentrations of carbon black fillers.

2.4.9 Electron Probe Microanalyzer (EPMA)

The principle of an electron microprobe is that if a solid material is bombarded by an accelerated and focused electron beam of sufficient energy it can liberate both matter and energy from the sample. These electron–sample interactions mainly

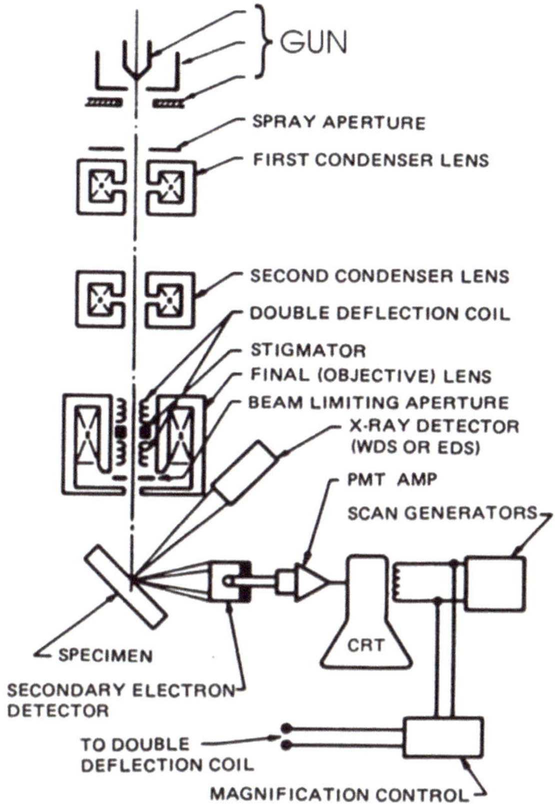

Fig. 2.19 Schematic diagram of a SEM

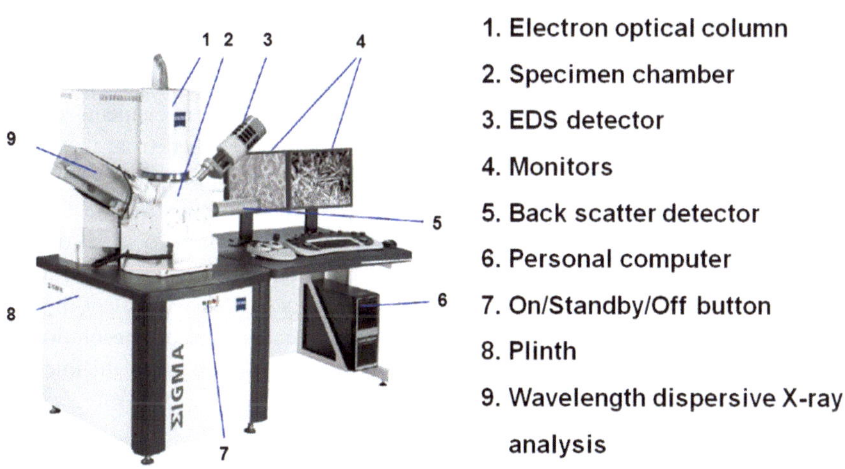

Fig. 2.20 Diagram showing the components of scanning electron microscope

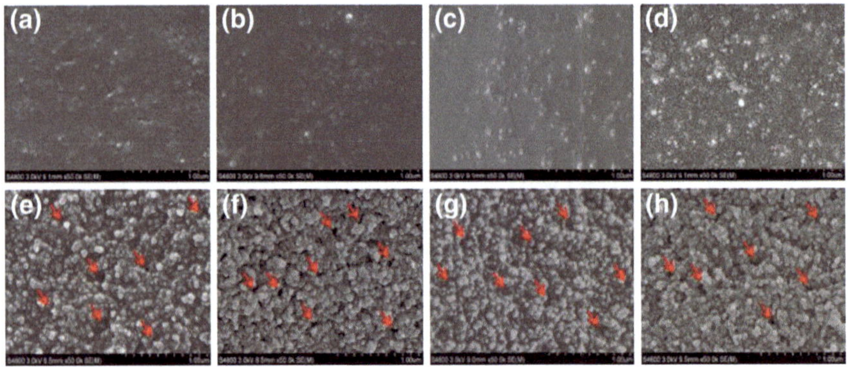

Fig. 2.21 SEM images of cryogenic fracture surfaces of the samples **a** NR40, **b** NR50, **c** NR60, **d** NR70; **e** CBG40, **f** CBG50, **g** CBG60, and **h** CBG70. *Arrows* show the pores in CBGs. "Adapted with permission from Gan S et al. Macromolecules, 2016, 49 (4), 1454–1463." Copyright (2014) American Chemical Society

liberate heat, but electrons and X-rays are also generated. Of most common interest in the analysis of materials are secondary and back-scattered electrons, which are useful for imaging a surface or obtaining an average composition of the material. X-ray generation is produced by inelastic collisions of the incident electrons with electrons in the inner shells of atoms in the sample; when an innershell electron is ejected from its orbit, leaving a vacancy, a higher-shell electron falls into this vacancy releasing some energy in the form X-rays. These X-rays are characteristic of the element and are quantized to determine the elemental composition. EPMA analysis is considered to be "nondestructive"; that is, X-rays generated by electron interactions do not lead to volume loss of the sample, so it is possible to reanalyze the same materials more than one time.

In EPMA, the elemental composition of the constituent nanomaterials is measured by recording wavelength dispersive spectra of the samples. During the EPMA analysis, the microvolume of the samples was exposed to a focused electron beam and collecting the X-ray photons generated by various constituent elements present in the nanocomposites. The sample composition can be identified from the wavelength dispersive spectra (WDS) recorded. EPMA can be used to detect the presence of fillers and constituents in NRNs with ppm sensitivity. Generally, electron beams with energy varying from 5 to 30 keV are usually used to carry out the EPMA measurements. Compared to the SEM-EDX analysis, the EPMA analysis provide much better results due to the superior ability of WDS to detect the elements with high sensitivity. The detector dead time and spectral resolution of EPMA are better than that of energy-dispersive spectrometer. Also the thickness of

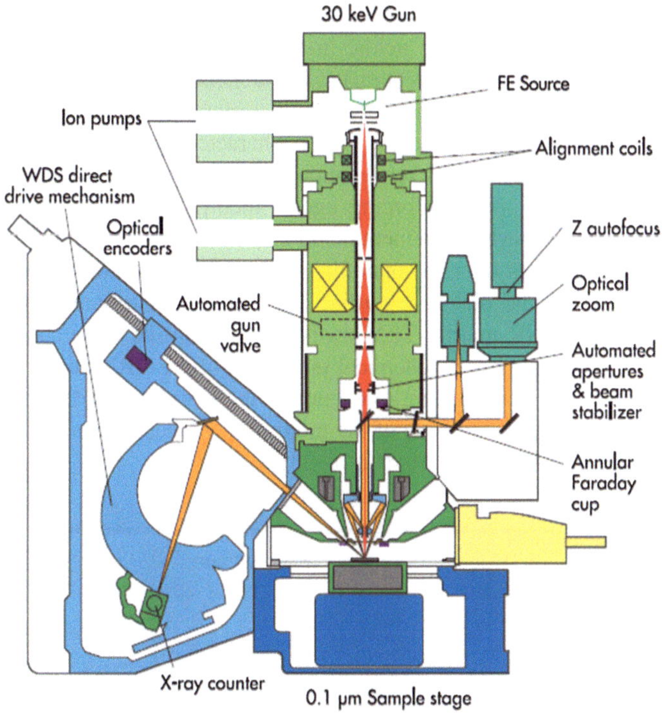

Fig. 2.22 Block diagram of electron probe microanalyzer

the nanofiller layers with size varying from nanometer to millimeter range can be measured using EPMA system with its flexible sample analysis abilities and magnification setup ranging from 40 to 400,000 (Fig. 2.22).

2.4.10 Transmission Electron Microscopy-Selected Area Electron Diffraction (TEM-SAED)

Transmission electron microscopy is widely used to study the surface morphology and topology of the nanomaterials. In TEM analysis, a beam of electrons are allowed to undergo transmission through a thin specimen which can transmit the electrons. During the transmission, the electrons will interact with the specimen and the image formed due to the result of these interactions is magnified and focused on to a fluorescent screen. The advanced image detection technique like CCD cameras is also is used for the digital image formation in TEM analysis. In comparison with the conventional light microscopes, since the electrons used in TEM analysis have small de Broglie wavelength, it can be used for very-high-resolution imaging purpose. Due to his high-resolution imaging ability, TEM analysis can reveal the fine details of even single atom or columns of atoms. TEM forms a major analysis method in a range of scientific fields, in both physical and biological sciences. Recent development in NRNs has attracted researchers to utilize the TEM technique to characterize the nanofillers and polymers incorporated into the nanocomposites during the manufacturing process. TEM characterization of nanofillers and nanocomposites provide crucial information regarding the topology and morphology of the nanofillers and reinforcing agents incorporated in the NR matrix. It includes the nature of surface interaction and influence on compounding on forming extra cross-linking sites over the nanofillers. The high-resolution TEM images of nanofillers forms a clear picture regarding the real time distribution of those filler materials inside the NR matrix. TEM analysis will help the researchers to extract information regarding mechanochemical properties of the NRNs by understanding the morphology and cross-linking dynamics of the nanofillers from the TEM pictures. Grain boundary interactions of the multiphase blends with nanofillers can be elucidated from the TEM analysis (Fig. 2.23).

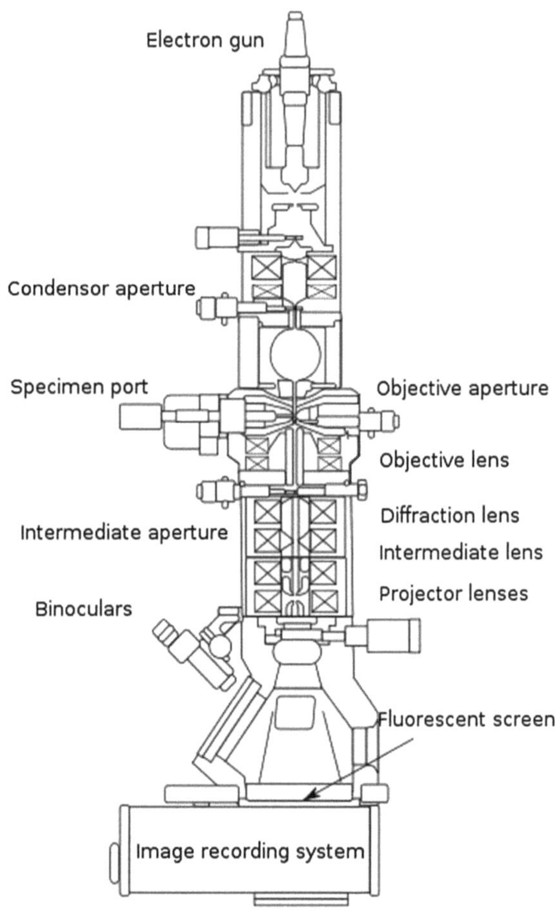

Fig. 2.23 Block diagram of basic TEM

Selected area electron diffraction is a crystallographic characterization method which is utilized hand to hand with TEM analysis. In the SAED analysis, the electrons are targeted on a typical spot in the specimen and allowed to transmit through it, and an aperture in the image plane is used to select the diffracted region of the specimen, giving site-selective diffraction analysis. The atoms in the electron transparent specimen will act as a diffraction grating to the electrons, and due to this, the transmitted electrons are diffracted and scattered to different directions and particular angles. This will form images of series of spots or rings on the screen of the TEM. We can generate the required condition for a specific kind of diffraction by tilting and rotating the samples. The diffraction pattern obtained by this kind can provide the information including crystallinity of the sample, crystal structure of the sample and orientation of the crystal planes of the samples, etc. The crystalline samples can be tilted to a low-index zone axes to identify the crystal structures and measure lattice parameters. SAED of nanocrystals gives ring patterns analogous to those from X-ray powder diffraction and can be used to identify texture and discriminate nanocrystalline from amorphous phases (Fig. 2.24).

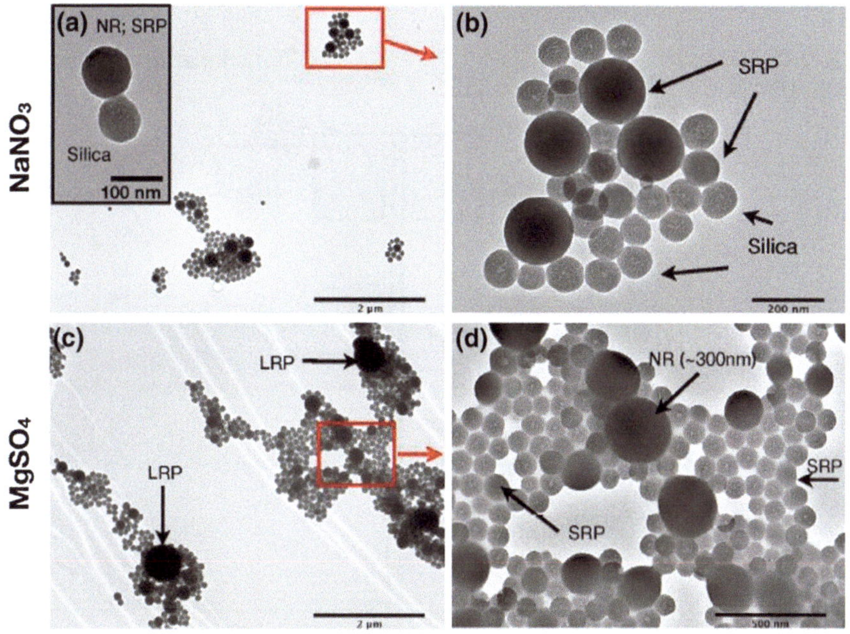

Fig. 2.24 TEM micrographs of NR-100 nm plain silica after 2 h of interaction in (**a**, **b**) $NaNO_3$ and **c**, **d** $MgSO_4$ suspensions at 100 mM ionic strength. **b** and **d** Magnified image of the *red rectangular* zone in (**a**) and (**c**), respectively. *Inset* in (**a**) magnified image of a 1:1 NR-plain silica particle interaction. "Adapted with permission from Chan AJ et al. Langmuir, 2015, 31 (45), 12437–12446." Copyright (2014) American Chemical Society

2.4.11 *Vibrating Sample Magnetometry (VSM)*

Vibrating sample magnetometry (VSM) is based on the Faraday's law which states that an electromagnetic force is generated in a coil when there is a change in flux linking the coil. In the measurement setup, a magnetic sample is moving in the vicinity of two pickup coils as given in Fig. 2.25. The oscillator provides a sinusoidal signal that is deciphered by the transducer assembly into a vertical vibration. The sample which is fixed to the sample rod vibrates with a given frequency and amplitude (60 Hz and ~1 mm, respectively). The sample is centered between the two pole pieces of an electromagnet that generates a magnetic field (H_0) of high homogeneity. Stationary pickup coils are mounted on the poles of the electromagnet. Their symmetry center coincides with the magnetic center of the sample. Hence, the change in magnetic flux originating from the vertical movement of the magnetized sample induces a voltage V_{ind} in the coils. This induced voltage is a measure of the magnetic moment of the sample. The VSM characterization technique is widely used to investigate the magnetic properties of the NRNs containing magnetically active filler nanoparticles such as Fe_2O_3, ZnO, and Gd_2O_3 (Fig. 2.26).

Fig. 2.25 Schematic representation of a VSM

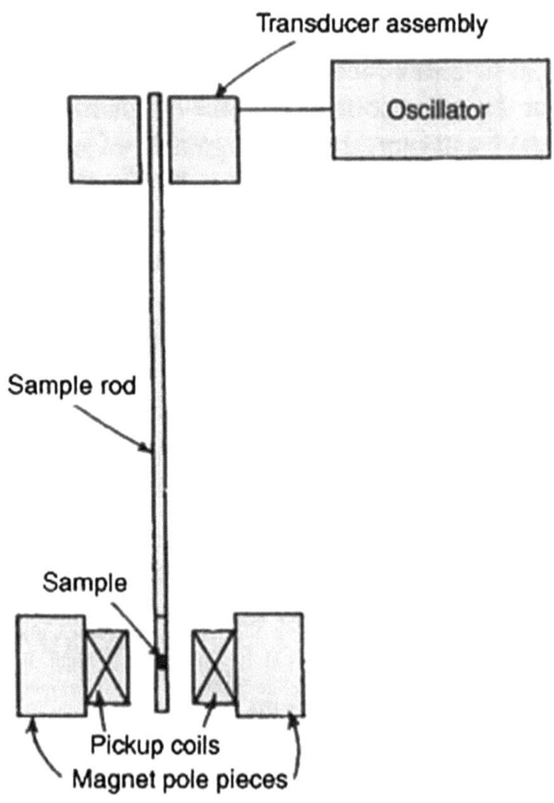

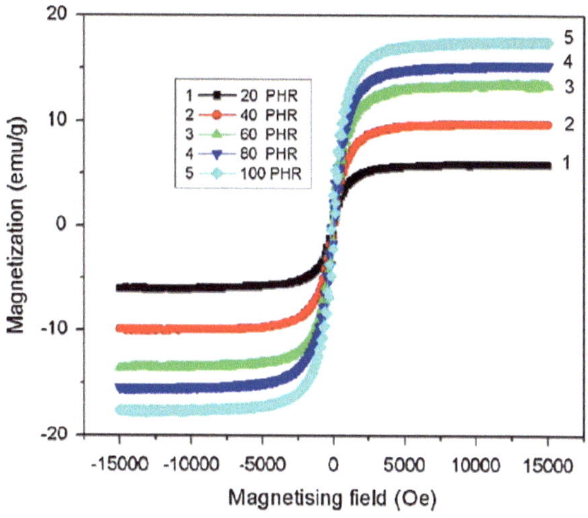

Fig. 2.26 Magnetic hysteresis of composites measured on VSM. The rubber composites show a ferromagnetic behavior and the saturation magnetization is found to increase with the nickel filler content. "Adapted from Jamal EMA et al. Materials Science and Engineering B 156 (2009) 24–31"

2.4.12 Fourier Transform Infrared Spectroscopy (FTIR)

Fourier transform infrared (FTIR) spectroscopy is one of the most common spectroscopic techniques used by organic and inorganic chemists. It is the absorption measurement of different IR frequencies by a sample positioned in the path of an IR beam. The crucial goal of IR spectroscopic analysis is to determine the chemical functional groups in the sample. Different functional groups absorb characteristic frequencies of IR radiation. IR spectroscopy is an important tool for identification and elucidation of NRNs. At temperatures above absolute zero, all the atoms in molecules are in continuous vibration with respect to each other. When the frequency of a specific vibration is equal to the frequency of the IR radiation focused on the molecule, the molecule absorbs the radiation. The resulting spectrum represents the molecular absorption and transmission, creating a molecular fingerprint of the sample. Like a fingerprint, no two unique molecular structures produce the same infrared spectrum. This makes infrared spectroscopy useful for numerous types of analysis (Fig. 2.27).

The FTIR technique is used to identify the functional groups present in the polymer blends and nanofillers in NRNs. The adsorption and surface interaction of natural rubber with various nanopolymers and nanofillers can be easily identified from the FTIR spectra by assigning the vibrational modes to the obtained FTIR spectra and comparing it with the FTIR modes of standard functional groups (Fig. 2.28).

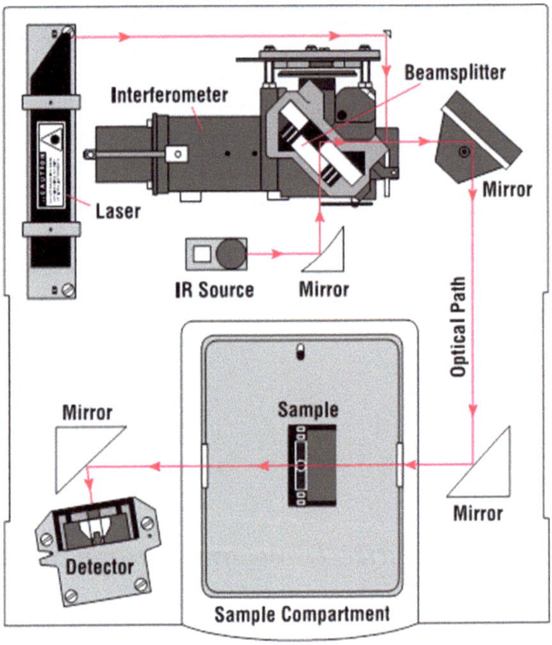

Fig. 2.27 Block diagram of an FTIR spectrometer

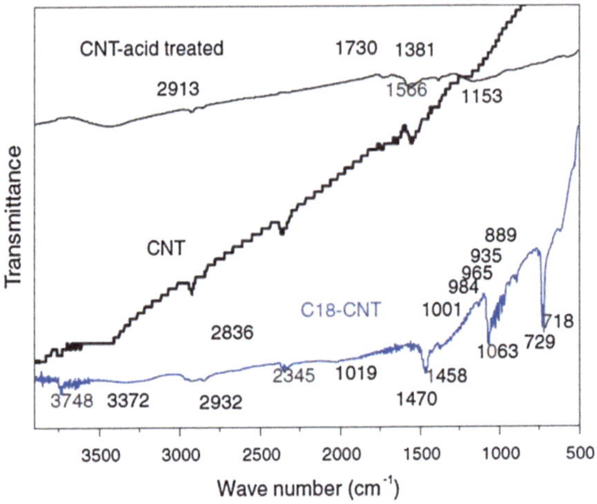

Fig. 2.28 FT-IR spectra of CNT, CNT treated with acid and C18-CNT. "Adapted from Thomas PS et al. J Mater Sci (2012) 47:3344–3349"

2.5 Novel Nanofillers for Developing NRNs and Their Applications

Fillers are generally defined as the class of materials or additives which are added into the NR matrix to reinforce and enhance the physicochemical and mechanical properties of the NR in a desired manner. To achieve the required material properties for industrial and technological applications, we have to add large amount of fillers into the NR matrix, which affects the transparency and brittleness and chemical properties of the natural rubber-based materials. Due to this type of issues associated with conventional fillers in the size range of micrometers, there is a research interest among scientist communities across the globe to develop new materials for filler applications. Recent trends in nanotechnology facilitated the development of nanomaterials for filler applications provided a tremendous upthrust in natural rubber industry. Nanomaterial-based fillers have unique properties including high surface to volume ratio, tunable surface energy and physicochemical properties which provide large surface interface area for the NR matrix to cross-link and interact with the filler. The monodispersed nanofillers can provide strong adhesion between the filler and NR matrix. The nanofiller-NR matrix interface can be engineered in such a way that the nanofiller provide optimum performance in reinforcement and enhancement in physicomechanical properties. Current trend in the nanofillers and overview of nanomaterials for the development of NRN are deliberated in the following sections.

2.5.1 Carbon Black

Carbon black is a material produced by the incomplete combustion of heavy petroleum products. Carbon black is a form of paracrystalline carbon that has a high surface area to volume ratio. The carbon black is mainly used as the reinforcing filler in tires and other rubber products. It is also commonly used in the aerospace industry for the development of elastomers for aircraft vibration control components and engine mounts. Also carbon black improves the mechanical properties such as tensile strength and wear resistance of the NR. The 70 % of the carbon black produced is used in automobile tires. Likewise, carbon black helps to improve the thermal conductivity of the rubber used in tire and drives the heat away from the tread and belt area. This helps to increase the life of tire and reduce thermal damages. The carbon blacks have chemisorbed oxygen complexes including carboxylic, lactonic, phenolic, and quinonic groups on their surfaces which contribute to the volatile nature of the material. Based on various parameters such as particle size, surface area and activity, structure and porosity, the carbon blacks are categorized into various types. These parameters have direct influence on the properties of the rubber compound. Generally, the mechanical properties such as modulus and tensile properties are found to be increasing with decreasing particle size due to this, the NRN shows high viscosity, electrical conductivity, and abrasion resistance. The particle size of the carbon black material is directly proportional to the scorch resistance properties. The Mooney viscosity and modulus of the carbon black–NRN mainly depend on the loading and structure of the carbon black in natural rubber matrix. Carbon black-based fillers reinforce the rubber compound and lowering the compound cost. The physicochemical interactions between the NR matrix and carbon black reinforces improve the NRN. Recent trends in NRNs show inquisitiveness to use carbon black-based hybrid fillers to improve the physicomechanical properties of the NRNs [29, 30].

Types of carbon black used in tires

Name	Abbrev.	ASTM design	Particle size (nm)	Tensile strength (MPa)	Relative laboratory abrasion	Relative road wear abrasion
Superabrasion furnace	SAF	N110	20–25	25.2	1.35	1.25
Intermediate SAF	ISAF	N220	24–33	23.1	1.25	1.15
High abrasion furnace	HAF	N330	28–36	22.4	1.00	1.00
Easy processing channel	EPC	N300	30–35	21.7	0.80	0.90
Fast extruding furnace	FEF	N550	39–55	18.2	0.64	0.72
High modulus furnace	HMF	N683	49–73	16.1	0.56	0.66
Semi-reinforcing furnace	SRF	N770	70–96	14.7	0.48	0.60
Fine thermal	FT	N880	180–200	12.6	0.22	–
Medium thermal	MT	N990	250–350	9.8	0.18	–

2.5.2 *Clay*

Montmorillonite (MMT)-, hectorite-, and rectorite-based natural clay fillers are attracting research interest due to their environmental friendly and non-polluting nature. MMT consists of the layered aluminosilicates stacked over one another. The interlayer distance between the aluminosilicates is in the order of 10–1000 nm [31]. Even though the electrostatic force between silicate layers limits the intercalation of clay-based fillers with the natural rubber matrix, they have been widely used as reinforcement material in the preparation of NRNs. Using various solvents, the interlayer spacing between the clay fillers can be expanded using a stepwise swelling process and improve the dispersion of the fillers. The concept of utilization of layered silicate fillers as reinforcing materials for the production of polymer nanocomposites was invented by Toyota research group. It is also to be considered that the physicochemical properties of clay fillers to exfoliate them and get good dispersion properties. The properties of the clay filler must be compatible with that of polymer and it must favor the steric repulsion of the chains to push the layers apart from each other. The strong interfacial interaction between the polymer and clay leads to the agglomeration due to the intercalation of clay fillers without the exfoliation of silicate layers and restricts the dispersion of fillers in the natural rubber matrix. In order to overcome this issue, the clay fillers are modified by exchanging the interlayer cations with organic cations to increase the organophilic properties and make them compatible with the natural rubber [32]. The exfoliated layers in the clay fillers provide unique characteristics for interfacial intercalation of natural rubber polymer chain which result the enhancement in reinforcement properties.

2.5.3 Carbon Nanotubes

Carbon nanotubes (CNTs) were discovered by Sumio Iijima in 1991. The sp2 hybridized carbon atoms present in the CNT-based fillers provide unique electrical and thermal conductivity to the NRNs. The physicochemical properties of the CNTs are influenced by the chirality and diameter of the carbon nanotubes. The CNTs are classified into two types as single-walled nanotubes and multiwalled nanotubes. The CNTs are generally used as reinforcing agents for high-performance rubber composites due to their high mechanical, thermal, electrical, and physical properties. NRNs of the CNTs can be prepared by solution mixing. Due to the low dispersion of CNTs in the polymer matrix, variety of methods are adopted to make CNT-NRNs. This method includes solution mixing in which the dispersion of CNTs mixed with monomer solution followed by polymerization of the monomers, dispersion of CNTs and dispersion of polymers are added together in a solvent followed by the removal of solvent by evaporation, and dispersed nanotubes are added to dispersed monomer followed by polymerization, etc. Melt mixing is a common and simple method, particularly useful for polymers like rubber. In melt processing, CNTs are mechanically dispersed into the natural rubber matrix using a high-temperature and high-shear force mixer or compounder. The shear forces help to break nanotube aggregates or prevent their formation. The disadvantage of this method is that the dispersion of CNTs in the rubber matrix is quite poor compared to the dispersion that may be achieved through solution mixing.

In addition, the CNTs must be lower due to the high viscosities of the composites at higher loading of CNTs. In the in situ polymerization method, the CNTs are dispersed in monomer followed by polymerization. A higher percentage of CNTs may be easily dispersed in this method and form a strong interaction with the natural rubber matrix. The high viscosity of the natural rubber inhibits the dispersion of CNTs with high aspect ratio into the natural rubber matrix. In order to achieve desired physicomechanical properties, we have to introduce some pretreatment techniques to functionalize the CNT fillers before compositing process. The conductivity of the rubber material can be improved by adding CNT fillers. The presence of CNT fillers in natural rubber matrix can be detected by Raman spectroscopic technique [33, 34]. The surface functionalization techniques are used to improve the dispersion of the CNTs. It can be classified into covalent and non-covalent functionalization. In the non-covalent functionalization, the ionic attachment of functional groups by neutralization of the acid-modified CNTs is performed by using metal oxides. As a result of these ionic attachments, ionic clusters are formed which inhibit the separation of the carbon nanotubes. Due to this reason, normally covalent functionalization is advisable for the preparation of CNT-NRN to avoid the inter particle attraction between the CNTs [35]. For the bulk production of CNT-NRN, different methods including spray drying and two-roll mill assisted solvent mixing are adopted [36].

2.5.4 Iron Oxide

Ferromagnetic α-Fe_2O_3 materials are economically viable, non-toxic, and thermodynamically stable nanomaterials which can be used as filler material for reinforcement and enhancing magnetic properties of NRNs. α-Fe_2O_3 is also called as hematite. Hematite nanomaterials have been studied for the wide range of applications because their magnetic properties greatly depend on their size and shape. The magnetic properties of α-Fe_2O_3 nanoparticles are of different shapes (spherical, rhombohedral. and circular) and can be prepared as powders by a various methods [37]. Ferromagnetic materials have different electrical, magnetic, electro-optical, and chemical properties depending on the different size variations in the nanometer range (1–100 nm). The nanomagnetic materials are characterized by different structural length scale, ranging from few interatomic distances to one micron. The magnetic moment m, in the nanomaterials, is originated from partially filled electrons in the inner shell of transition and rare earth materials. The magnetic moments are from the contribution of both spin S and orbital's L. Magnetic NRNs are having great research interest in the field of development of magnetic data storage media, drug delivery, magnetic sensors, actuators, etc. [38–41]. The magnetic-natural rubber nanoblends are prepared by mixing iron oxide nanoparticles with the natural rubber matrix followed by vulcanization process [42, 43].

The desired ferromagnetic properties can be imparted to the NRN by mixing the iron oxide nanofillers with specific size and magnetic properties. The elastomeric iron oxide-NRN can be used for flexible application platform for high voltage alternating current applications [44–46]. The magnetic and electrical properties of nanostructured iron oxide particles depends on the size of the nanoparticles. So by tuning the size of the iron oxide fillers, particular magnetic properties can be achieved [47–51]. (Ni-Fe_2O_3-NR) Jong-Ryul Jeong et al. (2004) synthesized γ-Fe_2O_3 nanoparticles using chemical coprecipitation technique through a typical pipette drop method (pipette diameter: 2000 μm) and a piezoelectric nozzle method (nozzle size: 50 μm). The size distribution of the γ-Fe_2O_3 nanoparticles prepared by the pipette drop method is from 5 to 8 nm. However, the nanoparticles made by the piezoelectric nozzle method show smaller and very narrow size distribution from 3 to 5 nm. Zero-field-cooled (ZFC)/field-cooled (FC) magnetization and magnetic hysteresis measurements were performed using superconducting quantum interference device (SQUID) magnetometer from 5 to 300 K to investigate the magnetic properties of the nanoparticles. The SQUID measurements revealed that the superparamagnetism of nanoparticles with the blocking temperature of 119.5 and 94.3 K for the nanoparticles made by the pipette drop method and the piezoelectric nozzle method, respectively.

2.5.5 *Zinc Oxide*

Zinc oxide (ZnO) is a white inorganic powder. It is nearly insoluble in water but soluble in acids or alkalies. ZnO is often called as II–VI semiconductor because zinc and oxygen belong to the 2nd and 6th groups of the periodic table, respectively. ZnO has a wide band gap (3.37 eV) and a large exciton binding energy (60 meV) and exhibits many potential applications in areas such as laser diodes, solar cells, gas sensors, optoelectronic devices, and rubber technology. ZnO has three crystal forms: the hexagonal wurtzite, the cubic zinc blende, and the cubic rock salt which is rarely observed. The wurtzite structure is the most commonly used as it has the highest stability under normal working conditions. Hexagonal and zinc blende structures do not display inversion symmetry and its properties are responsible for the piezoelectricity and pyroelectricity of ZnO. This semiconductor material has several favorable features such as: good transparency, high electron mobility, wide band gap, and strong room temperature luminescence. ZnO is widely used in rubber industry as curing agent and activator for sulfur vulcanization [52]. The surface adsorption of accelerators, sulfur, and fatty acids in natural rubber matrix on the ZnO nanoparticles assists in the formation of intermediate complexes which solubilizes the insoluble accelerators to form the catalyst which enhances the vulcanization process and cure characteristics of the NRNs [53, 54]. The nanosized zinc oxide can easily form hydroxyl complexes over the surface act as template for the formation of reactive intermediates which improves the dynamics of vulcanization reaction (Fig. 2.29).

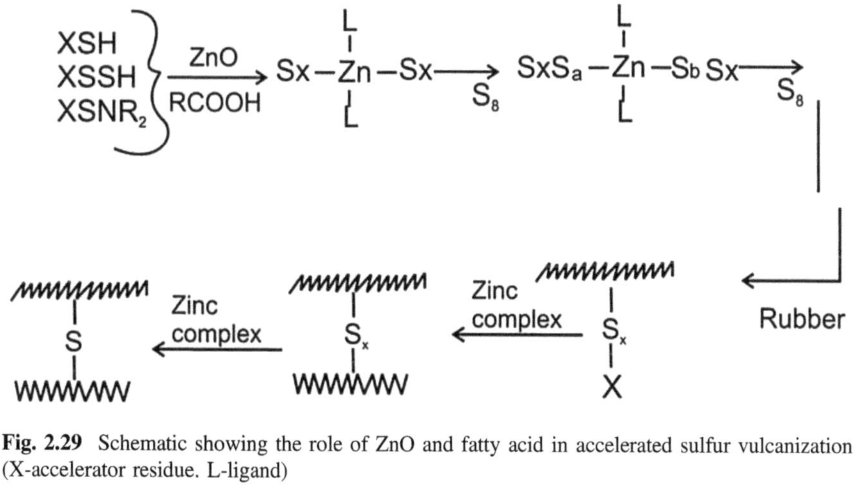

Fig. 2.29 Schematic showing the role of ZnO and fatty acid in accelerated sulfur vulcanization (X-accelerator residue. L-ligand)

The optimal concentration required for the catalytic conversion of accelerator to the zinc–accelerator sulfur complex was identified by Duchacek [52]. It is found that the optimum zinc oxide content to minimize reversion was slightly greater than this minimum. The high surface to volume ratio of nanosized ZnO provides high surface interface area, for chemisorption of accelerator molecules enhances the chemical kinetics of vulcanization reaction by increasing the formation of intermediate efficiency of accelerator–zinc complex [56]. Zinc oxide nanoparticles help to increase the cross-linking ability of monomers to form natural rubber matrix. ZnO nanoparticles act as sulfur transporting agent to include the sulfur atoms into the hydrocarbon chains. The ZnO nanoparticle-based fillers show better physicomechanical property than the conventional bulk ZnO nanofillers. It also helps to improve the reinforcing index, flex resistance, cross-linking density, and tensile properties of the NRNs [57].

2.5.6 *Nickel*

Nickel metal nanoparticles are used as magnetic fillers in natural rubber to incorporate the magnetic properties of nickel nanoparticles to the elastomeric composite of natural rubber. The electrical conductivity of the NRN increases with nickel loading concentration. Girgis et al. observed that the optimal loading concentration of 30 phr of nickel provides better rheological and electrical properties compared to that of iron [58]. M.R. Anantharaman et al. observed that the nickel nanoparticle can change the dielectric and magnetic properties of rubber simultaneously. The natural rubber–nickel nanocomposites can be used as flexible magnetic shielding materials for safety applications. The nanoengineered elastomers of nickel metal nanoparticles can be used for application in fabrication of low-frequency electronic peripherals. The dielectric permittivity of nickel-NRNs can be tuned by controlling the loading of nickel metal nanoparticles in the natural rubber matrix. The elastic modulus and magnetic properties of the nickel-NRN were proportional to the weight percentage loading of nickel nanoparticles in the rubber matrix. The dielectric permittivity of nickel-rubber nanocomposite was inversely proportional to the temperature, and the heterogeneous nature of the composite material attributes an interfacial polarization which results a change in dielectric permittivity with frequency as well as filler concentration. It is observed that the high volume expansion of elastomeric rubber indulges the dielectric permittivity of the nickel-loaded NRNs [59].

2.5.7 *Silica*

Monodisperse silica (SiO_2) is a promising alternative for carbon black, which invites industrial attention due to its light color and glossy nature. It is well acknowledged that the rolling resistance of tire can be improved by adding SiO_2 nanofillers. The hydrophilic and polar nature of silica is not compatible with the apolar polymer matrix like natural rubber. Due to this discordancy between silica filler and natural rubber matrix, the surface functionalization of silica is inevitable for the better intercalation of natural rubber matrix with silica filler [60]. Sol-gel technique is commonly used method to prepare nanostructured SiO_2 particles. In sol-gel method, an alkoxy precursor of silicon is dissolved in an organic solvents like propanol or ethanol followed by the addition of slightly acidic catalyst to initiate the polymerization reaction by the polycondensation of water molecules to form SiO_2 networks. The obtained gel was heated to remove the organic solvent to form silica xerogel. The xerogel was calcined above 800 °C to obtain the SiO_2 nanoparticles. Debapriya et al. observed that during the in situ polymerization of silica in natural rubber matrix, the cross-linking density of rubber composite increases with the loading of SiO_2 [61] (Fig. 2.30).

R. Scotti et al. demonstrated that the alkylthiol and alkyl polysulfide functionalized silica form anisotropic silica fillers which provide better interfacial surface interaction between silica fillers and rubber polymer matrix. The surface functionalization using thiol and polysulfide groups provides enhancement in filler–filler interaction and filler–natural rubber polymer matrix interaction due to their ability in inducing in situ formation of anisotropic SiO_2 [62]. S. Chuayjuljit et al. observed that the differential microemulsion polymerization technique can be adopted for the synthesis of NRN using polystyrene coated silica nanoparticles. The study revealed

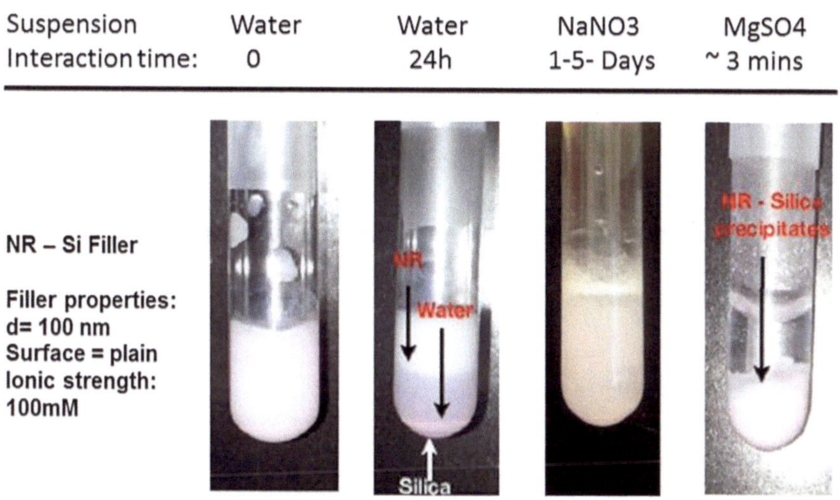

Fig. 2.30 Photographs of the mixed solutions of NR-100 nm plain silica in water, $NaNO_3$, and $MgSO_4$ suspension at 100 mM ionic strength. Aggregation was observed only in the presence of ions. Similar results can be obtained for NR-30-nm plain silica suspensions. NR concentration: ~0.7 wt%; volume percent ratio (dry silica/dry NR): 20 %. "Adapted with permission from Chan AJ et al./Langmuir, 2015, 31 (45), 12437–12446." Copyright (2014) American Chemical Society

that the optimized amount of polystyrene coated silica fillers reinforces the physicomechanical properties of the NRN. The prepared NRN exhibited excellent tensile strength, improved elasticity of rubber molecules near the glass transition temperature values and flammability due to the enriched mechanical and functional properties [63]. Nanosilica fillers also help to achieve better cure characteristics and tear strength through the interfacial interaction between filler and rubber matrix [64].

2.5.8 *Graphene*

Single-layer graphene is a purely two-dimensional material. Its lattice consists of regular hexagons with a carbon atom at each corner. The bond length between the adjacent carbon atoms is 1.42 Å and the lattice constant, a, is 2.46 Å (Fig. 2.31). The first report on this material appeared several decades ago, even before the name graphene had been coined, but it took the pioneering work done by the Manchester research group in 20041 to spark the present explosion of interest in this material [65]. At present, the most popular approaches for the graphene preparation are mechanical exfoliation, growth on metals, and subsequent graphene transfer to insulating substrates, and thermal decomposition of SiC to produce so-called epitaxial graphene on top of SiC wafers. Exfoliation is the most popular for laboratory use, and it is well suited for the fabrication of graphene-NRNs due to their feasibility in solution-based vulcanization processes. The delocalized electrons present in the graphene network give electrical conductivity to this material in addition to the in-plane σ C–C bond which provides extraordinary mechanical stability and strength. Graphene-based nanofillers and reinforcing agents provide large surface

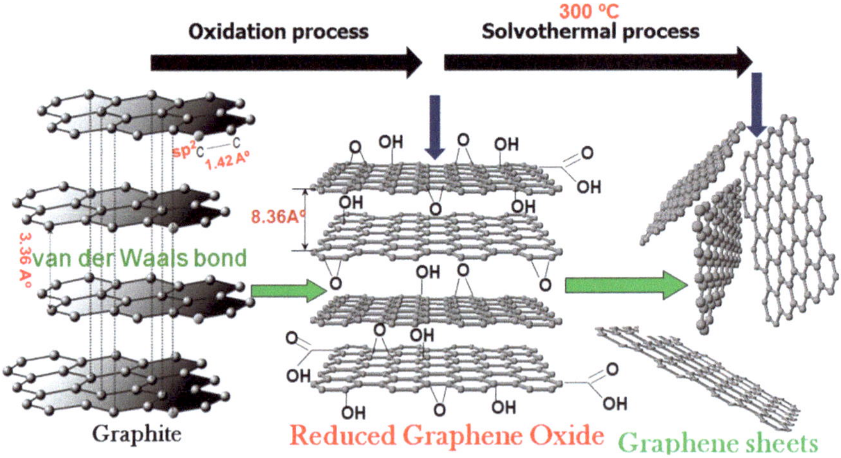

Fig. 2.31 Schematic of solvothermal formation of graphene sheets

interface interaction between graphene and rubber polymer matrix, and it also facilitates the intercalation of natural rubber matrix with the two-dimensional network of carbon atoms. These strong interfacial interactions make them mechanically stable and highly flexible without losing the electrical and electronic conductivity which is preferred for the development of NRNs for the flexible electronics applications.

2.5.9 Gold and Silver Metal Nanoparticles

Gold nanoparticles (GNPs) are generally purple pink color, and the color is tuned by varying the size of the Au nanoparticles. GNPs are generally prepared by reducing gold precursors through solution process. Surface plasmon resonance and size-dependent photoluminescent properties of nanosized gold particles are of particular interest due to their huge potential in membrane-based sensors. The Au nanoparticle-incorporated flexible natural rubber composites can replace conventional sensor modules in analytical instruments. GNPs show an optical quenching upon change in the pH of the elastomer, it is found that the GNPs tend to agglomerate when the pH is changed and this agglomeration bases the reduction in surface plasmon resonance. The photoluminescent properties of natural rubber can be mainly attributed to the organic acid components of latex. The GNPs incorporated into the natural rubber matrix interact with these organic acid complexes which influence their photoluminescent properties. Job et al. demonstrated that the natural rubber can be used as a substrate to reduce the Au atoms present in the gold precursor materials and form GNPs. This in situ reduction process result the formation of metallic nanoclusters on the surface of natural rubber substrate [66].

Silver nanoparticles (SNPs) are considered for wide range of applications such as development of antimicrobial, electrical, thermal, and optical functional materials. SNPs have glossy faint greenish color with optical and electronic properties suitable for industrial applications. In comparison with the broad optical absorption band of natural rubber at 239 nm, SNPs have optical absorption band around 425 and 484 nm. SNPs also exhibit surface plasmon resonance, and the band gap of SNPs can be tuned by varying particle size. SNP is the only material with tunable surface plasmon resonance which can be regulated into any wavelength in the visible spectrum. The SNP-NRNs have huge potential in the development of flexible antimicrobial coatings, elastomeric optical transducers, flexible electronic components, and membrane-based sensors [67] (Fig. 2.32).

SNPs can be synthesized by adopting different methods such as chemical reduction, photoreduction, and microwave-assisted reduction process. The common methods used for the synthesis of silver nanoparticles include Turkevich method, Lee-Meisel method, and Creighton method [55, 68, 69]. In Turkevich method, $AgNO_3$ is used as silver source and reduction of this metal source produces silver nanoparticles in a narrow distribution of particle sizes but in Meisel method, a broad range of silver nanoparticles can be produced through the chemical reduction

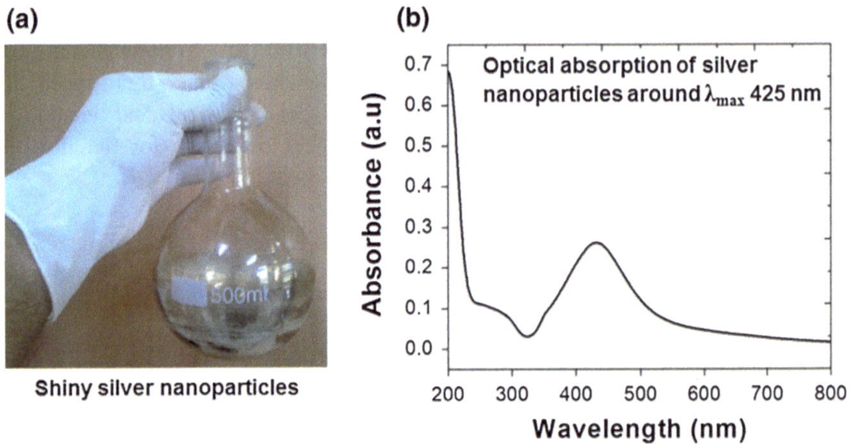

Fig. 2.32 **a** Picture showing the silver nanoparticles coated inside the glass bottle 32, **b** UV–Vis absorption spectrum of silver nanoparticles

method. In Creighton method, $NaBH_4$ is used as the reducing agent to reduce $AgNO_3$ metal source, and the SNPs synthesized by this method have very narrow size distribution with 10 nm average particle size [56].

2.6 Summary and Outlook

In summary, this chapter provides an introduction to NRNs and various strategies for fabricating NRNs. A general out look of literature has been provided followed by the major outcomes of recent research trends. Different types and synthesis methods for preparing NRNs include skim latex mixing, latex/melt intercalation method, freeze-drying and in situ non-aqueous sol-gel method. It is followed by the schematic representations of various physicochemical characterization techniques including XRD, XPS, XRF, UV–Vis spectroscopy, photoluminescence spectroscopy, Raman spectroscopy, TGA-DTA, SEM-EDX, EPMA, TEM-SAED, VSM, and FTIR. Those techniques can be used for analyzing various composite properties such as structural, thermal, optical, morphological, magnetic, and size distribution of the constituent nanofillers. A brief idea of currently available novel nanofillers including iron oxide, carbon black, graphene, silver, and gold nanoparticles which are utilized for the successful development of NRNs was described. In addition, by changing the physicochemical properties of the nanofillers, the composite properties can be tuned with desired nanofunctionalities in order to employ them in various applications such as flexible electronic applications, antimicrobial coating, membrane-based sensors have been elaborately deliberated to get a clear notion.

References

1. Bendahou A, Kaddami H, Dufresne A (2010) Investigation on the effect of cellulosic nanoparticles morphology on the properties of natural rubber based nanocomposites. Eur Polymer J 46:609–620
2. Jinrong W et al (2013) Vulcanization kinetics of graphene/natural rubber nanocomposites. Polymer 54:3314–3323
3. Suil G et al (2008) Preparation and properties of natural rubber composites reinforced with pretreated carbon nanotubes. Polym Adv Technol 19:1543–1549
4. Thomas PS et al (2012) Electrical properties of natural rubber nanocomposites: effect of 1-octadecanol functionalization of carbon nanotubes. J Mater Sci 47:3344–3349
5. Nakason C, Kaesaman A, Supasanthitikul P (2004) The grafting of maleic anhydride onto natural rubber. Polym Testing 23(1):35–41
6. Bokobza L (2012) Multiwall carbon nanotube-filled natural rubber: electrical and mechanical properties, eXPRESS. Polym Lett 6(3):213–223
7. Il-Jin K et al (2010) Effect of nano zinc oxide on the cure characteristics and mechanical properties of the silica-filled natural rubber/butadiene rubber compounds. J Appl Polym Sci 117:1535–1543
8. Abraham E et al (2013) Physicomechanical properties of nanocomposites based on cellulose nanofibre and natural rubber latex. Cellulose 20:417–427
9. Kumari P, Unnikrishnan G (2013) Thermal properties of compatibilized and filled natural rubber/acrylonitrile butadiene rubber blends. J Therm Anal Calorim 114:67–75
10. Ajesh KZ et al (2014) Rheological behaviour of clay incorporated natural rubber and chlorobutyl rubber nanocomposites. RSC Adv 4:58047–58058
11. Jamal EMA et al (2009) Synthesis of nickel–rubber nanocomposites and evaluation of their dielectric properties. Mater Sci Eng B 156:24–31
12. Flavio C et al (2014) Organic acids and protein compounds causing the photoluminescence properties of natural rubber membranes and the quenching phenomena from Au nanoparticle incorporation. Luminescence 29:1047–1052
13. Bakar NHHA, Ismail J, Bakar MA (2007) Synthesis and characterization of silver nanoparticles in natural rubber. Mater Chem Phys 104:276–283
14. Zakaria MZ, Ahmad SH (2013) Investigation on thermal conductivity and mechanical properties of thermoplastic natural rubber filled with alumina and boron carbide nanocomposites. Energy Environ Eng J 4(1):11–14
15. Dong-Ah L et al (2012) Natural rubber/fluoroelastomer blended composites using colloid stabilization-destabilization method. Macromol Res 20(7):673–681
16. Kaltseis R et al (2014) Natural rubber for sustainable high-power electrical energy generation. RSC Adv 4:27905
17. Quitmann D et al (2013) Solvent-sensitive reversible stress-response of shape memory natural rubber. ACS Appl Mater Interfaces 5:3504–3507
18. Venkatanarasimhan S, Raghavachari D (2013) Epoxidized natural rubber–magnetite nanocomposites for oil spill recovery. J Mater Chem A 1:868
19. Chen Y, Yuan D, Xu C (2014) Dynamically vulcanized biobased polylactide/natural rubber blend material with continuous cross-linked rubber phase. ACS Appl Mater Interfaces 6:3811–3816
20. Parulekar Y, Mohanty AK (2006) Biodegradable toughened polymers from renewable resources: blends of polyhydroxybutyrate with epoxidized natural rubber and maleated polybutadiene. Green Chem 8:206–213
21. Kong I et al (2010) Magnetic and microwave absorbing properties of magnetite–thermoplastic natural rubber nanocomposites. J Magn Magn Mater 322:3401–3409
22. Sookyung U et al (2014) Influence of modifying agents of organoclay on properties of nanocomposites based on natural rubber. Polym Testing 33:48–56

23. Hernandez M et al (2012) Overall performance of natural rubber/graphene nanocomposites. Compos Sci Technol 73:40–46
24. Potts JR et al (2012) Processing–morphology–property relationships and composite theory analysis of reduced graphene oxide/natural rubber nanocomposites. Macromolecules 45:6045–6055
25. Alex R, Nah C (2006) Preparation and characterization of organoclay-rubber nanocomposites via a new route with skim natural rubber latex. J Appl Polym Sci 102:3277–3285
26. Abdollahi M et al (2011) Structure and properties of natural rubber/butadiene rubber (NR/BR) blend/sodium montmorillonite nanocomposites prepared via a combined latex/melt intercalation method. Polym Sci Ser A 53(12):1175–1181
27. Pojanavaraphan T, Magaraphan R (2008) Prevulcanized natural rubber latex/clay aerogel nanocomposites. Eur Polym J 44:1968–1977
28. WahbaL et al (2014) A novel non-aqueous sol–gel route for the *in situ* synthesis of high loaded silica–rubber nanocomposites. Soft Matter 10:2234–2244
29. Thaptong P et al (2014) Properties of natural rubber reinforced by carbon black-based hybrid fillers. Polym-Plast Technol Eng 53(8):818–823
30. Bhattacharya M, Bhowmick AK (2010) Synergy in carbon black-filled natural rubber nanocomposites Part I: mechanical, dynamic mechanical properties and morphology. J Mater Sci 45:6126–6138
31. Mondragon M et al (2009) Injection molded thermoplastic starch/natural rubber/clay nanocomposites: morphology and mechanical properties. Carbohydr Polym 77:80–86
32. Rezende CA et al (2010) Natural rubber-clay nanocomposites: mechanical and structural properties. Polymer 51:3644–3652
33. Kueseng K, Jacob KI (2006) Natural rubber nanocomposites with SiC nanoparticles and carbon nanotubes. Eur Polymer J 42:220–227
34. Kueseng P, Sae-oui P, Rattanasom N (2013) Mechanical and electrical properties of natural rubber and nitrile rubber blends filled with multi-wall carbon nanotube: effect of preparation methods. Polym Testing 32:731–738
35. Sui G et al (2008) Preparation and properties of natural rubber composites reinforced with pretreated carbon nanotubes. Polym Adv Technol 19:1543–1549
36. Thomas PS et al (2012) Electrical properties of natural rubber nanocomposites: effect of 1-octadecanol functionalization of carbon nanotubes. J Mater Sci 47:3344–3349
37. Zhang X et al (2013) Synthesis, optical and magnetic properties of α-Fe_2O_3 nanoparticles with various shapes. Mater Lett 99:111–114
38. Qiaoling LI, Zhang C (2010) Preparation of Fe_2O_3 microtubules and the effect of a surfactant on their properties. J Ceram Process Res 11(3):331–334
39. Chirita M, Grozescu I (2009) Fe_2O_3—nanoparticles, physical properties and their photochemical and photoelectrochemical applications. Chem bull "POLITEHNICA" Univ (Timisoara) 54(68): 1–8
40. Jong-Ryul J et al (2004) Magnetic properties of γ-Fe_2O_3 nanoparticles made by coprecipitation method. Phys Status Solidi (b) 241(7):1593–1596
41. Suber L et al (1998) Structural and magnetic properties of α-Fe_2O_3 nanoparticles. Appl Organomet Chem 12:347–351
42. Akbar S et al (2004) Synthesis of Fe_2O_3 nanoparticles by new Sol-Gel method and their structural and magnetic characterizations. Ar Xiv:cond-mat/0408480
43. Teng X, Yang H (2004) Effects of surfactants and synthetic conditions on the sizes and self-assembly of monodisperse iron oxide nanoparticles. J Mater Chem 14:774–779
44. Sahoo SK et al (2010) Characterization of γ- and α-Fe_2O_3 nano powders synthesized by emulsion precipitation-calcination route and rheological behaviour of α-Fe_2O_3. Int J Eng Sci Technol 2(8):118–126
45. Ma J et al (2010) α-Fe_2O_3: hydrothermal synthesis, magnetic and electrochemical properties. J Phys Chem C 114(24):10671–10676
46. Liu L et al (2006) Surfactant-assisted synthesis of α-Fe_2O_3 nanotubes and nanorods with shape-dependent magnetic properties. J Phys Chem B 110(31):15218–15223

47. Wu C et al (2006) Synthesis of hematite (α-Fe2O3) nanorods: diameter-size and shape effects on their applications in magnetism, lithium ion battery, and gas sensors. J Phys Chem B 110 (36):17806–17812
48. Liao L et al (2008) Morphology controllable synthesis of α-Fe_2O_3 1D nanostructures: growth mechanism and nano device based on single nanowire. J Phys Chem C 112(29):10784–10788
49. Wang X et al (2009) Fast preparation, characterization, and property study of α-Fe_2O_3 nanoparticles via a simple solution-combusting method. J Phys Chem C 113(17):7003–7008
50. Cao SW, Zhu YJ (2008) Hierarchically nanostructured α-Fe2O3 hollow spheres: preparation, growth mechanism, photocatalytic property, and application in water treatment. J Phys Chem C 112(16):6253–6257
51. El-Nashar DE, Mansour SH, Girgis E (2006) Nickel and iron nano-particles in natural rubber composites. J Mater Sci 41:5359–5364
52. Duchacek V (1978) Effect of thiourea on thiuram-accelerated sulfur vulcanization and its significance for vulcanization mechanism. J Appl Polym Sci 22(1):227–237
53. Barnard D et al (1970) Rubber natural. Encyclopedia of polymer science and technology: plastics, resins, rubbers, fibers 12:178
54. Sahoo S et al (2008) Synthetic zinc oxide nanoparticles as curing agent for polychloroprene. Polym Polym Compos 16(3):193–198
55. Lee PC, Meisel DJ (1982) Adsorption and surface-enhanced Raman of dyes on silver and gold sols. J PhysChem 86:3391–3395
56. Creighton JA, Blatchford CG, Albrecht MG (1979) Plasma resonance enhancement of Raman scattering by pyridine adsorbed on silver or gold sol particles of size comparable to the excitation wavelength. J Chem Soc Farad Trans II 75:790–798
57. Sabura BPM (2010) Studies on the use of nano zinc oxide and modified silica in NR, CR and SBR. http://hdl.handle.net/10603/1406
58. El-Nashar DE, Mansour SH, Girgis E (2006) Nickel and iron nano-particles in natural rubber composites. J Mater Sci 41:5359–5364
59. Jamal EMA et al (2009) Synthesis of nickel–rubber nanocomposites and evaluation of their dielectric properties. Mater Sci Eng B 156:24–31
60. Sarkawi SS et al (2013) Morphology of silica reinforced natural rubber: the effect of silane coupling agent. Presented at the fall 184th technical meeting of the rubber division of the American Chemical Society, inc. Cleveland, Ohio, 7–10 Oct. ISSN 1547-1977
61. Debapriya D et al (2014) Effect of sol-gel-derived nano-silica on the properties of natural rubber-poly butadiene rubber-reclaim rubber ternary blends/silica nanocomposites. Polym-Plast Technol Eng 53:1131–1141
62. Scotti R et al (2012) Rubber–silica nanocomposites obtained by in situ sol–gel method: particle shape influence on the filler–filler and filler–rubber interactions. Soft Matter 8:2131–2143
63. Chuayjuljit S, Boonmahitthisud A (2010) Natural rubber nanocomposites using polystyrene-encapsulated nanosilica prepared by differential microemulsion polymerization. Appl Surf Sci 256:7211–7216
64. Lay M et al (2013) Effect of nanosilica fillers on the cure characteristics and mechanical properties of natural rubber composites. Adv Mater Res 626:818–822
65. Novoselov KS et al (2004) Electric field effect in atomically thin carbon films. Science 306:666–669
66. Flavio C et al (2014) Organic acids and protein compounds causing the photoluminescence properties of natural rubber membranes and the quenching phenomena from Au nanoparticle incorporation. Luminescence 29:1047–1052
67. BakarNHH Abu, Ismail J, Abu Bakar M (2007) Synthesis and characterization of silver nanoparticles in natural rubber. Mater Chem Phys 104:276–283
68. Turkevich J, Stevenson PC, Hillier J (1951) A study of the nucleation and growth processes in the synthesis of colloidal gold. Disc Faraday Soc 11:55–75
69. Lee PC, Meisel DJ (1982) Adsorption and surface-enhanced Raman of dyes on silver and gold sols. Phys Chem 86:3391–3395

Chapter 3
Nitrile Rubber Latex Blends: Preparation, Characterization and Applications

Garry L. Rempel and Hui Wang

Abstract Latex blending of desired rubbers with the objective of obtaining an optimized combination of properties is an important research topic in the field of rubber science and technology. This chapter aims to provide an overview of the recent advances toward the development of the nitrile rubber latex-based blends, starting from a general introduction of different types of rubbers commonly used in the rubber latex blending process. Different blend combinations between the nitrile rubber latex and other types of rubber latexes or nanofillers are then reviewed. This is followed by a summary of techniques used for the preparation of rubber latex blends. Next, the characterization of rubber latex blends is presented. Finally, the applications of nitrile rubber latex blends are summarized.

Keywords Nitrile rubber latex · Carboxylated nitrile rubber · Blend · Preparation · Characterization · Applications

Nomenclature

ATR	Attenuated Total Reflection
CBS	Cyclohexyl Benzthiazyl
CCD	Charged-Coupled Device
CR	Chloroprene Rubber
DCA	Dicholoroacetic Acid
DLS	Dynamic Light Scattering
DMTA	Dynamic Mechanical Thermal Analyzer
DRC	Dry Rubber Content
DSC	Differential Scanning Calorimetry

G.L. Rempel (✉)
Department of Chemical Engineering, University of Waterloo, 200 University Ave. West, Waterloo, ON N2L 3G1, Canada
e-mail: grempel@uwaterloo.ca

H. Wang (✉)
Institute of Chemical and Engineering Sciences, 1 Pesek Road, Jurong Island, Singapore 627833, Singapore
e-mail: Wang_Hui@ices.a-star.edu.sg

G. Marković and Visakh P.M. (eds.), *Rubber Nano Blends*, Springer Series on Polymer and Composite Materials, DOI 10.1007/978-3-319-48720-5_3

DTG	Derivative Thermogravimetric Analysis
EPDM	Ethylene Propylene Diene Terpolymer
FTIR	Fourier Transform Infrared Spectroscopy
GR-N	Government Rubber-Nitrile
HNBR	Hydrogenated Nitrile Rubber
HXNBR	Hydrogenated Carboxylated Nitrile Rubber
MBTS	Dibenzthiazyl Disulphide
MWCNT	Multiwalled Carbon Nanotube
Na-MMT	Sodium Montmorillonite
NBR	Acrylonitrile–Butadiene Rubber
NR	Natural Rubber
OMLS	Organically Modified Layered Silicate
PCN	Polymer–Clay Nanocomposites
Phr	Parts Per Hundred Rubber
PVC	Polyvinyl Chloride
SBR	Styrene–Butadiene Rubber
SDBS	Sodium Dodecyl Benzene Sulphonate
SEM	Scanning Electron Microscopy
SNRL	Skim Natural Rubber Latex
SR	Synthetic Rubber
TEM	Transmission Electron Microscopy
TG	Thermogravimetry
Tg	Glass Transition Temperatures
TGA	Thermogravimetric Analysis
TMQ	2,2,4-Trimethyl-1,2-Dihydroquinoline
TMTD	Tetramethyl Thiuram Disulphide
XNBR	Carboxylated Nitrile Rubber
XRD	X-Ray Diffraction

3.1 Historical Background

Nitrile–butadiene rubber (NBR) is one of the most commonly used commercialized and mass-produced rubbers among various types of rubbers that have been manufactured [1]. The creation of NBR occurred during and between the two world wars, and a patent for this new oil-resistant rubber was awarded to three German chemists Eduard Tschunkur, Helmut Kleiner and Erich Konrad on April 26, 1930 [2]. NBR was originally called Buna-N (a name coined from its constituents butadiene sodium Na and nitrile) and was later produced in the USA during World War II as GR-N (Government Rubber-Nitrile) [3]. As time passed, this group of acrylonitrile–butadiene rubbers eventually was well known as nitrile rubbers. It should be mentioned that prior to the development of nitrile rubber, Eduard Tschunkur had invented another important synthetic rubber with German chemist Walter Bock—styrene–

Fig. 3.1 Inventors of NBR and SBR: Walter Bock (*left* for SBR), Eduard Tschunkur (*right* for SBR and NBR) [2]

butadiene rubber (SBR). The patent for the copolymerization of butadiene and styrene was awarded on June 21, 1929, and up to today SBR has represented about one-half of the total world production for the synthetic rubbers [4] (Fig. 3.1).

3.2 Introduction

3.2.1 Latex Rubber Characteristics

Latex refers generically to a stable dispersion of small polymer particles in an aqueous medium. Rubber latices may be natural such as natural rubber (NR) or synthetic rubber (SR). Most of SRs are prepared by emulsion, miniemulsion, microemulsion, or dispersion polymerization techniques. Latex products are used extensively as alternatives to organic solvent-based systems to produce adhesives, sealants, and coatings because of concerns over environmental pollution and toxicity of solvents as well as for practical reasons, such as ease of use.

Latex blending is a technique carried out with the objective of obtaining a combination of properties inherent in the individual polymers. The level of compatibility between the components of the mixture is greatly affected by the domain size of the dispersed phase. The particle size of the rubber latex may span a wide range from several tens of nanometers to hundreds of micrometers. A smaller dispersed phase size could indicate a higher level of blend compatibility of the system. Consequently, the final structure, physical properties, and mechanical properties of the blend can be enhanced with a higher level of compatibility. In a multitude of situations, diene rubbers, such as SBR, butadiene rubber (BR), or NR, are generally used to reduce the compound costs. It is beneficial to the physical properties of the product by selecting compatible polymers as blending components to form the blends during the mixing process. In the assessment of polymer blends, caution has to be taken into account that a diminished level of physical properties could appear in comparison with a similar formulation based on the pure parent polymers [5]. Regardless of circumstance, all blending scenarios require a well-adjusted mixing procedure.

A number of rubber blends spanning many types of diene rubbers are currently used in the rubber industry, and each rubber added carries its own benefits to improve the desired characteristics of the resulting blend. NR enhances the building tack, low-temperature flexibility, elasticity, and reduces cost. BR features its excellent ability in improving low-temperature brittleness. SBR has a predominant benefit of reducing cost and lowering the crystallization hardening as well. NBR is primarily used for improved oil resistance and for a better energy uptake in a microwave cure. Ethylene propylene diene terpolymer (EPDM) allied with chloroprene rubber (CR) can be used in EPDM vulcanizates to achieve a certain level of oil resistance. CR can improve the adhesion of EPDM rubber to reinforcing the substrates. In blends where CR is dominant, EPDM rubber can acquire the price reduction and better ozone-resistance [5].

3.2.2 Category of Rubbers Commonly Used in the Rubber Blend

Acrylonitrile–Butadiene Rubber (NBR)

NBR, a complex family of unsaturated copolymers of acrylonitrile and butadiene, is produced by emulsion polymerization, as shown in Fig. 3.2. In some NBR brands, small amounts of a third monomer are added in order to enhance the adhesive or elastomeric properties. NBRs are special rubbers. Its specific property is resistance

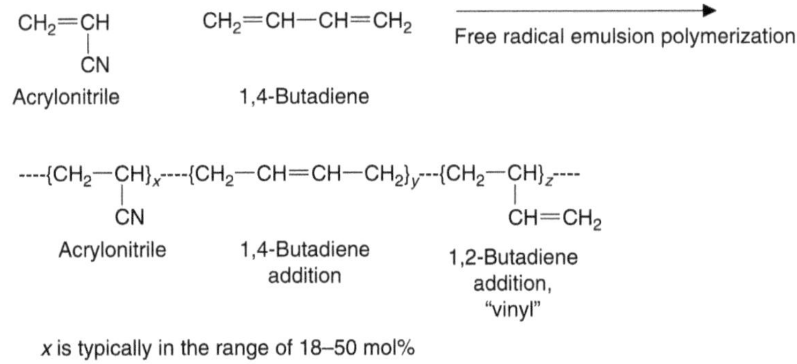

Fig. 3.2 Polymerization and compositional structure of NBR [6]

to non-polar solvents, fats, oils motor fuel, and chemicals. NBR is a type of statistical copolymer containing butadiene and nitrile units, which provides room to control the proportions of the constitutional units in balance with the properties for various end-use applications. In comparison with other commonly used rubbers, NBR has excellent compression set, outstanding tear resistance, and abrasion resistance. A greater content of acrylonitrile promotes higher strength, greater resistance to swelling by hydrocarbon oils, and lower permeability, which, however, comes at the cost of lower flexibility at lower temperatures [3]. The amount of acrylonitrile in the final copolymer often ranges from 18 to 50 % in order to promote certain qualities of the rubber. In particular, a higher percentage of acrylonitrile would be used to obtain a high-strength rubber while a lower percentage of acrylonitrile would be used to obtain a more flexible rubber at lower temperatures. The main drawback of NBR is poor ozone and weather resistance, which makes it unsuitable for applications where these qualities would be required. NBR has an operating temperature ranging from −30 to +100 °C, within which NBR can withstand most of the severe applications. The high oil resistance and durability of nitrile rubber enable it to be a valuable material in latex blend processing.

Hydrogenated Nitrile Rubber (HNBR)

NBR can also be modified through hydrogenation reactions that promote the desired advantageous properties such as the mechanical properties and heat resistance. The subsequent formation of the hydrogenated nitrile rubber is often represented as HNBR (Fig. 3.3), which is known for the elimination of the original deficiencies in nitrile rubber. The hydrogenation reaction is usually carried out via

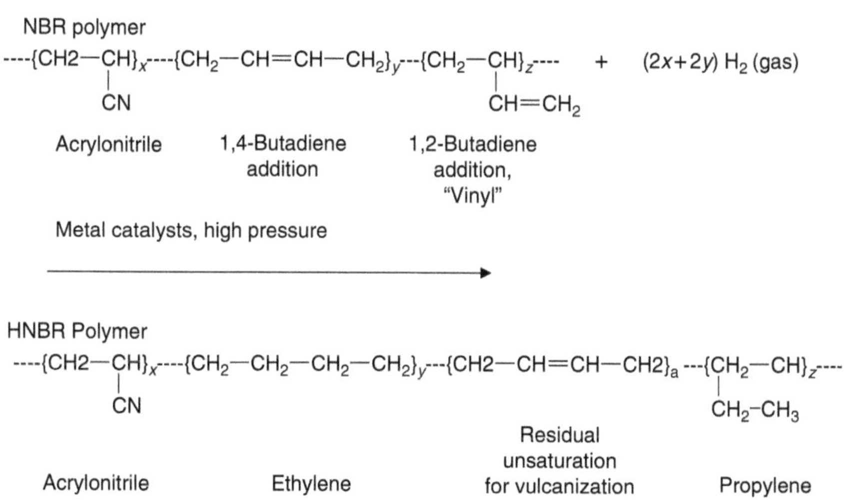

Fig. 3.3 Catalytic hydrogenation of NBR to prepare HNBR and the polymer microstructure of HNBR [8]

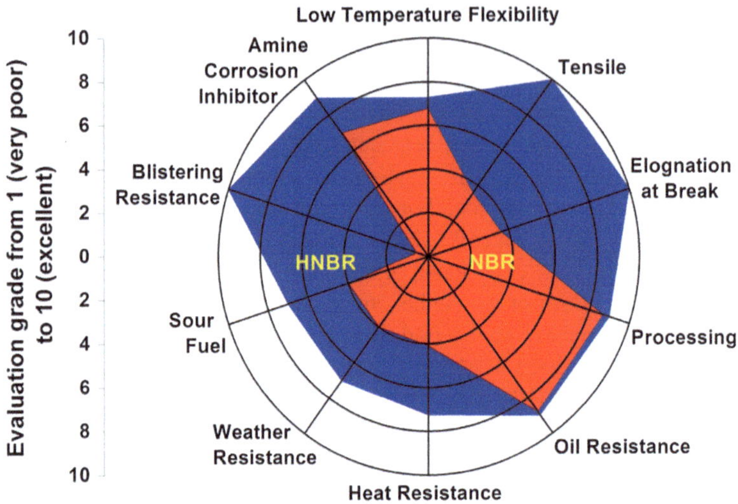

Fig. 3.4 HNBR versus NBR property comparison [1]

catalysis in the presence of the high pressure hydrogen, as shown in Fig. 3.3. HNBR can be regarded as a tetrapolymer, which is constituted by the acrylonitrile, ethylene formed by the hydrogenation of the 1,4-butadiene segments, propylene formed by the hydrogenation of the 1,2-butadiene segments, and residual butadiene units. HNBR retains the oil and fuel resistance properties and flexibility at lower temperatures of NBR and obtains a heat and oxidation resistance like EPDM rubber [7]. A detailed comparison of properties between NBR and HNBR can be seen in Fig. 3.4.

Ozone resistance of HNBR represents a significant improvement over that of NBR [9]. However, the drawback of using HNBR is the increased cold flow upon hydrogenation. Cold flow is the deformation of a product at room temperature as an external force is applied continuously over time. Despite the increased cold flow, HNBR has a wider operating temperature ranging from −40 to 165 °C with minimal degradation over long periods of time. HNBR is also subject to the similar effects of the NBR with respect to the content of the constitutional units. Furthermore, a higher concentration of acrylonitrile will lead to an increase in resistance to heat and petroleum-based oils and fuels and a decrease in the low temperature performance while a lower concentration of acrylonitrile will do the opposite. HNBR is a high-performance rubber due to its higher strength, improved fuel, heat, and ozone resistance and other properties, which make it valuable in very harsh environments where NBR is not suitable.

Carboxylated Nitrile Rubber (XNBR)

Carboxylated nitrile rubber (XNBR) is a terpolymer consisting of acrylonitrile, butadiene, and an unsaturated monomer containing carboxyl groups, such as the acrylic or methacrylic acids [10]. Lateral carboxyl functionalities –COOH can create extra curing sites and facilitate the use of curing agents to react with these carboxyl groups with the objective of ionic bond formation. The most common curing agents used to create ionic cross-links are based on salts and oxides of multivalent metals such as zinc oxide (ZnO), magnesium oxide (MnO), and calcium oxide (CaO) [10]. Carboxylated nitrile rubber is a polar and highly reactive polymer that possesses excellent compatibility with polar or non-polar resin materials [11]. Carboxylic nitrile rubber also possesses a higher degree of interaction with fillers to enhance the properties of the XNBR. XNBR has superior strength and improved damping properties in comparison with NBR. All those characteristics create numerous possibilities for XNBR as a matrix polymer. XNBR is often reserved to applications that are subject to extreme stress and wear [12].

Hydrogenated Carboxylated Nitrile Rubber (HXNBR)

In a similar way to NBR, carboxylated nitrile rubber can also be modified through a hydrogenation reaction to enhance the physical and mechanical properties of the carboxylated nitrile rubber (Fig. 3.5). The hydrogenated XNBR is often denoted as HXNBR. Introducing a carboxylic acid group attached to the backbone of HNBR to form HXNBR is expected to further improve HNBR's superior mechanical strength, adhesion properties, and abrasion resistance. HXNBR is mainly produced by LANXESS with the brand name Therban® XT [13]. Therban XT-based compounds exhibit unusually high tear strength at high temperatures when compared with regular Therban HNBR (LANXESS HNBR brandname) and other synthetic elastomers. In addition, Therban XT showed excellent adhesion properties to metals, plastics, natural, and synthetic fibers over a broad range of temperatures, especially in the higher temperature range [14].

The physical properties of HXNBR exceed the already high levels of strength and durability shown by HNBR. Figure 3.6 shows a comparison of some of the key

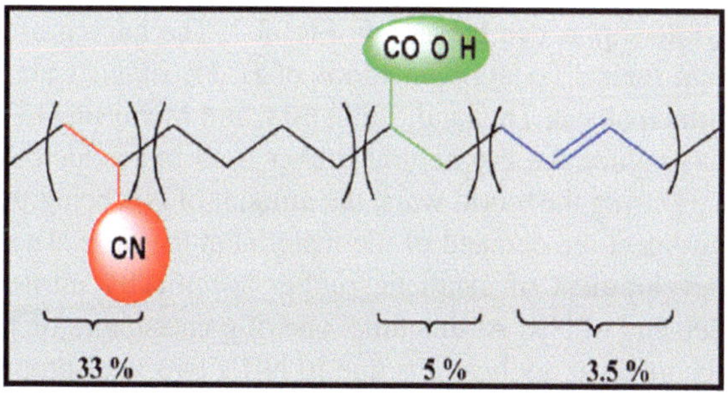

Fig. 3.5 Polymer microstructure of HXNBR [14]

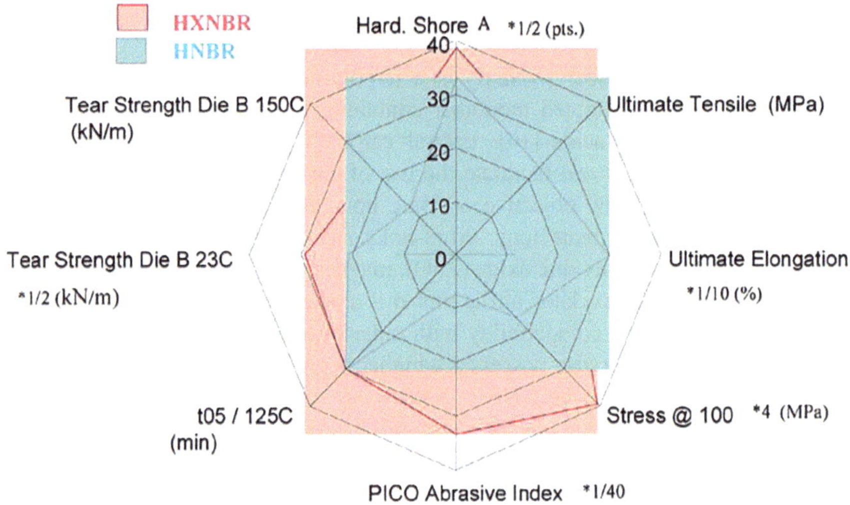

Fig. 3.6 HXNBR versus HNBR properties comparison [14]

physical properties of Therban XT versus regular Therban HNBR in a typical HNBR peroxide compound formulation. Despite the improved qualities of HXNBR, this type of rubber is often limited to extreme scenarios, where extreme performance is required, due to the high cost required to produce the rubber. This cost is associated with the extra steps and chemicals such as the precious metal catalysts required to produce the HXNBR. HXNBR is an ideal rubber for applications where extreme durability and strength are prioritized over cost. It is an efficient way to decrease the cost by blending the Therban XT with other rubbers such as regular Therban. Meanwhile, the level of active groups (COOH) is lowered via the blending process, which therefore can improve the HXNBR processibility when mixing with active ingredients such as zinc diacylate.

Natural Rubber (NR)

Natural rubber (NR) is the only type of rubber that can be extracted and refined from existing organisms. NR latex can be found in the *Hevea brasiliensis* tree as a white sap as can be seen in Fig. 3.7. The habitat of the rubber tree is the Amazon rain forest. Today, plantations of *H. brasiliensis* are also found in Southeast Asia, principally in Thailand, Indonesia, and Malaysia [15]. This latex is then coagulated to produce the dry natural rubber used in products spanning multiple industries.

During the world wars, the amount of NR being produced was limited and could not meet the demand of the material at the time. The demand for NR expedited the development of synthetic rubber specifically nitrile rubber to alleviate the large demand of NR. At this time, one of greatest uses of NR was to create rubber gloves for medical applications due to NR's two vital properties: chemical resistance and

Fig. 3.7 *Hevea Brasiliensis*, the rubber tree [16]

abrasion resistance. Today, NR is generally known for its excellent physical properties including low compression set, high tensile strength, high abrasion, and tear resistance [17], which allows NR to have much broader applications. The advantages that set NR apart from most other synthetic rubbers includes its good surface friction properties, excellent bonding capabilities to metal substrates, and good vibration dampening characteristics. The biggest detractor for the use of NR is its poor resistance to petroleum oils, which could prevent its use in the oil industry. Furthermore, NR has poor ozone and UV resistance, which makes this type of rubber unsuitable for outdoor use. The operating temperature range of NR is approximately −51 to 104 °C, which indicates that NR is capable of remaining flexible at lower temperatures.

Styrene–Butadiene Rubber (SBR)

Styrene–butadiene rubber (SBR) is an alternative synthetic rubber to NBR with similar properties to NR. SBR is a combination of styrene and butadiene, which can be obtained by two processes: from solution anionic polymerization (S-SBR) or as a free radical emulsion polymerization (E-SBR) [18]. The E-SBR is more widely used in industrial production. SBR is often compared to NR due to their similar properties. SBR is particularly useful due to its good resistance to brake fluids and water. The resistance of brake fluids makes SBR very useful in the automotive industry and more specifically the tire industry. However, SBR is not effective in terms of weather resistance and petroleum oil resistance. In addition, SBR is limited by its poor tear strength without reinforcement with fillers such as carbon black and low resilience especially at higher temperatures [19]. SBR can be used at lower temperatures with an operating temperature ranging from −46 to 100 °C. The amount of styrene rubber used in a blend mixture is often decreased when a higher degree of heat resistance is required [4]. SBR becomes harder as it ages due to the increased amount of polymer chain interlinking as a result of oxidation. This will considerably prevent the long-term use of SBR in most cases due to the decreased flexibility of aged SBR. SBR is the least expensive rubber among almost all types of rubbers, which enables it an excellent candidate for blend process for purposes of reducing cost. SBR is an excellent alternative for NBR in some cases within the automobile industry and works admirably as a NR replacement due to their similar qualities.

Chloroprene Rubber (CR)

Chloroprene rubber (CR), also known as neoprene, is a family of synthetic rubbers that are produced by free radical polymerization of chloroprene. Neoprene is a versatile rubber that is known for its excellent tensile strength, resilience, oil and flame resistance and degradation by oxygen and ozone over a wide temperature range [20]. Neoprene is restrained from its moderate water resistance and ineffectiveness in solvent environments. The main limitation to neoprene is the cost associated with producing this rubber which often limits its use to special properties required for certain applications. Neoprene has many desirable rubber characteristics and thus has been widely used to prepare the blends with other rubbers.

Polyvinyl Chloride (PVC)

Polyvinyl chloride is a synthetic polymer produced from the polymerization of vinyl chloride, which is the third-most widely produced synthetic plastic polymer. PVC is characterized by its lightweight and rigid form. However, PVC can also be created in a softer and more flexible form after being mixed with a plasticizer such as phthalate esters. Most PVC products can be heated and mixed with a plasticizer in order to keep the plastic flexible at low temperatures, wherein the desire for flexibility is over rigidity [21]. PVC possesses a wide array of applications ranging from raincoats and shower curtains to window frames and indoor plumbing.

3.3 Nitrile Rubber Latex-Based Blends

3.3.1 NR and NBR Latexes Blend

Blending NR and NBR would aim to obtain the oil resistance of nitrile rubber while retaining the durability or NR. The ratio of NR to NBR would determine the presence of each rubber's prominent qualities within the final mixture of latex. A higher ratio of NBR to NR would promote the oil resistance characteristics of the latex blend while a lower ratio would promote the mechanical properties and heat buildup of the latex blend. Moonprasith et al. [22] carried out a study on the blending of NR and NBR latexes and investigated their morphology and mechanical properties in various NR/NBR dry rubber ratios: 100:0, 75:25, 50:50, 25:75, and 0:100. In addition, the oil swelling properties of NR and NBR solid blends were studied, and the effects of different constitutional ratios on NR and NBR were examined. It was found that the rubbers were uniformly mixed both in the latex form and as a result the thin film is formed. Morphology of the blend films was characterized by scanning electron microscopy (SEM). It was shown that the particles in dispersed phase (i.e., blended film) retained their original sizes and shapes, and phase separations between NR and NBR were in the range of submicrometer to a few micrometers (which is relatively small in comparison with a melt blend). With respect to the oil swelling test for solid blends, the NR/NBR latex blends were dried, mixed with vulcanizing agents, and subsequently vulcanized. Mechanical and chemical properties of the vulcanizates were thus examined. As expected, the oil swelling and hardness properties of these blends increased with an increase in NBR content. Nevertheless, the tensile properties of the blends are generally improved on increasing the NR component because of the good mechanical properties of NR. The blend ratio has a significant effect on the oil-resistant and tensile properties of NR/NBR blends. With 75, 50, and 25 % NBR content, better oil resistance is achieved than for pure natural rubber. The 50:50 NR/NBR has a poor oil resistance and tensile strength, which is likely to be affected by phase formation. Blends of NR and NBR have become very important within the rubber industry due to their combined oil resistance and superior mechanical properties.

3.3.2 HXNBR and PVC Latex Blend

Yue and coworkers prepared a new type of self-cross-linking HXNBR/PVC composite film by using the latex blend method. They further carried out an investigation of the properties of self-cross-linking HXNBR/PVC composite film [7]. HXNBR latex was first obtained by the hydrogenation of XNBR latex via the diimide hydrogenation route. The hydrogenation degree was 95 % on the basis of the calculation according to FTIR spectroscopy. Blending of HXNBR and PVC emulsions was then prepared with different blend ratios of XNBR and PVC in a round flask at 40 °C under constant stirring. The HXNBR/PVC composite film was prepared from their mixed emulsions through a series of operations. The HXNBR/PVC composite film can be self-cross-linked without any additional cross-linking agent. The self-cross-linking HXNBR latex has good film-forming properties, but PVC has poor film-forming properties because the film-forming temperature of PVC is higher than room temperature. When the PVC content was lower than 30 %, the composite film showed no cracks on the surface of the film and homogeneous and transparent films were obtained. At 20 % of PVC content, the composite film showed excellent mechanical properties. Dynamic mechanical thermal analysis (DMTA) showed that there were two glass transition temperatures (T_{g}) in the composite film. PVC can also improve the thermal stability of a stronger rubber in HXNBR and the formulated HXNBR/PVC composite film with a 20 % PVC content has the best thermal stability. Conventionally, HNBR/PVC blends were prepared by the mechanical blend method or the solution blend method and the two components in the resulting blends were miscible [23]. The latex blend pathway generally could make dispersions of two rubber components more uniform than the mechanical and solution blend methods.

3.3.3 Nitrile Rubber Latex Blends with Nanofillers

These blends of natural rubber and nitrile rubber can be further enhanced through the addition of clay or layered silicates. This type of blend can be prepared through different intercalation methods, either individually or combined, such as latex shear blending methods [24]. Each method promotes certain improvements within the latex blend, which, however, often leads to shortcomings in other properties. For example, the shear blending method can yield poor intercalation and dispersion morphology of silicate layers [25]. A common source of layered silicate used for creating these types of latex blends is sodium montmorillonite (Na-MMT). In general, latex blending requires the mixing of rubber latex and a clay aqueous dispersion under certain conditions to enhance good dispersion, and then co-coagulating the mixture by adding an electrolyte [26]. The formation of the nanocomposite in the latex stage may occur through two potential routes. One route involves aqueous dispersions of swelling clay and the penetration of the rubber molecule into the intergalleries of the clay. This route would promote a good level of dispersion due to the shorter path required to transfer. The second route requires vigorous mixing and involves an aqueous clay dispersion occurring where the latex particles are in-between separated clay layers [26]. These separated clay layers have structures similar to those of intercalated/exfoliated polymer–clay nanocomposites. Wei and coworkers prepared an elastomeric NBR/silicate nanocomposite with a mostly exfoliated and partially intercalated coexisting structure by a modified latex shear blending process aided with ball milling, as shown in Fig. 3.8. Compared to neat NBR, the tensile mechanical properties and tear strengths of NBR/silicate nanocomposites were found to increase with the amount of layered silicates. Specifically, the tensile and tear mechanical properties of the NBR/layered silicates increase by 200 and 60 %, respectively. Additionally, the produced NBR/silicate nanocomposites display higher thermal stabilities than neat rubber.

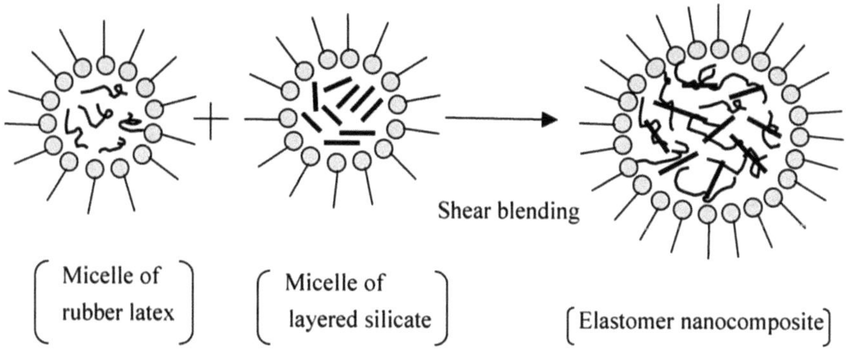

Fig. 3.8 Latex blending process for the preparation of NBR/silicate nanocomposites [25]

3.4 Preparation Techniques of Nitrile Rubber Latex-Based Blends

Blending is a commonly used method in the rubber latex industry to bring the desirable traits of one rubber to another rubber. For example, NR has excellent mechanical properties and heat buildup characteristics but has poor resistance to oil [22]. In comparison, nitrile rubber has a strong resistance to oils, ozone, and chemicals. The commonly used method for the preparation of nitrile rubber latex-based blend is to simply mix the nitrile rubber latex with other types of rubber latexes or nanosized fillers and then to blend them with the assistance of high shear forces, which can be imparted by vigorous agitation, sonication, and other mechanical equipments [27]. Latex shear blending is often advantageous due to its simplicity, lower cost, and its environmental safety [26]. Therefore, the latex shear blending method has been used in a large number of works that involved the preparation of polymer latex-based blends.

3.5 Characterization of Nitrile Rubber Latex-Based Blends

3.5.1 X-ray Diffraction (XRD)

X-ray crystallography, or X-ray diffraction (XRD), is a tool used for identifying the atomic and molecular structure of a crystal, in which the crystalline atoms cause a beam of incident X-rays to diffract into many specific directions [28]. By measuring the angles and intensities of these diffracted beams, a crystallographer can create a three-dimensional picture of the density of electrons within the crystal. From this electron density, the mean positions of the atoms in the crystal

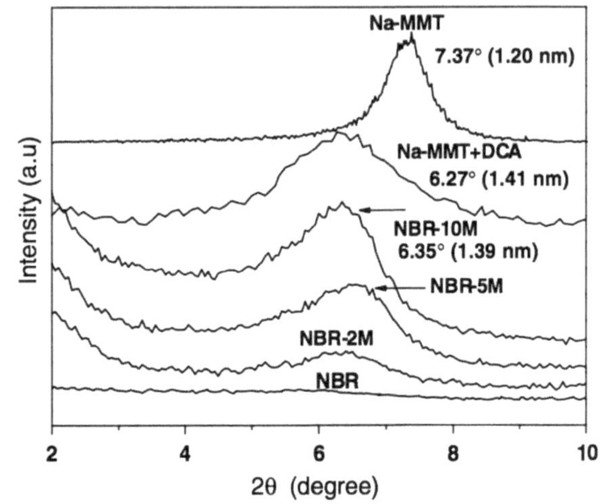

Fig. 3.9 Wide angle XRD diffraction for Na-MMT, NBR, and NBR/Na-MMT nanocomposites with varying proportions of Na-MMT (Na-MMT/dicholoroacetic acid is also included for comparison [26])

can be identified, as well as their chemical bonds, patterns, and various other information. The structure of polymer–clay nanocomposites (PCN) has commonly been analyzed using XRD and transmission electron microscopy (TEM). XRD is the most widely used technique to identify intercalated structures due to the periodic arrangement of the silicate layers both in the pristine and the intercalated states. The repetitive multilayer structure is well preserved in intercalated structures, allowing the interlayer spacing for determination. Intercalation of polymer chains between the clay layers usually increases the interlayer spacing, which results in shifting of the XRD peak toward lower angles. A representative example of a measurement for XRD can be seen in Fig. 3.9, which shows the diffraction patterns of 001 plane in montmorillonite (Na-MMT), dicholoroacetic acid (DCA)-treated Na-MMT, NBR, and NBR/Na-MMT nanocomposites with various Na-MMT loadings [26]. As shown in Fig. 3.9, the interlayer spacing of Na-MMT was observed at $2\theta = 7.37°$ (d spacing calculated from Bragg's law = 1.20 nm), whereas DCA-treated Na-MMT showed the peak at $2\theta = 6.27$ (d_{001} = 1.41 nm). For NBR/Na-MMT samples, the peak appeared at $2\theta = 6.35°$ (d_{001} = 1.39 nm) with a decrease in peak intensity compared to pristine Na-MMT. This indicates a small gallery expansion of about 0.2 nm in the layered clay in Na-MMT/DCA and NBR/Na-MMT samples. Therefore, the structures of these polymer–clay nanocomposites can be understood.

3.5.2 Fourier Transform Infrared Spectroscopy (FTIR)

FTIR is a technique used to obtain an infrared spectrum of absorption, emission, photoconductivity or Raman scattering of solid, liquid, or gas materials [29]. A FTIR spectrometer simultaneously collects high spectral resolution data over a

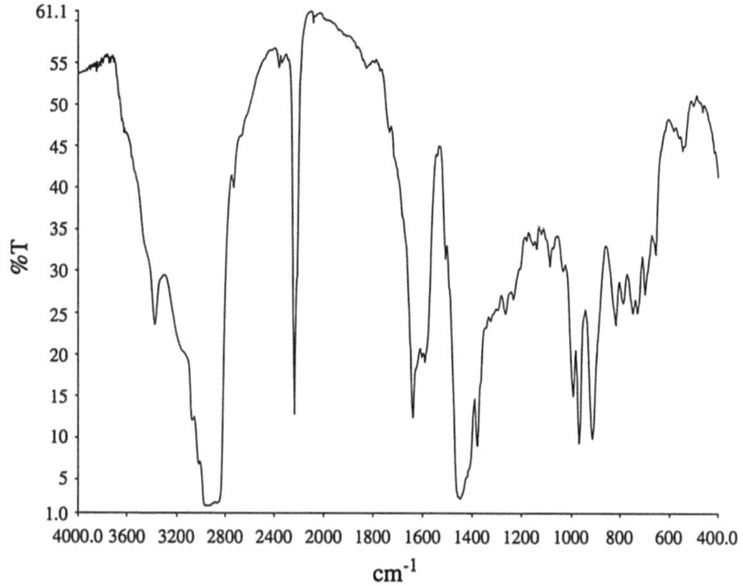

Fig. 3.10 FTIR spectra of acrylonitrile–butadiene rubber (film casting) [30]

wide spectral range. This represents a significant advantage over a dispersive spectrometer which measures intensity over a narrow range of wavelengths at a time. The term Fourier transform infrared spectroscopy originates from the fact that a Fourier transform (a mathematical process) is required to convert the raw data into the actual spectrum.

Bandyopadhyay and coworkers performed a study on the applications of FTIR on the characterization of NBR [30]. Figure 3.10 depicts the FTIR spectra of a NBR film sample. The wavelengths (measured in cm^{-1}) are connected to different substances within the spectrum. For example, the nitrile stretching can be found at 2238 cm^{-1} and the C–H stretching (out of plane) of butadiene double bond can be found at 969 cm^{-1}. The ratio of nitrile stretching to the C–H stretching of butadiene double can be used to characterize the nitrile rubber such as for the determination of the acrylonitrile content. FTIR can be used to characterize nitrile rubber-based blends and to determine the individual contents within the blend.

Joseph and coworkers carried out research on the incorporation of the dispersed multiwalled carbon nanotube (MWCNT) into the XNBR latex with the assistance of sodium dodecyl benzene sulphonate (SDBS) via the latex shear blending technique [31]. Figure 3.11 shows the infrared spectrum of pristine MWCNT and XNBR/MWCNT composite, respectively. MWCNT does not show any distinct peaks in the FTIR spectrum. The FTIR spectrum of the composite shows characteristic peaks of NBR at 2238 cm^{-1} (CN group), 2925 and 2854 cm^{-1} (CH_2 group). This demonstrates the absence of a chemical bond between nanotube and NBR in the composite.

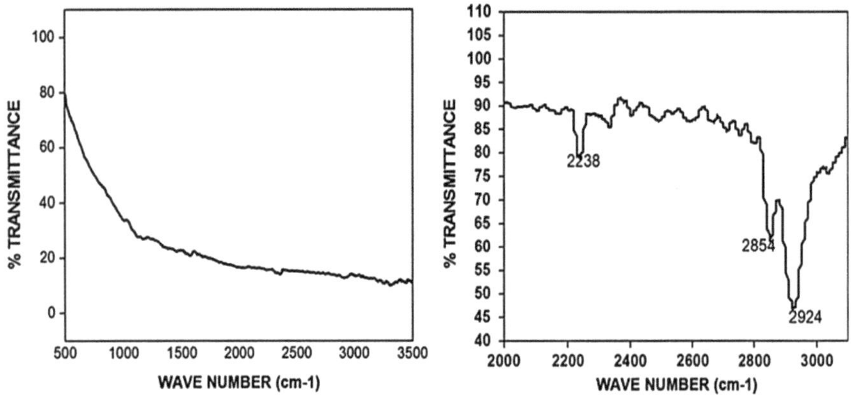

Fig. 3.11 FTIR spectra of MWCNT (*left*) and FTIR spectra of XNBR-MWCNT composite (*right*) [31]

3.5.3 Dynamic Light Scattering (DLS)

DLS is a technique in physics that can be used to determine the size distribution profile of small particles in suspension or polymers in solution [32]. In the scope of DLS, temporal fluctuations are usually analyzed by means of intensity or photon auto-correlation function (also known as photon correlation spectroscopy or quasi-elastic light scattering). When the incident light hits the small particles, the light scatters in all directions (Rayleigh scattering) as long as the particles are small comparable to the wavelength. If the light source is a laser, which is monochromatic and coherent, the scattering intensity fluctuates over time. This fluctuation is due to the fact that the small molecules in solutions are undergoing Brownian motion. Therefore, the distance between the scattering in the solution is constantly changing with time. This scattered light then undergoes either constructive or destructive interference by the surrounding particles, and within this intensity fluctuation, information is contained about the time scale of movement of the scattering materials. Sample preparation either by filtration or centrifugation is critical to remove dust and artifacts from the solution.

Alex and Nah carried out a study focusing on NR/NBR/organoclay nanocomposites, and the particle size of rubber latex was determined using a particle size analyzer based on the DLS technique [33]. The DLS method allows for deeper analysis of the properties of natural rubber and nitrile rubber nanocomposites. Figure 3.12 presents the particle size distributions of the skim NR latex (SNRL), creamed SNRL and NBR latex nanoparticles. The creaming process of the NR can enables the SNRL to coagulate easily to a consolidated mass. In addition, the proteins adsorbed on natural rubber particles reduce after the creaming treatment. As shown in Fig. 3.12, the particle size varied from 94 to 630 nm for pure SNRL, and this distribution changed from 79 to 578 nm for the creamed skim latex. The particle size recorded for the NBR latex was comparatively lower. There was a

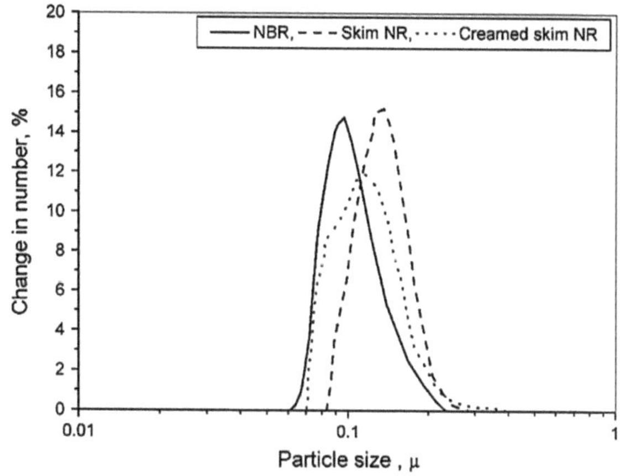

Fig. 3.12 Particle size distribution of skim NR latex (SNRL), creamed SNRL, and NBR latex [33]

considerable size reduction for skim latex particles after creaming as more than 14 % of the particles recorded a size lower than the smallest detected particles of pure skim rubber latex. This was below the average particle size of fresh natural rubber latex since larger particles were known to readily separate into concentrated fraction [34]. The particle size of fresh natural rubber latex from *H. brasiliensis* was shown to have a bimodal distribution with a mean diameter of 1.07 μm and in some cases a lower size range of 0.08–0.75 μm [35]. The particle size of SNRL was reduced to a size very close to that of NBR latex after creaming, and therefore the creamed SNRL was expected to exhibit excellent dispersion when compared to NBR latex, which thus confirms the advantages of using skim latex in blending with synthetic rubber latexes.

3.5.4 Thermogravimetric Analysis (TGA)

TGA is a method of thermal analysis in which changes in physical and chemical properties of materials are measured as a function of increasing temperature (at a constant heating rate), or as a function of time (at constant temperature and/or at constant mass loss) [36]. TGA can provide information about physical occurrences, including vaporization, sublimation, absorption, adsorption, and desorption. Similarly, TGA can provide information about chemical phenomena including chemisorption, desolvation (especially dehydration), decomposition, and solid–gas reactions (e.g., oxidation or reduction).

3.5.5 Microscopy

Transmission Electron Microscopy (TEM)

TEM is a microscopy technique in which a beam of electrons is transmitted through an ultrathin specimen while interacting with the specimen as it passes through it. An image is formed from the interaction of the electrons transmitted through the specimen, which is then magnified and focused onto an imaging device such as a fluorescent screen on a layer of photographic film, or detected by a sensor such as a charge-coupled device camera [37]. The major advantage of TEM is that it is capable of imaging at a significantly higher resolution than optical microscopes, which enables the possibility to examine finer details to the level of a single column of atoms. TEM has become an essential analysis method in a wide range of scientific fields.

TEM can be used to detail the structure of nitrile rubber based blends. Figure 3.13 presents the TEM micrographs of the dispersion of nanosized layered silicates in a NBR matrix. The NBR/silicate nanocomposites were prepared by latex shear blending of nitrile latex with surfactant-treated hydrophilic-layered silicates [25]. A few multilayer bundles in the case of NBR containing 3 and 5 wt% layered silicates can be found in Fig. 3.13a, b. In both cases, the exfoliated layered silicates, where the interlayer *d*-spacing was larger than 3.2 nm, can be observed. In Fig. 3.13c, the dispersion of silicates in NBR can be found to actually adopt a bimodal structure, consisting of both intercalated and exfoliated states. This reveals that a higher silicate loading would lead to poorer dispersion and more aggregated bundles in the rubber film. A critical concentration of 7.5 wt% layered silicate in NBR/silicate nanocomposites can be utilized to exert the exfoliated nanocomposite with the use of high shear force generated in the ball milling process.

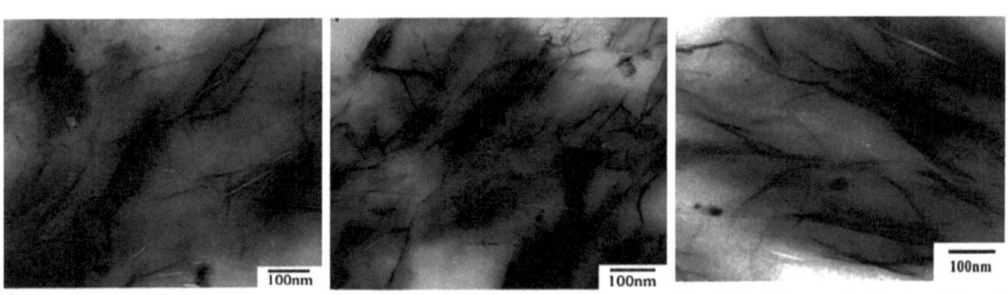

(a) NX-3 (layered silicates:3phr) **(b) NX-5 (layered silicates:5phr)** **(c) NX-7.5 (layered silicates:7.5phr)**

Fig. 3.13 TEM micrographs of NBR/silicate nanocomposites containing 3, 5 and 7.5 phr layered silicates [25]

Scanning Electron Microscopy (SEM)

A scanning electron microscope (SEM) is a type of electron microscope that produces images of a sample by scanning it with a focused beam of electrons. The electrons interact with atoms in the sample thereby producing various signals that contain the information of the surface topography and composition of the sample. The electron beam is generally scanned in a raster scan pattern, and the beam's position is combined with the detected signal to produce an image. SEM can achieve a very high resolution of better than 1 nm. Specimens can be observed in high vacuum, in low vacuum, in wet conditions (in environmental SEM), and at a wide range of cryogenic or elevated temperatures [38]. The most common SEM mode is to detect the secondary electrons emitted by atoms excited by the electron beam. The number of secondary electrons that can be detected depends on factors such as the angle at which the beam meets the surface of the specimen, more specifically concerning the topography of specimen. By scanning the sample and collecting the secondary electrons with a special detector, an image displaying the topography of the surface is thus created. Nah and coworkers utilized the SEM techniques to observe the tear morphology of a torn surface of a NBR/Na-MMT blended composite and pure NBR matrix, as shown in Fig. 3.13 [26]. In general, the tearing energy of the nanocomposite increased with an increase in the loading level of clay fillers. The tear morphology of torn surface of NBR-M10 (NBR/Na-MMT composite with 10 wt% Na-MMT) (Fig. 3.14a) showed a tear pattern clearly different from those of commonly used composites, which normally shows the presence of a cross-hatched pattern with numerous webs and steps of different sizes [39]. The tear pattern of the nanocomposites showed a series of tear ridge parallel to the tearing direction, whereas the unfilled NBR showed a simple torn surface without complexity (Fig. 3.14b). The tear propagated in a particular fashion resulting in tilting of tear lines with respect to the tearing direction. The uniformly dispersed clay layers and small tactoids alter the tear path along their length depending on the orientation of clay layers. This leads to more resistance for tear propagation and therefore a higher tearing energy.

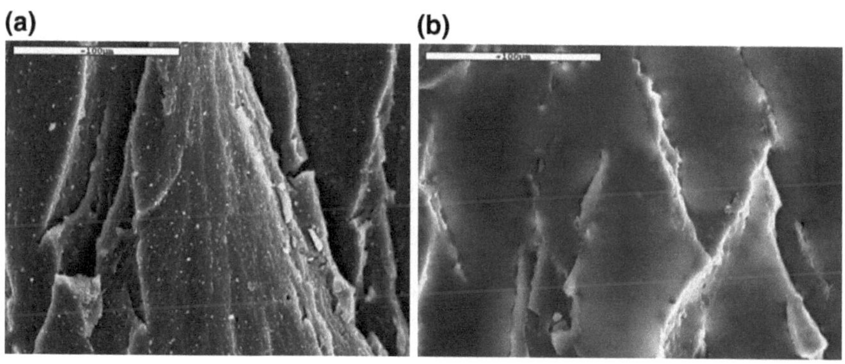

Fig. 3.14 SEM image of torn surface of **a** NBR/Na-MMT composite with 10 wt% Na-MMT. **b** Pure NBR [26]

3.6 Applications of Nitrile Rubber Latex-Based Blends

NBR/NR blends are prominent within the food industry due to their unique oil and fat resistance [40]. These resistances allow for NR/NBR blends to follow health regulations necessary in more sensitive industries. The dairy industry in particular benefits from the use of NBR and NR blends due to strict regulations surrounding dairy production. NBR/NR blends are commonly used in the manufacturing of milking inflations. The regulations on milking inflations are strict due to the fact that this tool makes direct contact with the milk extracted from the cow.

Blends of NBR and PVC can be found in a wide array of applications due to its special combination oil resistance and durability. PVC/NBR blend is perhaps the oldest commercial blend since its introduction 60 years ago [41]. PVC is miscible with NBR (23–45 % acrylonitrile content) at all composition ranges. PVC can improve the ozone and chemical resistance as well thermal aging of NBR for certain applications. The ratio of PVC-to-NBR content can be adjusted in order to promote the desired qualities of the rubber blend. Namely, a higher ratio of PVC to NBR would promote ozone and chemical resistance of the rubber while a lower ratio would increase the role of nitrile rubber as a plasticizer. PVC/NBR blends can be produced for applications such as wire and cable insulation, food containers, and pond liners.

SBR is commonly used as a component in rubber blends, which can improve their resistance to abrasion, crack initiation, and heat [42]. In contrast, NBR does not crystallize when stretched and so requires reinforcing fillers for optimum tensile strength, abrasion resistance, and tear resistance. SBR/NBR blends reduce the overall oil resistance but increases the resistance to polar liquids in proportion to the SBR content. The blending of SBR with NBR is mostly used in applications where oil seals and break fluids are present.

References

1. Wang H, Yang L, Rempel GL (2013) Homogeneous hydrogenation art of nitrile butadiene rubber: a review. Polym Rev 53:192–239
2. Hofmann's Successors. LANXESS. Retrieved from http://lanxess.cn/en/about-lanxess-china/100-years-synthetic-rubber/history-of-synthetic-rubber-cn/hofmann-successors-china/
3. Nitrile rubber (NBR) (2015) Encyclopaedia Britannica. Retrieved from http://www.britannica.com/technology/nitrile-rubber
4. Styrene-butadiene rubber (SBR) (2015) Encyclopaedia Britannica. Retrieved from http://www.britannica.com/science/styrene-butadiene-rubber
5. Musch R, Magg H (2008) 1 Polychloroprene. In: Klingender RC (ed) Handbook of specialty elastomers. CRC Press, Boca Raton, p 15
6. Keller RW (2008) Hydrogenated nitrile rubber (Chapter 3). In: Klingender RC (2008) Handbook of specialty elastomers. CRC Press, Boca Raton, p 94
7. Han Y, Su L, Mao L, Zhang L, Yue D (2014) Self-cross-linking hydrogenated nitrile-butadiene rubber latex/polyvinyl chloride emulsion composite film and its properties. Polym Plast Technol Eng 53:306–311

8. Keller RW (2008) Hydrogenated nitrile rubber (Chapter 3). In: Klingender RC (2008) Handbook of specialty elastomers. CRC Press, Boca Raton, p 95
9. Wang H, Rempel GL (2015) Aqueous-phase catalytic hydrogenation of unsaturated polymers. Catal Today 247:117–123
10. Laskowska A, Zaborski M, Boiteux G, Gain O, Maniukiewicz W (2014) Ionic elastomers based on carboxylated nitrile rubber (XNBR) and magnesium aluminum layered double hydroxide. eXPRESS. Polym Lett 8:374–386
11. Mousa A, Heinrich G, Simon F, Wagenknecht U, Stöckelhuber KW, Dweiri R (2012) Carboxylated nitrile butadiene rubber/hybrid filler composites. Mater Res 15:671–678
12. Krynac. LANXESS. Retrieved from http://lanxess.in/en/industries-products-india/industries-india/other-india/krynac-india/
13. Therban XT. LANXESS. Retrieved from http://therban.com/en/technical-information/carboxylated-technology/
14. Guo S, von Hellens W, Campomizzi EC, Ferrari L (2003) Therban XT boosts abrasion resistance, adhesion. Rubber Plast News 1–5
15. Greve H-H (2012) Rubber, 2. Natural. Ullmann's encyclopedia of industrial chemistry. In: Ullmann's encyclopedia of industrial chemistry, vol 31, 8th edn. VCH, Weinheim, Germany, p 584
16. Kew Royal Botanical Gardens. *Hevea Brasiliensis* (rubber tree). Retrieved from http://www.kew.org/science-conservation/plants-fungi/hevea-brasiliensis-rubber-tree
17. Greve H-H (2012) Rubber, 2. Natural. Ullmann's encyclopedia of industrial chemistry. In: Ullmann's encyclopedia of industrial chemistry, vol 31, 8th edn. VCH, Weinheim, Germany, pp 583–594
18. Threadingham D, Obrecht W, Wieder W, Wachholz G, Engehausen R (2012) Rubber, 3. Synthetic rubbers, introduction and overview. In: Ullmann's encyclopedia of industrial chemistry, vol 31, 8th edn. VCH, Weinheim, Germany, pp 597–616
19. Obrecht W, Lambert JP, Happ M, Oppenheimer-Stix C, Dunn J, Krüger R (2012) Rubber, 4. Emulsion rubbers. In: Ullmann's encyclopedia of industrial chemistry, vol 31, 8th edn. VCH, Weinheim, Germany, pp 623–646
20. Neoprene CR (2015) Encyclopaedia Britannica. Retrieved from http://www.britannica.com/science/neoprene
21. Fischer I, Schmitt WF, Porth H-C, Allsopp MW, Vianello G (2014) Poly(vinyl chloride). Ullmann's encyclopedia of industrial chemistry, pp 1–30
22. Moonprasith N, Suchiva K, Tongcher O (2006) Blending in latex form of natural rubber and nitrile latices: a preliminary study of morphology and mechanical properties. In: Material technical seminar, Thailand
23. Sotiropoulou DD, Avramidou OE, Kalfoglou NK (1993) Property-composition dependence of hydrogenated poly (butadiene-co-acrylonitrile)/poly (vinyl chloride) blends. Polymer 34:2297–2301
24. Paul DR, Robeson LM (2008) Polymer nanotechnology: nanocomposites. Polymer 49:3187–3204
25. Hwang WG, Wei KH, Wu CM (2004) Preparation and mechanical properties of nitrile butadiene rubber/silicate nanocomposites. Polymer 45:5729–5734
26. Kader MA, Kim K, Lee YS, Nah C (2006) Preparation and properties of nitrile rubber/montmorillonite nanocomposites via latex blending. J Mater Sci 41:7341–7352
27. Sirisinha C, Baulek-Limcharoen S, Thunyarittikorn J (2001) Relationships among blending conditions, size of dispersed phase, and oil resistance in natural rubber and nitrile rubber blends. J Appl Polym Sci 82:1232–1237
28. Zhang JP, Liao PQ, Zhou HL, Lin RB, Chen XM (2014) Single-crystal X-ray diffraction studies on structural transformations of porous coordination polymers. Chem Soc Rev 43:5789–5814
29. Dettinger U, Egelhaaf HJ, Brabec CJ, Latteyer F, Peisert H, Chassé T (2015) FTIR study of the impact of PC [60] BM on the photodegradation of the low band gap polymer PCPDTBT under O_2 environment. Chem Mater 27:2299–2308

30. Chakraborty S, Bandyopadhyay S, Ameta R, Mukhopadhyay R, Deuri AS (2007) Application of FTIR in characterization of acrylonitrile-butadiene rubber (nitrile rubber). Polym Testing 26:38–41
31. Nair KP, Thomas P, Joseph R (2012) Latex stage blending of multiwalled carbon nanotube in carboxylated acrylonitrile butadiene rubber: mechanical and electrical properties. Mater Des 41:23–30
32. Mailer AG, Clegg PS, Pusey PN (2015) Particle sizing by dynamic light scattering: non-linear cumulant analysis. J Phys Condens Matter 27:145102
33. Alex R, Nah C (2006) Studies on natural rubber/acrylonitrile butadiene rubber/organoclay nanocomposites. Plast Rubber Compos 35:219–225
34. Ohya N, Tanaka Y, Wittisuwannakul R, Koyama T (2000) Activity of rubber transferase and rubber particle size in Hevea latex. J Rubber Res 3:214–221
35. Singh AP, Wi SG, Chung GC, Kim YS, Kang HS (2003) The micromorphology and protein characterization of rubber particles in *Ficus carica, Ficus benghalensis* and *Hevea brasiliensis*. J Experim Bot 54:985–992
36. Haines PJ (2012) Thermogravimetry. In: Thermal methods of analysis: principles, applications and problems. Springer Science & Business Media, Berlin, pp 22–62
37. Proetto MT, Rush AM, Chien M, Baeza PA, Patterson JP, Thompson MP, Olson NH, Moore CE, Rheingold A, Andolina CM, Millstone JE, Howell S, Browning ND, Evans JE, Gianneschi NC (2014) Dynamics of soft nanomaterials captured by transmission electron microscopy in liquid water. J Am Chem Soc 136:1162–1165
38. Hughes L, Hawes C, Monteith S, Vaughan S (2014) Serial block face scanning electron microscopy-the future of cell ultrastructure imaging. Protoplasma 251:395–401
39. Nah C, Ryu HJ, Han SH, Rhee JM, Lee MH (2001) Fracture behaviour of acrylonitrile-butadiene rubber/clay nanocomposite. Polym Int 50:1265–1268
40. Lewan MV (1998) NR/NBR blends—compounding for food contact applications. In: Tinker AJ, Jones KP (eds) Blends of natural rubber. Chapman & Hall, London, pp 94–105
41. Perera MCS, Ishiaku US, Ishak ZAM (2000) Thermal degradation of PVC/NBR and PVC/ENR50 binary blends and PVC/ENR50/NBR ternary blends studied by DMA and solid state NMR. Polym Degrad Stab 68:393–402
42. Noriman NZ, Ismail H (2013) Effect of carbon black/silica hybrid filler on thermal properties, fatigue life, and natural weathering of SBR/recycled NBR blends. Int J Polym Mater Polym Biomater 62:252–259

Chapter 4
Polyurethane Rubber-Based Nanoblends: Preparation, Characterization and Applications

Zizi I. Abdeen

Abstract To improve the mechanical, thermal, flame, moisture and chemical resistance, of Ealstomer polyurethane (TPU) (Thermoplastic), the TPU is blended with different types of the nanoparticles, such as, the nanoclay and carbon nanofibers (CNFs) to form nanocomposites. Different examples of TPU material based on polyester or polyether pre-polymers, were manufactured, that contains a linear segmented block composed of soft segments which is aliphatic polyester polyols or aliphatic polyether polyols are coupled with hard segment formed from aromatic di-isocyantes and hard diols urethane linkages. The structures of the different blends of TPU-nanofillers samples were characterized by FTIR spectroscopy, XRD and the particle diameter of CNTs, clay and TPU-clay blending samples was determined by using transmission electron microscopy (TEM). The XRD characterization showed the shifting position of the diffraction peak of the nanoclay powder intercalated in the polymer matrix. This emphasizes that the melt blending is the most promising and practical method to be used in industry, where, solvents are not required, to produce nanocomposites based on usual compounding devices, such as, extruders or blends. So it was clear that the TPU-nanofillers blend showed an improved thermal resistance, mechanical properties and physical properties.

Keywords Ealstomer polyurethane · Thermoplastic · Nanoblends · Nanoclay · Carbon nanofibers

4.1 Introduction

Thermoplastic polyurethane (TPU) was discovered in the late 1930s as part of Germany's World War II research. It is a thermoplastic elastomer (TPE), which combines the mechanical and physical properties of rubber with the advantages of

Z.I. Abdeen (✉)
Polymers Laboratory, Petrochemicals Department, Egyptian Petroleum Research Institute (EPRI), 1-Ahmad Alzomar St, Nasr City, Cairo 11727, Egypt
e-mail: ziziabdeen@yahoo.com

G. Marković and Visakh P.M. (eds.), *Rubber Nano Blends*, Springer Series on Polymer and Composite Materials,
DOI 10.1007/978-3-319-48720-5_4

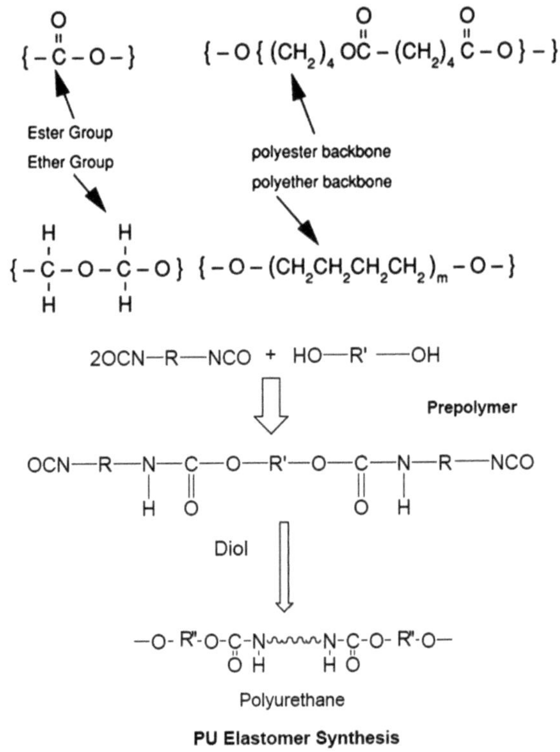

Fig. 4.1 Schematic representation of TPU preparation

thermoplasticity and processability. Other examples of TPEs include polyethylene and polypropylene. Also, the definition of a thermoplastic rubber or elastomer (TPE) is "a polymer blend or compound which, above its melt temperature, exhibits a thermoplastic character that enables it to be shaped into a fabricated article and which, within its design temperature range, possesses elastomeric behavior without cross-linking during fabrication". This process is reversible and the products can be reprocessed and remolded. Polymer blends have been of great interest particularly in the last two decades or so. The factors that have increased this interest are cost and time associated with the development of new polymers and the ability to tailor properties by blending, which may result in new, desirable, and, in some cases, unexpected synergistic effects on the properties [1, 2]. Elastomer polyurethane (TPU) (Thermoplastic) is a linear segmented block composed of hard and soft segments. The soft segment polyurethane is aliphatic polyester polyols or aliphatic polyether polyols are coupled with hard segment formed from aromatic di-isocyanate and hard diols urethane linkages [3] (Fig. 4.1).

In general, since urethane rubber has a specific fluidity before cross-linking, it is possible to cast said rubber into a desired shape under atmospheric pressure or a low pressure, and urethane rubber can be used in a broad application, for example, solid tire, roller, and machine parts. However, urethane rubber is poor in cut–crack resistance and therefore the urethane rubber is limited in the application of products and the use condition. Many researchers have been published on synthesis and modification of polyurethanes. Zhen et al. [4] improved of mechanical properties of polyurethane modification with silica; there is an observed increase in the modulus of elasticity, hardness, and compressive strength of the polymer-impregnated silica. Gite and coworkers [5] have been studied chemical and mechanical properties of polyurethane coating using isophorone di-isocyanate and polyethylene glycol (PEG) of different ratio. The field synthesis nanoparticles polyurethane elastomer modification was also studied by some research groups [6–8]. TPU is an outstanding contributor of polymer blends. When added to polyvinyl chloride (PVC), TPU improves compression set, abrasion resistance, and low-temperature flexibility. When added to polycarbonate or acrylonitrile-butadiene-styrene (ABS), a TPU resin with a nominal flexural modulus of 18,000 produces compounds with flexural modulus values up to 150,000 psi. In this way, specialized compounds can be made that improve the properties of polycarbonate or ABS or other plastics. Blending the thermoplastic with elastomeric modifiers is a successful way to improve the toughness characteristics. Many works have documented the effect of various parameters of the modifiers on the toughness response including rubber particle size [9–11], rubber concentration [12], and interparticle distance [11, 13]. Typical rubber particle sizes for effective toughening should be in nanoscale range. Rubber particles with microsize have been found to be relative inefficient, although they might be active in crack bridging. Ceasar G. et al., has used TPUR (Thermoplastic polyurethane rubber) as rubber phase for nylon-6 modification and their nanocomposites with sodium montmorillonite and organically modified montmorillonite are discussed [14]. TPUR has excellent resistance to oxidation with ozone and oxygen which is lacking in conventional elastomers such as natural rubber or synthetic poly isoprene, poly butadiene, and styrene butadiene rubber [SBR] [14]. The Polymer alloys and blends (PAB's) have been utilized in the plastics industries because of their unique and combination of characteristics. The PABs provide improvement in impact strength, chemical resistance, flame resistance, heat distortion temperature (HDT) etc., and also lower cost [optimum] depending upon the type of the polymeric materials used in the PAB's [2]. Hence many plastic alloys' blends are commercially available. The blends of nylon-6 with TPUR in different proportions have been reported [15]. The blending of the nylon 6—TPUR with nanofiller montmorillonite will show the improved on the mechanical and thermal properties [14]. Polyurethane when filled with silica nanoparticles in a weight fraction of 20 % shows an increase in the strain to failure of 500 % compared to pure polyurethane [16]. Also, the formulations with the blending of nanoclays and carbon nanofibers in a TPU were melt-compounded using twin screw extrusion [17]. The TPU blending nanocomposites (TPUNs) are proposed to replace Kevlar-filled ethylene-propylene-diene-monomer rubber. Koo et al. reported in a series of review papers [18–20] on using polymeric composites as ablative thermal protection systems (TPSs) for a variety of military and aerospace applications. The introduction of inorganic nanomaterials as additives into polymer systems has resulted in polymer nanostructured materials exhibiting multifunctional, high-performance polymer characteristics beyond what traditional polymer composites possess. Multifunctional features attributable to polymer blend nanocomposites (PNCs) consist of improved thermal and flame resistance, moisture resistance, decreased permeability, charge dissipation, chemical resistance, and enhanced thermal and electrical conductivity [21–24].

4.2 Preparation of Polyurethane Rubber Nanoblends

4.2.1 Materials

Polymer Resin

Thermoplastic elastomer (TPE) is a family of rubber-like materials that can be processed and recycled like thermoplastics, unlike conventional vulcanized rubbers. There are different examples of TPU material that they are based on polyester or polyether pre-polymers, the latter has better hydrolytic stability and low-temperature performance. The polyesters are typically derived from succinate or adipate monomers combined with ethylene, propylene, butane, or dimers thereof based diols. Succinates can be derived from sugars creating interest because of their bio-origin. The succinate and adipate polyester–polyurethane were compared and found to be similar, with slightly higher glass transition temperature, increased interaction between hard and soft phases and a slightly lower abrasion resistance [6]. They were manufactured by Dow Plastics CO. They have the following properties: outstanding abrasion resistance, low-temperature flexibility and impact resistance, resistance to fuels, oils, and most nonpolar solvents, hydrolytic stability, range of hardness, low-compression set, high compressive strength, low gas/vapor permeability, relatively high moisture vapor transmission rates, easy process ability with conventional extrusion and thermoplastic molding techniques, and low extractable levels.

Nanoparticles

There are different types of the nanoparticles, for examples, the nanoclay and carbon nanofibers (CNFs). The montmorillonite (MMT) nanoclay, that is a natural MMT modified with an organic modifier, is an example for the nanoclay [25]. Achieving exfoliation of organo montmorillonite in various continuous phases is a function of the surface treatment of the MMT clays and the mixing efficiency of the dispersing equipment. Surface treatment of MMT is conveniently accomplished with the exchange of inorganic counterions, for example, sodium and so forth, with quaternary ammonium ions. The example for the second type of nanoparticles is CNFs that are a form of vapor-grown carbon fibers, which are a discontinuous graphitic filaments produced in the gas phase from the pyrolysis of hydrocarbons [26]. In properties of physical size, performance improvement, and product cost,

CNFs complete a continuum bounded by carbon black, fullerenes, and single-wall to multiwall carbon nanotubes on one end and continuous carbon fiber on the other end. CNF has fiber diameters about 100–200 nm and fiber lengths about 30–100 m.

4.2.2 *Preparation of Polyurethane Rubber (TPUR) or TPU Nanoblends*

The different blends were prepared by melt blending TPU and nanoclay or CNF using twin screw extruder which is configured for a wide variety of materials. The energy profile of the screw is adjusted to optimally meet the needs of the target product. For the nanoclay, a relatively long residence time is preferred for the screw profile allowing high shear characteristics for complete separation of the nanoclay platelets with the polymer matrix being uniformly dispersed into the clay galleries [14, 17] (Figs. 4.2 and 4.3). For the CNF, the screw was configured to allow separation of the CNF without fracturing them. Separate volumetric feeders were used for the base polymer and the nanoparticles. The TPU, nanoclays, and carbon nanofiber were dried in a desiccant drier overnight before blending [14, 17]. Injection molded specimens of each blend were prepared and examined by transmission electron microscopy (TEM). The TPU/nanoparticles blends were prepared in the different ratios of them. Where, the nanoparticles are added in the ratio % by weight according to the needs of the target product.

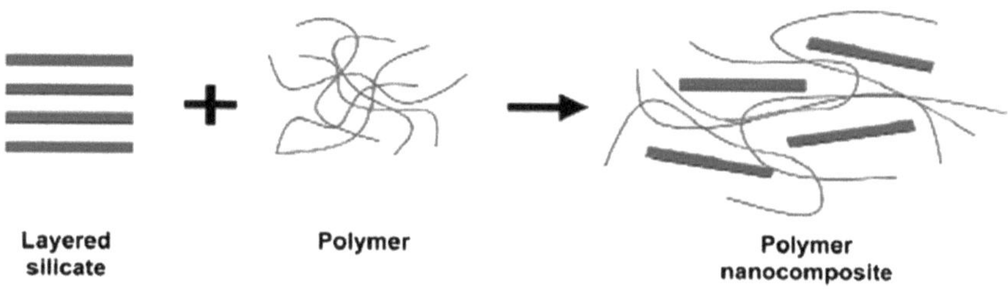

Fig. 4.2 Schematic representation of TPU-nanofiller preparation

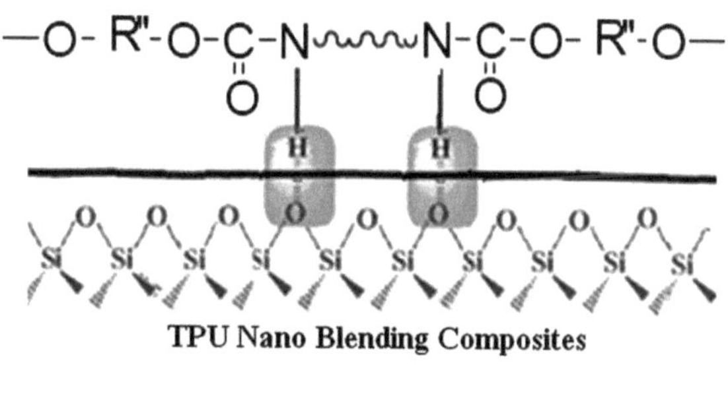

Fig. 4.3 Schematic illustration of polyurethane elastomeric nanoblending composite based on clay ($R=$ ⟨○⟩-CH_2-⟨○⟩)

4.3 Characterizations of Polyurethane Rubber Nanoblends

Thermoplastic polyurethane (TPU) elastomers reinforced with nano-sized particles were obtained by a blending technique. A TPU was blended with nanofiller in different ratios (wt%). The TPU nonoblending was characterized with improved multifunctional behaviors. Each of the TPU-clay and TPU-CNF nano blending has a higher density than the neat TPU. The specific heat of the TPU-clay is substantially higher than the neat TPU, while the TPU-CNF is lower than the neat TPU. The thermal conductivities of both the TPU-clay and TPU-CNF are higher than the neat TPU, with the carbon-based TPU-CNF substantially more thermally conductive than the neat TPU. The thermal stability was improved due to the inorganic nature of the nanoclays. Higher Thermal stability was achieved with 5 % nanoclays. The thermal stability of the organoclay is determined by the thermal stability of the onium ion treatment. It is reported in the literature that the TGA curve for montmorillonite shows the onset of decomposition begins above 200 °C for the onium ion on montmorillonite [27–29]. However, the onset temperature does not consider the kinetic of decomposition and do not have data on the kinetics of the onium ion decomposition. Taking into consideration of the thermal stability of the organoclay, extrusion is still an acceptable process to make many nanocomposites. As nanoparticles' concentration was increasing, the tensile strength and tensile modulus were increasing. In the case of clay, the 5 % formulation has higher tensile strength and modulus. Generally, higher tensile modulus obtained with nanoclay has more compatibility due to the polar interaction of the –OH groups present in the organic modifier. If further clay concentration will increased the tensile strength and flexural modulus, but the impact strength will be drastically reduced due to saturation of the nanoclay concentration. Hence optimum nanoclay was used. For TPU–CNTs, the optimum result of tensile strength and Young's modulus is recorded CNTs at 2 %, however, elongation at break decreased with increasing the amount of CNTs [30].

The structures of TPU, nanofiller, and TPU-nanofiller samples were characterized by FTIR spectroscopy Fig. 4.4. It is clear that the TPU has a strong intensity of

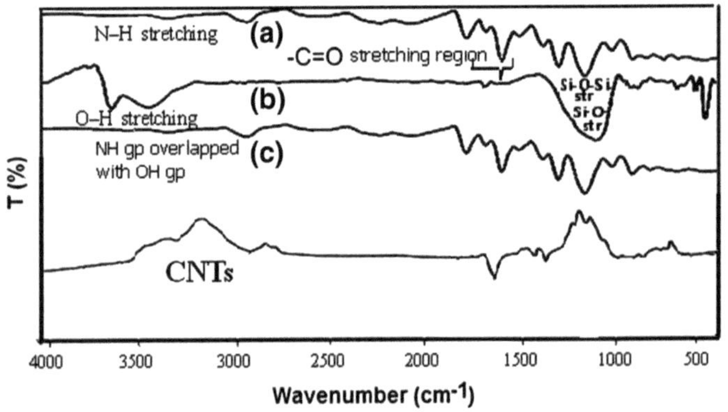

Fig. 4.4 FTIR spectra of TPU (**a**), clay (**b**), TPU-nanofiller (**c**), and CNTs

the bands from 900 to 1300 cm^{-1}, indicating the presence of the NCO/OH with absence of the band around 2250 cm^{-1} of the isocyanate groups (–NCO), referring to completing the reaction and consuming all amount of the MDI. From the FTIR spectra of nanofiller and TPU-nanofiller, the appearance of adsorption peaks for the TPU at 3290, around 2877, 1704, 1515, and 1079 cm^{-1} corresponds to the N–H stretching group, $–CH_2–$ stretching, carbonyl urethane stretching, CHN vibration, and C–O stretching, respectively. Also, the spectra of TPU and TPU-nanofiller refer to all the characteristic bands of TPU staying unmoved with improvement in bands at 1704 and 1515 cm^{-1}, as a result of increase in the formation of urethane linkages in TPU-nanofiller. The band at 1297 cm^{-1} was for amide group that was formed between nanofiller and the TPU matrix [27]. Also, the characteristic bands of clay at 3628 and 3428 cm^{-1} for stretching vibrations of hydroxyl groups, and at 1112 and 462.41 cm^{-1} correspond to Si–O stretching and bending vibration, respectively [27–29]. Figure 4.4 shows the FTIR spectra of CNTs bands at 1000–1300 cm^{-1} denoted the absorption of stretching vibration of C–O bonds. The other peaks at 1616.49 and 1386 cm^{-1} can be assigned to the nitro groups on the surface of carbon materials, as a result of the efficiency of the modification with nitric acid that was demonstrated by a significant increase in the C–O and –NO functional groups [30]. It may be observed from the figure that there are common characteristic peaks at 3230–3330 cm^{-1} for all samples. These peaks are attributed to the bonded OH stretching because of the presence of oxygen-containing groups, such as hydroxyl (OH) and carboxylic acid (COOH).

By using TEM analysis, the particle diameter of CNTs, clay, and TPU-clay blending samples was determined and is shown in Fig. 4.5a–c. Figure 4.5 shows the characteristic TEM image of CNTs. This micrograph showed the surface of purified CNTs was very clean and the most of the impurity phases, such as amorphous carbon and graphitic nanoparticles, were removed by the purification. The diameter of the CNTs was 20–30 nm. The lattice fringe of the organoclay silicate layers was observed in the TPU matrix; it was noticed that, at 5 wt% filler loading, the clay layers were nearly adjoining to each other with some interlayer spacing larger than the one of raw clay. The particle size of clay is 10.80–26.70 nm and that of TPU-clay is 14.47 nm; this is evidence that this organoclay is intercalated in the TPU matrix. At higher organoclay content of 5 wt% bentonite, the layered filler is arranged in intercalated layers. In this case, the effective entry of TPU molecules between the organically modified inter-lamellar spacing could not be achieved to cause an exfoliation of the silicate layers in TPU [27].

The XRD characterization is commonly based on the observation of the shifting position of the diffraction peak of the nanoclay powder intercalated in the polymer matrix [27–29]. It is shown that the reflections belong to 001 plane, natural bentonite clays (d-spacing = 1.24 nm, $2\theta = 7.12°$) Fig. 4.6. The comparative curves obtained for the blending nanocomposite TPU and nanofiller represent an amorphous halo in the TPU-nanofiller sample. Three diffraction peaks in the range $2\theta = 18–23°$ can be noticed. The incorporation of nanofiller reduces the peak intensity of nanofiller, and also a shift by 2θ is seen in Fig. 4.4c for TPU- nanofiller. The XRD of TPU-nanofiller shows that the diffraction peak of TPU-nanofiller

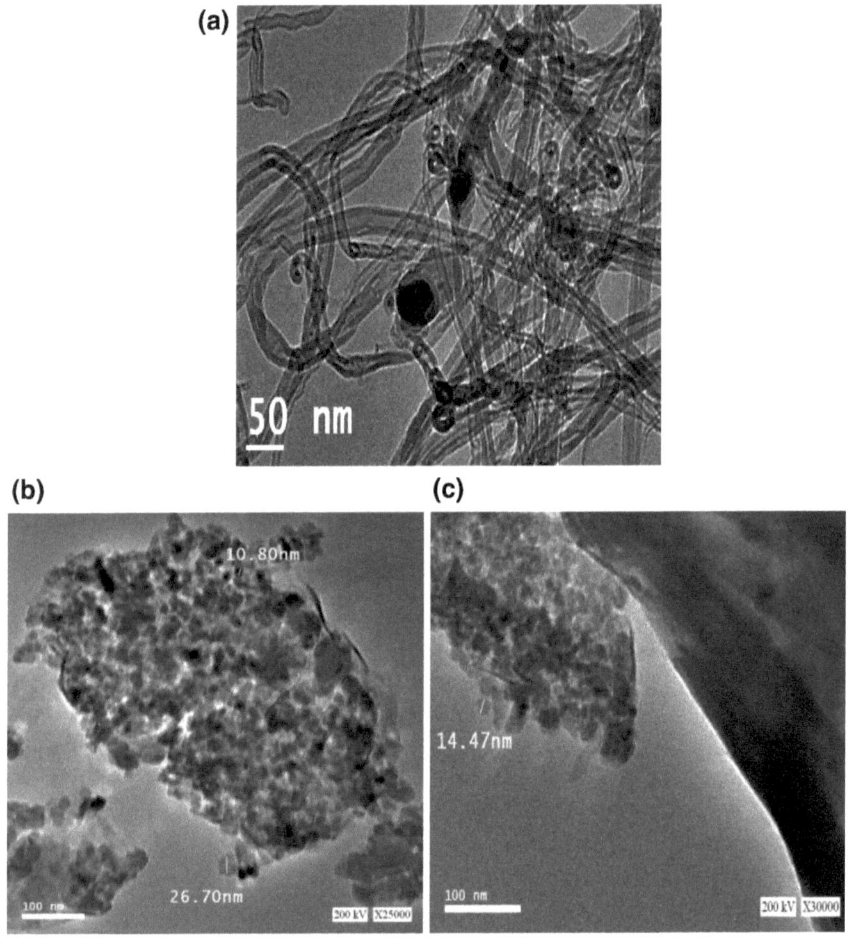

Fig. 4.5 TEM micrographs of CNTs (**a**), clay (**b**), and TPU-clay blending (**c**)

appears at $2\theta = 21.3°$, when compared to $2\theta = 18.48°$ for TPU, suggesting that the nanofiller is highly intercalated. The characteristic diffraction peak of CNTs appeared clearly at $2\theta = 26.2°$, implying that its d-spacing resembles the one of pristine graphite. The CNTs sheets shows a peak of diminished intensity at 23.4° indicating that graphite has been efficiently intercalated within the polymer matrix.

4.3.1 Melt Intercalation

Melt blending is the most promising and practical method to be used in industry. Since solvents are not required, this can be applied to polymer processing industry to produce nanocomposites based on usual compounding devices, such as extruders

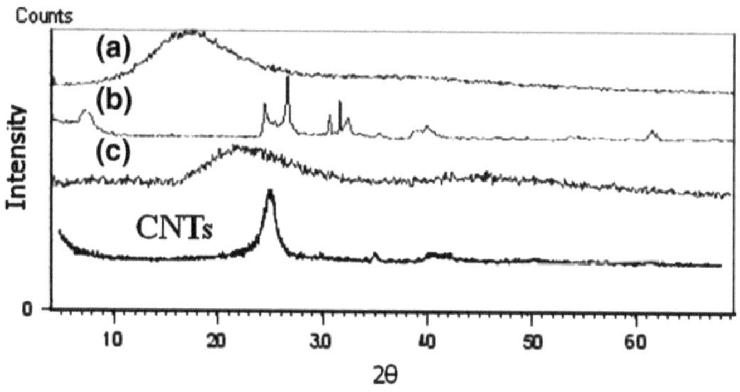

Fig. 4.6 X-ray diffraction (XRD) of TPU (**a**), clay (**b**), TPU-nanofiller (**c**), and CNTs

Fig. 4.7 Schematic view of preparation methods for polymer silicates nanocomposites [31]

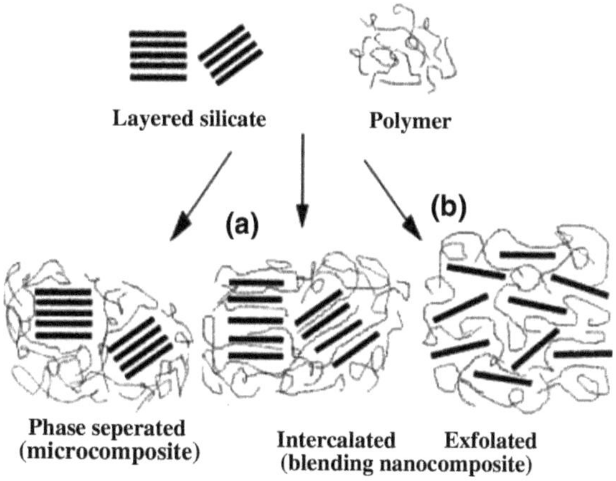

or blends [32]. Direct blending, due to simplest, economical, and environmentally friendly technique, is the mainstream for the fabrication of polymer silicates nanocomposite [32, 33]. Usually during extrusion, the clay is mixed with the molten polymer, whose chains penetrate inside the galleries of clay layers inducing its intercalation as shown in Figs. 4.7 and 4.8 [34]. According to the compatibility between layers' surface and polymer, either intercalated or exfoliated nanocomposites can be obtained, depending on the degree of layers separation (Fig. 4.2). This method commonly requires "trial-and-error-based" experiments to test different process conditions to optimize dispersion [35]. Processing equipment and processing conditions have important role since the combination of shear stresses, residence time, and temperature improve clay dispersion [36]. Other important parameter for effective clay dispersion is the polymer molecular weight. Polymers with high molecular weight enhance clay dispersion since these can transfer shear stress to the clay. Nevertheless, normally lower levels of exfoliation are achieved when compared with in situ polymerization.

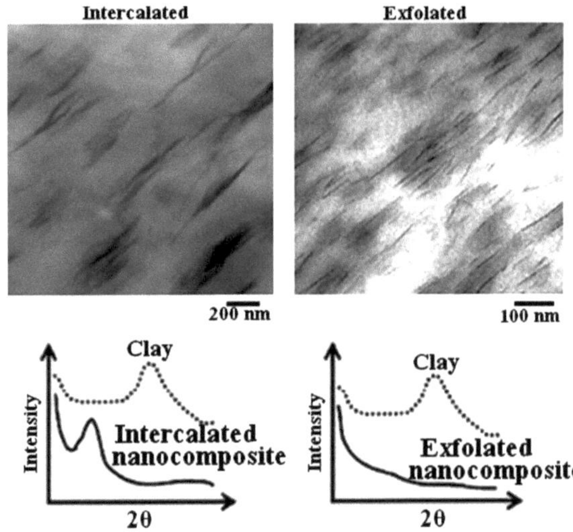

Fig. 4.8 TEM micrograph and WAXS of intercalated and exfoliated organoclay in polymer [34]

Carbon nanotubes (CNTs) are recognized as the stiffest and strongest man-made materials known to date [37]. CNTs are at least 100 times stronger than steel, but only one-sixth as heavy, so nanotube fibers could strengthen almost any material. Nanotubes can conduct heat and electricity far better than copper. CNTs are already being used in polymers to control or enhance conductivity and are added to anti-static packaging. Recognized by high flexibility, lower mass density, and large aspect ratio (typical > 1000) and extremely high tensile modules and strength. A carbon nanotube is a tube-shaped material, made of carbon, having a diameter measuring on the nanometer scale. CNTs are unique because the bonding between the atoms is very strong and the tubes can have extreme aspect ratios. A carbon nanotube can be as thin as a few nanometers yet can be as long as hundreds of microns. Carbon nanotubes have many structures, differing in length, thickness, and number of layers. The characteristics of nanotubes can be different depending on how the graphene sheet has rolled up to form the tube causing it to act either metallic or as a semiconductor. The graphite layer that makes up the nanotube looks like rolled-up chicken wire with a continuous unbroken hexagonal mesh and carbon molecules at the apexes of the hexagons. There are many different types of carbon nanotubes, but they are normally categorized as either single-walled (SWNTs) or multi-walled nanotubes (MWNTs). A single-walled carbon nanotube is just like a regular straw. It has only one layer, or wall. Multi-walled carbon nanotubes are a collection of nested tubes of continuously increasing diameters. They can range from one outer and one inner tube (a double-walled nanotube) to as many as 100 tubes (walls) or more. Each tube is held at a certain distance from either of its neighboring tubes by interatomic forces. Additionally, single-walled carbon nanotubes (SWCNTs) can transport current for long length without significant lost of strength making them more conductive than copper [38]. Ajayan and coworkers [39] reported for the first time polymer nanocomposites using CNTs as filler. Since then, many studies related with processing and resulting mechanical and/or electrical properties of polymer/CNTs nanocomposites have been published. Nevertheless, to obtain polymer/nanotubes nanocomposites with good properties depends on many parameters, such as carbon nanotube production, length, diameter, and chirality. Moreover, the dispersion of carbon tubes in polymeric matrices is recognized as the main difficulty. The failure of tubes dispersion means phase separation and consequently poor mechanical and electrical properties. Then, the outstanding CNTs properties, as high strength and stiffness, could not yet be fully transferred to the final products [37, 40]. Aiming to improve the dispersion of CNTs on a polymer matrix, research has been performed on its functionalization. The CNTs functionalization is divided in two main approaches: covalent or noncovalent modifications [41, 42]. The noncovalent modification covers physical interactions, van der Waals forces, and charge-transfer interactions that can be used to wrap polymers around the nanotubes. Covalent modification involves the chemical reaction of CNTs surface and the incorporation of new species.

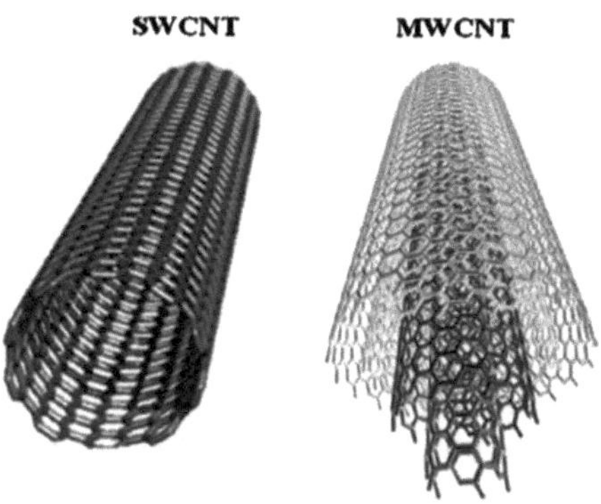

4.4 New Challenges of Polyurethane Rubber Nanoblends

Thermoplastic elastomers, TPEs, have been used to replace the traditional elastomers due to their high mechanical efficiency, easy processability, and low cost. In an increasingly broad range of applications (e.g., automobile industry and military, medical component, electronics, and civil construction), the demands increase as well as the need for materials with improved and multifunctional properties. The combination of a polymer with filler gives rise to a new polymer system with improved performance and added functionalities as a result of the combination of the existing properties. A TPU was blended with nanofiller in different ratios (wt%). The TPU nanoblending was characterized with improved multifunctional behaviors. They have improved thermal and mechanical properties, making them to be in the automotive industry for under-the-hood applications. Unfortunately, a few commercial successes have been achieved with these new materials but more is to come. The next decade might see the emergence of TPU blended with nanofiller in our everyday life. TPU-nanofillers blending has been developed to enhance materials' properties for high-temperature applications. Thermoplastic polyurethane elastomer nanofiller is a novel class of insulation materials developed by Air Force Research Laboratory (AFRL) that is lighter, exhibits better ablative performance and insulation characteristics, and possesses a more cost-effective manufacturing process than the current material, Kevlar-filled ethylene-propylene-diene-monomer rubber (EPDM).

4.5 Recent Advances on Polyurethane Rubber Nanoblends

The recent advance is an obtaining homogeneous TPU-nanofillers blending with improved materials properties and is especially suitable for commercial CNTs and clay materials and further up-scaled production.

4.6 Industrial Values and Applications

Polymer blends are classified as incompatible, compatible, and miscible. Polymer blends suitable for application must be compatible; though they may be miscible, often miscibility is not desired because properties are averaged. It was found that the TPU-nanofillers blend showed an improved thermal resistance, mechanical properties, and physical properties. So, the feasibility of using TPU-nanofillers blending is clearly demonstrated for rocket motor insulation, rocket nozzle ablatives, carbon/carbon composites, and damage tolerant high-performance epoxy fiber-reinforced composite systems. All the studies demonstrated that the TPUNs developed primarily for military usage as internal insulation materials for solid rocket motor which can be extended for commercial usage as flame retardant materials. TPU has many applications including automotive instrument panels, caster wheels, power tools, sporting goods, medical devices, drive belts, footwear, inflatable rafts, and a variety of extruded film, sheet, and profile applications. Further, TPU can be blended with nanofillers for use in numerous applications where greater structural integrity is required, such as for automotive body side moldings. Where, TPU's properties of abrasion resistance, paint ability, high impact strength, and good low-temperature flexibility can all become enhanced, and the blend also demonstrates a good fuel and oil resistance and high melt flow characteristics. In addition, when TPU was added to polyvinyl chloride (PVC), TPU improves compression set, abrasion resistance, and low-temperature flexibility. When added to polycarbonate or acrylonitrile-butadiene-styrene (ABS), a TPU resin with a nominal flexural modulus of 18,000 produces compounds with flexural modulus values up to 150,000 psi. In this way, specialized compounds can be made that improve the properties of polycarbonate or ABS or other plastics. Moreover, polyurethane rubber nanoblends is an industrial urethane rubber compound designed to meet the most stringent requirements of industrial production applications. It is a very hard rubber that offers superior tear and tensile strength as well as impact and abrasion resistance. It is suitable for a variety of industrial applications including making concrete stamping pads, mold liners, and molds for casting concrete. It is also good for making ball mill liners, industrial rollers and belts, and rubber mechanical parts.

4.7 Conclusions

TPU rubber nanoblends or nanocomposites prepared by blending method are directly applicable in industry. A clays and CNTs have been successfully used in the synthesis of TPU-nanofiller blending. These fillers made improvements in mechanical and physical properties of these products to be in industry. Also, these nanoblends materials have enhancement in thermal and mechanical properties might even extend the use of them in the automotive industry for under-the-hood applications. As nanofiller concentration increases, the tensile strength, tensile modulus, and modulus were increasing. The elongation at break decreases. Optimum mechanical properties were obtained with 5 % clay filled TPU-clay blends. The XRD shows the intercalation of clay, and thermal stability was improved for both types of nanoblending composites. If further clay concentration will increase, the tensile strength and modulus may increase but the impact strength will be drastically reduced. If we obtained the thermo physical properties making these nano-blending used in a numerical model to predict the material behavior at a higher heat flux, simulate a solid rocket motor environment.

References

1. Paul DR, Bucknall CB (2000) Polymer blends: formulation and performance. Wiley, New York
2. Utracki LA (1989) Polymer alloys and blends: thermodynamics and rheology. Hanser, Munich
3. Tatjana R, Saulute B, Krzysztof P, Jan P (2006) Application of polyurethane-based materials for immobilization of enzymes and cells: a review. Chemija 17:74–89
4. Yang ZG, Zhao B, Qin SL, Hu ZF, Jin ZK, Wang JH (2004) Study on the mechanical properties of hybrid reinforced rigid polyurethane composite foam. J Appl Polym Sci 92:1493–1500
5. Gite VV, Mahulikar PP, Hundiwale DG, Kapadi UR (2004). Polyurethane coating using trimer of isophorone diisocyanate. J Sci Ind Res 63:348–354
6. Sonnenschein MF, Guillaudeu SJ, Landes BG, Wendt BL (2010) Comparison of adipate and succinate polyesters in thermoplastic polyurethanes. Polymer 51:3685–3692
7. Robert CK (2008) In: Pascal F (ed) Handbook of specialty elastomers (Chapter 4). CRC Press, Raton, pp 134–154
8. John S (1997) In: Arcella V, Ferro R (ed) Modern fluoropolymers (Chapter 2). Wiley, West Sussex, pp 71–90

9. Borggreve RJM, Gaymans RJ, Schuijer J (1987) Impact behaviour of nylon-rubber blends: 5. Influence of the mechanical properties of the elastomer. Polymer 30(1):71–77. ISSN 0032-3861
10. Margolina A, Wu S (1988) Percolation model for brittle-tough transition in nylon/rubber blends. Polymer 29(12):2170–2173. ISSN 0032-3861
11. Wu S (1988) A generalized criterion for rubber toughening: the critical matrix ligament thickness. J Appl Polym Sci 35(2):549–561. ISSN 0021-8995
12. Baldi F, Bignotti F, Tieghi G, Riccó T (2006) Rubber toughening of polyamide 6/organoclay nanocomposites obtained by melt blending. J Appl Polym Sci 99(6):3406–3416. ISSN 0021-8995
13. Jiang W, Yuan Q, An L, Jiang B (2002) Effect of cavitations on brittle-ductile transition of particle toughened thermoplastic. Polym Commun 43(4):1555–1558. ISSN 0032-3861
14. GunaSingh C, Soundararajan S, Palanivelu K (2013) Studies on mechanical, thermal properties and characterization of nanocomposites of nylon-6—thermoplastics poly urethane rubber [TPUR] blend. IOSR J Appl Chem 4(1):65–75
15. Palanivelu K, Sivaraman P, Sharma SK, Verma SK (2003) Int J Plast Technol 7:133
16. Petrović ZS, Javni I, Waddon A, Bánhegyi G (2000) Structure and properties of polyurethane–silica nanocomposites. J Appl Polym Sci 76:133–151
17. Ho WK, Koo JH, Ezekoye OA (2010) Thermoplastic polyurethane elastomer nanocomposites: morphology, thermophysical, and flammability properties. J Nanomater 2010:11 (Article ID 583234)
18. Koo JH, Ho DWH, OA Ezekoye (2006) A review of numerical and experimental characterization of thermal protection materials—part I: numerical modeling. In: Proceedings of the 42nd AIAA/ASME/SAE/ASEE joint propulsion conference, AIAA Paper 2006–4936, vol 8. AIAA, Reston, pp 5895–5945
19. Koo JH, Ho DWH, Bruns MC, Ezekoye OA (2007) A review of numerical and experimental characterization of thermal protection materials—part II: properties characterization. In: Proceedings of the AIAA/ASME/ASCE/AHS/ASC structures, structural dynamics and materials conference, AIAA Paper 2007-2131. AIAA, Reston, pp 5136–5175
20. Ho DWK, Koo JH, Bruns MC, Ezekoye OA (2007) A review of numerical and experimental characterization of thermal protection materials—part III: experimental testing. In: Proceedings of the 43rd AIAA/ASME/SAE/ASEE joint propulsion conference, AIAA Paper 2007-5773, vol 8. AIAA, Reston, pp 7524–7598
21. Pinnavaia TJ, Beall GW (eds) (2000) Polymer-clay nanocomposites. Wiley, New York
22. Koo JH (2006) Polymer nanocomposites: processing, characterization, and applications. McGraw-Hill, New York
23. Morgan AB, Wilkie CA (eds) (2007) Flame retardant polymer nanocomposites. Wiley, Hobken
24. Bhattacharya SN, Gupta RK, Kamal MR (2008) Polymeric nanocomposites: theory and practice. Hanser Gardner, Cincinnati
25. Koo JH (2006) Polymer nanocomposites: processing, characterization, and applications. McGraw-Hill, New York, pp 19–26
26. Luo ZP, Koo JH (2007) Quantifying the dispersion of mixture microstructures. J Microsc 225 (2):118–125
27. Abdeen Z (2014) Enhanced recovery of Pb^{2+} ions from aquatic media 4 by Using polyurethane composite as adsorbent. Environ Process. doi:10.1007/s40710-014-0048-0
28. Dimitry IHO, Abdeen IZ, Ismail AE, Saad LGA (2011) Studies of particle dispersion in elastomeric polyurethane/organically modified montmorillonite nanocomposites. Int J Green Sci Nanotechnol 3:197–212
29. Dimitry IHO, Abdeen IZ, Ismail AE, Saad LGA (2010) Preparation and properties of elastomeric polyurethane organically modified montmorillonite nanocomposites. J Polym Res 17:801–813
30. Hesheng X, Mo S (2005) Preparation and characterization of polyurethane carbon nanotubes composites. Soft Matter 1:386–394

31. Oriakhi CO, Farr IV, Lerner MM (1997) Thermal characterization of poly (styrene sulfonate)/ layered double hydroxide nanocomposites. Clays Clay Miner 45:194–202
32. Nguyen QT, Baird DG (2006) Preparation of polymer–clay nanocomposites and their properties. Adv Polym Tech 25:270–285
33. Kalaitzidou K, Fukushima H, Drzal LT (2007) A new compounding method for exfoliated graphite-polypropylene nanocomposites with enhanced flexural properties and lower percolation threshold. Compos Sci Technol 67:2045–2051
34. Youssef AM (2013) Polymer nanocomposites as a new trend for packaging applications. Polym Plast Technol Eng 52:635–660
35. Kumar AP, Depan D, Singh Tomer N, Singh RP (2009) Nanoscale particles for polymer degradation and stabilization—trends and future perspectives. Prog Polym Sci 34:479–515
36. Giraldi ALFM, Bizarria MTM, Silva AA, Velasco JI, d'Ávila MA, Mei LHI (2008) Effects of extrusion conditions on the properties of recycled poly(ethylene terephthalate)/nanoclay nanocomposites prepared by a twin-screw extruder. J Appl Polym Sci 108:2252–2259
37. Spitalsky Z, Tasis D, Papagelis K, Galiotis C (2010) Carbon nanotube–polymer composites: chemistry, processing, mechanical and electrical properties. Prog Polym Sci 35:357–401
38. Dürkop T, Kim BM, Fuhrer MS (2004) Properties and applications of high mobility semiconducting nanotubes. J Phys Condens Matter 16:R553
39. Ajayan P, Stephan O, Colliex C, Trauth D (1994) Aligned carbon nanotube arrays formed by cutting a polymer resin—nanotube composite. Science 265:1212–1214
40. Andrews R, Weisenberger MC (2004) Carbon nanotube polymer composites. Curr Opin Solid State Mater Sci 8:31–37
41. Singh P, Campidelli S, Giordani S, Bonifazi D, Bianco A, Prato M (2009) Organic functionalisation and characterisation of single-walled carbon nanotubes. Chem Soc Rev 38:2214–2230
42. Choudhary V, Gupta A (2011) Polymer/carbon nanotube nanocomposites. In: Dr. Yellampalli S (ed) Carbon nanotubes—polymer nanocomposites. InTech

Chapter 5
Chlorosulfonated Rubber-Based Nanoblends: Preparation, Characterization and Applications

Gordana Marković, Milena Marinović-Cincović, Vojislav Jovanović, Suzana Samaržija-Jovanović and Jaroslava Budinski-Simendić

Abstract In this chapter, the preparation, characterization, and applications of chlorosulphonated polyethylene (CSM) rubber based nano blends have been investigated. CSM rubber based nanoblend compounds have been prepared on a laboratory-size (300x600 mm) two-roll mixing mill maintained at 40±5 °C. Curing was performed at 160 °C up to the optimum cure time (t), which was determined from the rheograph using a Monsanto R-100 model. The mechanical properties, namely tensile strength (TS), tensile modulus at 100 % and 300 % elongation (M100 and M300), elongation at break (Eb%) and hardness have been followed up as a function of nano fillers and irradiation dose (dose rate of 10 kGy h^{-1} and total absorbed dose of 100, 200, 300, and 400 kGy) as well as blend composition. Cross-linking density has been determined using swelling measurements. A scanning electron microscope has been used for morphological study. Thermal stability of samples has been determined using thermogravimetric (TG) measurements. Dynamical mechanical properties were studied on a mechanical analyzer. The assignment of the principal bands in chlorosulphonated polyethylene (CSM) rubber based nano blends has been carried out by FTIR ATR spectra. Elastomeric materials based on the chlorosulfonated polyethylene (CSM) rubber nanoblend have a quite good crack resistance, wet grip, and weather resistance. On the other hand, they show a good resistance to interactions with alcohol's as well as strong acids and bases. This type of material also exhibits excellent ultraviolet light and oxygen stability.

G. Marković (✉)
Tigar, Nikole Pašića 213, 18300 Pirot, Serbia
e-mail: gordana1markovic@gmail.com

M. Marinović-Cincović
Institute of Nuclear Science Vinča, University of Belgrade, Belgrade, Serbia

V. Jovanović · S. Samaržija-Jovanović
Faculty of Natural Science and Mathematics, University of Priština, Kosovska Mitrovica, Serbia

J. Budinski-Simendić
Faculty of Technology, University of Novi Sad, Novi Sad, Serbia

G. Marković and Visakh P.M. (eds.), *Rubber Nano Blends*, Springer Series on Polymer and Composite Materials,
DOI 10.1007/978-3-319-48720-5_5

Keywords Chlorosulphonated polyethylene rubber based nano blends · Preparation · Characterization · Applications

5.1 Introduction

Blends of rubbers are of technological and commercial importance, since they allow the user to access properties of the final blended and vulcanized rubber that is not accessible from a single, commercially available rubber alone. These potentially improved properties include chemical, physical, and processing benefits. In reality, all blends show compositionally correlated changes in all of these properties compared to the blend components. The technology of rubber blends is largely focused on the choice of individual rubbers and the creation of the blends to achieve a set of final properties [1]. Miscible blends of rubbers that consist of a single rubber phase with microscopically uniform cross-linking and distribution of reinforcing agents reflect a compositionally weighted average of the microscopic properties of the constituents. Miscible blends are commonly used though they have been very rarely recognized. Analysis of such blends, particularly after vulcanization, is difficult. The current analytical techniques are only slightly more capable than the classical techniques of selective precipitation of the components of an unvulcanized rubber blend from solution [2]. Immiscible blends show additional, more complex changes due to a microscopically inhomogeneous phase structure of the two component rubbers. The two separate phases typically have differences in the retention of the fillers and plasticizers as well as vulcanization in the presence of the curative. Changing the properties of rubbers by uneven distribution of fillers and vulcanization is, however, the more common use of blends of immiscible rubbers [3].

Many reviews [4–6] are focused on immiscible blends of rubbers. Isobutylene-based rubbers include butyl rubber and halogenated butyl rubber, and due to their impermeability and resistance to heat and oxidation, these polymers find application in tire innerliners, innertubes, curing bladders and envelopes, and other specialty applications where air retention and resistance to heat and oxidation are desired [7]. Chlorosulfonated polyethylene (CSM) is an important rubber which has been frequently used in many applications such as sheeting cable and geomembrane due to its outstanding resistance to deterioration by heat, oils, ozone, and oxidation [8]. The blending together of isobutylene-based rubbers (IIR) with chlorosulfonated polyethylene (CSM) is intended to produce a vulcanizate with the best properties from each component, i.e., CSM's impermeability and IIR and CIIR's good strength properties and resistance to oil [9]. Chlorobutyl (CIIR) and bromobutyl (BIIR) rubbers are manufactured similarly to butyl rubber, but with an added subsequent halogenation step. Due to the nearly equivalent reactivity ratios of the monomers and low concentration of isoprene, the isoprenyl units in these polymers are randomly distributed along the polymer chain. Commercially, chlorobutyl rubber (CIIR) and bromobutyl rubber (BIIR) are the most important derivatives of butyl rubber. Their predominant application is in tire innerliners with the performance of the modern radial tire being impacted by the amount of halobutyl rubber used in the innerliner compound. Typical performance requirements of the tire innerliner are air retention, adhesion to tire carcass compound, resistance to fatigue and cracking, heat resistance, high tensile, and tear strength [10].

Silica has been used as non-black reinforcing filler in the rubber industry for a long time. Reinforced rubber blends are suitable materials for industrial practice [11]. Obviously, in such systems, both components (filler and rubber) have the reactive groups for the additional cross-linking reaction to take place [12]. Besides the so-called dispersion forces, there are a variety of other interactions between particles. A key one among them is hydrogen bonding, which tends to be a significant force in the case of fumed silica, where hydrogen bonding between surface silanol groups takes place. Many authors have investigated the role of filler networking in the elastic properties of elastomer composites [13]. The surface of hydrated or precipitated silica is highly polar and hydrophilic because of the presence of numerous silanol groups. The evaluation of the reinforcing effect is usually based on the determination of the processability of elastomers and the dynamic mechanical properties of cross-linked materials [14].

Among many other particulate fillers, the use of carbon black (CB) is very important in rubber technology for obtaining elastomers with good dynamic mechanical as well as ultimate properties. New materials are also considered as fillers, whether for more economical reasons or rather to impart some desirable characteristics [15, 16] of CB as one of the major inputs in tire industry, has some extraordinary features. The surface specific area of CB is considered to be one of the most critical parameters considered for achieving a high degree of reinforcement. This characteristic property supplies the necessary adsorption sites to the polymer chains for wetting. As polymer chains may also enter into the voids of the filler agglomerates and aggregates, the rheological properties (especially the viscosity) of the filled polymer melts are affected by both the so-called physical entrapment and the physical cross-links, created by a strong physisorption of the chains on the filler surface. These effects may also play an important role in the reinforcement of cross-linked elastomeric matrices. One of the most important parameters for the performance of a composite material is the degree of adhesion (phase bonding) at the matrix–filler interface [17]. In this study, fourteen different SBR/CSM formulations were produced by mechanical mixing and vulcanized. The mechanical properties such as the hardness, the modulus, the tensile strength, and the abrasion loss were investigated. On the other hand, some physical characteristics were investigated via the vulcanization test.

The main interest was related to the effect of biogenic silica (BSi) as a filler on the deformation mechanisms and mechanical properties of composites based on the polymer matrix. Diatomaceous comes as a biogenic silica (BSi) from the word diatom, which is the single-celled aquatic plant. Diatoms are living phytoplanktons which form an important part of marine and freshwater. The silicate shell, termed "frustule," supports a fleshy body. The fossil beds are skeletons sunk to the bed of lake or sea after the body died and formed deposits there. Diatoms are found in a great variety of forms [18]. The skeletal remains of diatoms with submicron-sized hole are described as "indescribable particle shape" far from any simple classification [19]. The specific diatom morphology and chemical resistance could be the base of using this material as filler in composites. There is little in the literature on this subject. Adhesion between matrix and filler phases in composites is important in crack propagation and composite failure. The study of the deformation and failure process, their relation to the structure of composite, and the strength of interactions between the polymer matrix and filler phases may provide useful information for the development of the new composite materials with diatoms.

5.2 Preparation

The rubber-based nano-blend compounds were prepared using a laboratory mixing roll mill of dimensions 400 × 150 mm at a speed ratio of the rollers $n_1/n_2 = 28/22$, at a roller temperature of 40–50 °C. The time of the preparation of the blends was ca. 20 min. Curing was performed at 160 °C up to the optimum cure time (t_{c90}), which was determined from the rheograph obtained using a Monsanto R-100 model. The addition of ingredients during mixing was carried following the same order and conditions of mixing. Two different types of rubbers were used IIR and CIIR blended with CSM at different ratios. IIR (Butyl 068), Mooney viscosity 51 (ML1+8 at 125 °C) with isoprene content of 1.15 mol% and CIIR (Chlorobutyl 1066), Mooney viscosity 38 (ML1+8 at 125 °C) with isoprene content of 1.95 mol % and halogen content of 1.26 mol% were supplied by Exxon Mobil, England. Their respective technical specifications are shown in Tables 5.1 and 5.2, and the compound formulations of the samples are given in Tables 5.3, 5.4, 5.5, and 5.6.

The compound formulation is given in Table 5.4.

Table 5.1 Analytical data for the fillers

Filler	BET (N_2)/m^2g^{-1}	DBP absorption/cm^3 100 g^{-1}	pH	Particle size
Nano-silica	160–200	220	6.7	15 nm
Micro-silica	/	130	10	28 μm

Table 5.2 Technical specification of isobutylene-co-isoprene (IIR), chlorinated isobutylene-co-isoprene, and chlorosulfonated polyethylene (CSM) rubbers

Elastomer	Grade	Mooney viscosity (ML1 +8 at 125 °C)	Isoprene (mol%)/sulfur content (mol%)	Halogen (mol%)	Halogen (mol%)
Isobutylene-co-isoprene (IIR) rubber	Butyl 068	51	1.15	–	–
Chlorinated isobutylene-co-isoprene rubber	Chlorobutyl 1068	38	1.95	Cl	1.28
Chlorosulfonated polyethylene (CSM) rubber	Hypalon 40S	46	1	Cl	34.5

5.3 Characterization

5.3.1 *Cure Characteristics*

The sheeted rubber compound was conditioned at 23 ± 2 °C for 24 h prior to cure assessment on a Monsanto moving die rheometer (model 100S, USA) at 160 °C. The compound formulations (Tables 5.3, 5.4, 5.5, and 5.6) are exprcssed in part per hundred parts of rubber (phr). All test specimens were compression-molded at 160 °C during the respective optimum cure time (t_{c90}) determined from the Monsanto rheometer. The scorch time, t_{s2}, is the time to 2 units of torque increase above minimum torque, and optimum cure time, t_{c90}, is the time to 90 % of maximum torque development calculated from the following expression:

$$M_{t90} = (M_h - M_l) \times 0.9 + M_l \tag{5.1}$$

where M_h is the maximum torque and M_{c90} is a optimum cure time determined from the cure traces generated at 160 ± 2 °C by oscillating disk rheometer curemeter at

Table 5.3 Formulation of the compounds and nano- and micro-silica particle-sized filled polychloroprene (CR) and chlorosulfonated polyethylene rubber blends

Compounds (phr[a])	1	2	3	4	5	6	7
CSM	50	50	50	50	50	50	50
CR	50	50	50	50	50	50	50
ZnO	2.5	2.5	2.5	2.5	2.5	2.5	2.5
Stearic acid	2	2	2	2	2	2	2
Nano-silica	0	10	20	35	0	0	0
Micro-silica	0	0	0	0	10	20	35
Naphthenic oil	0	3.3	6.7	10	3.3	6.7	10
MgO	2	2	2	2	2	2	2
ETU	0.5	0.5	0.5	0.5	0.5	0.5	0.5
TMTD	1	1	1	1	1	1	1
S	0.5	0.5	0.5	0.5	0.5	0.5	0.5
Diethylene glycol	0	0.3	0.6	1.1	0.3	0.6	1.1

[a]Part per hundred parts of rubber

Table 5.4 Blend formulations (phr) of IIR and CIIR with CSM (cure system: 5.0 phr zinc oxide, 2.0 phr stearic acid, 0.5 phr MBTS, 2 phr TMTD, and 2 phr sulfur)

CSM	IIR/CIIR
100	0
80	20
60	40
40	60
20	80
0	100

Table 5.5 Blend combination of NR and CNR with CSM rubber

CSM (phr)	NR/CNR (phr)[a]
100	0
80	20
60	40
40	60
20	80
0	100

Cure system: zinc oxide 5; stearic acid 2; sulfur 1.5; tetramethylthiuram disulfide (TMTD) 2; magnesium oxide 4
[a]Part per hundred parts of rubber

Table 5.6 Formulation of the cross-linked systems based on acrylonitrile butadiene rubber (NBR), polychloroprene rubber (CR), chlorosulfonated polyethylene rubber (CSM), and nano- and micro-particle-sized SiO_2-filled NBR/CSM and CR/CSM cross-linked systems

Compounds/phr[a]	1	2	3	4	5	6	7	8	9
NBR[b]	100	–	50	50	50	–	–	–	–
CSM[c]	–	100	50	50	50	–	50	50	50
CR[d]	–	–	–	–	–	100	50	50	50
ZnO	5	5	2.5	2.5	2.5	5	2.5	2.5	2.5
Stearic acid	2	2	2	2	2	2	2	2	2
SiO_2^e	–	–	–	35	–	–	–	35	–
SiO_2^f	–	–	–	–	35	–	–	–	35
Naphthenic oil[g]	–	–	–	11	11	10	10	10	10
MgO[h]	–	4	2	2	2	–	2	2	2
ETU[i]	–	–	–	–	–	1	0.5	0.5	0.5
TMTD [j]	2.5	2	2.5	2.5	2.5	–	1	1	1
CZ [k]	2		1	1	1	–	–	–	–
S	0.5	1	0.75	0.75	0.75	–	0.5	0.5	0.5
Diethylene glycol	–	–	–	1.1	1.1	–	–	1.1	1.1

[a]Part per hundred parts of rubber, [b]Butadiene–acrylonitrile copolymer rubber–Europren N3345, [c]Chlorosulfonated polyethylene rubber—Hypalon 40S, [d]Polychloroprene rubber—Baypren 110, [e]White filler (15 nm)—Ultrasil VN 3, [f]White filler (28 μm)—Celite 281, [g]Naphthenic oil—TEPHIL NAPHTHENE P. O. 100, [h]Magnesium oxide—Anscor P, [i]Ethylene thiourea—ETU, [j]Very fast accelerator—tetramethylthiuram disulfide (TMTD), [k]Low fast accelerator—N-cikloheksil-2-benzothiazole sulfenamide(VULKACIT CZ)

an angular displacement of ±3 and a frequency of 1.7 Hz [20]. The cure rate index (CRI) is the measure of rate of vulcanization based on the difference between optimum cure time of vulcanization t_{c90} and the scorch time t_{s2}. It can be calculated from the relation:

$$\mathrm{CRI} = \frac{1}{t_{c90} - t_{s2}} \times 100 \tag{5.2}$$

5.3.2 Mechanical Properties

The sheets were cut, marked according to the time and temperature determined from the oscillating disk rheometer, and vulcanized in clean polished molds in a press at 160 °C. The vulcanized sheets were cut into dumbbell-shaped specimens (five replicates from each sample) for the evaluation of the mechanical properties using an electronic tensile testing machine (Zwick 1425, Germany) at a speed of 500 mm/min. Samples of at least 0.12 mm in thickness with flat surface were cut for hardness test. The measurement was carried out according to ASTM D 2240 using durometer of model 306L type. The tensile properties were determined using the tensile testing machine type Zwick 1445 according to ASTM D-412. The average value of the mechanical properties was calculated using at least three samples. A cross-head speed of 50 mm/min was used, and the tests were performed at 25 °C.

5.3.3 Hardness Measurements

Samples of at least 0.12 mm in thickness with flat surface were cut for hardness test. The measurement was carried out according to ASTM D 2240 using durometer of model 306L type. The unit of hardness is expressed in (Shore A).

5.3.4 Swelling Measurements and Cross-linking Density Determination

The investigated rubber composites were obtained as small sheets (1 g) weighed accurately and immersed in a 25 cm^3 benzene solvent in closed simple bottles, at 25 °C. After 72 h, the samples were removed from the solvent and weighed in a glass-stopper bottle after removing the surface fluids by blotting with a filter paper. The volume swelling in percentage was calculated using the following equation [21]:

$$R_v = 1 + [(w - w_0) - 1]\rho_r/\rho_s \quad (5.3)$$

$$R_w = w/w_0 \quad (5.4)$$

$$v_2 = 1/R_v \quad (5.5)$$

where R_v and R_w are the volume and mass ratio of rubber samples in the swollen gel; w and w_0 are weights of the sample before and after swelling, respectively; ρ_r and ρ_s are densities of the rubber and the solvent, respectively; and v_2 is the volume fraction of the sample in the swollen gel. The elastic modulus (E) values were obtained from dumbbell samples by stress–strain measurements [22] . The values of E were obtained from the initial slope of the curves in neo-Hooke's stress–strain relation σ versus (α and α^{-2}) in the equation below:

$$\sigma = E\left(\alpha - \alpha^{-2}\right) \tag{5.6}$$

where α is the elongation ratio (L/L_0).

The E modulus is related to cross-linking density by the well-known equation from the kinetic theory of elasticity:

$$E = g\nu RT \tag{5.7}$$

where the factor $g < 1$, n is the concentration of cross-linking density, R is the universal gas constant, and T is the absolute temperature.

$$\nu = \rho/M_c \tag{5.8}$$

$$M_c = \rho RT/E \tag{5.9}$$

where M_c is the average molecular weight between two cross-links. The Flory–Huggins interaction parameter, w, was calculated for the rubber composites in a benzene solvent, according to the formula below:

$$x = \left[\nu V_1\left(\nu_2^{1/3} - 0.5\nu_2\right) + \ln(1 - \nu_2) + \nu_2\right]/\nu_2^2 \tag{5.10}$$

where V_s is the molar volume of the benzene solvent. Cross-linking density was determined from the swelling data. The samples were allowed to swell in the benzene solvent, and the equilibrium uptake was noted. The molecular weight between the cross-link M_c was calculated using the following equation:

$$M_c = -2V_1\rho_r V_r^{1/3}/\ln(1 - V_r) + V_r + 2x^2V_1^2 \tag{5.11}$$

where M_c is the molecular weight of the polymer between two cross-links; ρ_r is the rubber composite density; V_s is the molar volume of benzene solvent; V_r is the volume fraction of rubber composites; and w is the filler–polymer interaction parameter. The Kraus model, used for the investigation of the filler–rubber interaction, shows how to quantitatively calculate the effect of particles that are either completely unbound or completely and permanently bound to the polymer. The Kraus equation is given in the equation [23]:

$$\frac{V_{r0}}{V_{rf}} = 1 - [m\theta/(1 - \theta)] \tag{5.12}$$

$$m = 3C\left(1 - V_{r0}^{1/3}\right) + V_{r0} - 1 \tag{5.13}$$

where V_{r0} and V_{rf} are the volume fraction rubber in the solvent-unfilled and solvent-filled rubber composites; y is the volume fraction of the filler in the rubber composites; the parameter m is obtained from the graphic V_{r0}/V_{rf} versus $\Theta/1 - \Theta$ and describes how the swelling of polymer matrix decreases when the filler content increases; and C is the constant characteristic for the filler, but it depends on the solvent.

5.3.5 Morphological Study Using a Scanning Electron Microscope

SEM images were obtained using a JEOL 5300 SEM operated at an acceleration voltage of 20 kV and equipped with energy-dispersive systems. All the studied samples were coated with a thin layer of gold in a nitrogen atmosphere.

5.3.6 Thermal Stability

Thermogravimetric (TG) measurements have been performed on the PerkinElmer TGS-2 instrument. Rubber samples were heated from 25 to 750 °C at a heating rate of 10 °C min^{-1} in the nitrogen atmosphere with the gas flow rate of 50 cm^3 min^{-1}. The mass of the samples ranged from 17 to 25 mg.

5.3.7 DSC Measurements of Rubber Blends

The thermal stability of pure CR/CSM rubber blend and CR/CSM rubber blends filled with nano- and micro-silica particles was investigated by non-isothermal thermogravimetric analysis (TG) using a PerkinElmer TGS-2 instrument. The measurements were conducted at heating rates of 10 °C min^{-1} in a dynamic nitrogen atmosphere (flow rate 25 cm^3 min^{-1}), in the temperature range from 25 to 750 °C. The average sample mass was about 8 mg.

5.3.8 Dynamic Mechanical Analysis (DMA)

Dynamic mechanical properties were studied on a mechanical analyzer Tritec 2000 (Triton Technology). Complex Young's modulus $E^* = E' + iE''$ (E' is a storage and E'' loss modulus) was measured in a single cantilever bending mode. Temperature dependencies of dynamic mechanical properties were measured at a heating rate of 2 °C/min in the temperature range from −50 to 150 °C at 5 frequencies (0.31; 1; 3.1; 10; and 31 Hz). Multifrequency mode was used to determine a dependence of E^* on frequency (f) at frequency $f = 1$ Hz. Cross-linking density can be obtained using DMA by calculating M_c, the (number) average molar mass of elastically effective chains (between cross-links), according to the following equation derived from the theory of rubber elasticity [24, 25]:

$$E = 3\rho RT/M_c$$

where E is the elastic modulus in the rubbery plateau (assimilated to the storage modulus E^* measured at 1 Hz), R the gas constant, T the absolute temperature, and q the material density at T.

5.3.9 FTIR Analysis

Fourier transform infrared spectra (FTIR) were recorded on a Bruker IFS-66 spectrometer with an attenuated total reflection (ATR) attachment. The internal reflection element (IRE) chosen was a 45° KRS-5. Samples for measurements were approximately 0.3-mm-thick sheets, prepared by compression molding between two Teflon films at a temperature of 100 °C.

5.4 New Challenges

5.4.1 Nano- and Micro-silica-Filled CR/CSM Rubber Blends

The nano-silica-filled CR/CSM rubber blends have higher values for cure characteristics than micro-silica particle-sized filled compounds. Nano-silica type activated the vulcanizing process through the promotion of hydrogen sulfide formation and the rupture of S–N linkage when heated with sulfonamides in rubber either in the presence or in the absence of other compounding ingredients. The strong peak from SiO–C at 1070 and 1079 cm^{-1} in the IR spectrum of nano- and micro-silica-filled CR/CSM rubber blend suggested the presence of interaction with silica which leads to increased stability of the filled CR/CSM polymer blends. According to the Kraus equation, it can be observed that the nano-silica-filled CR/CSM rubber blend has higher Vr_0/V_{rf} values and better rubber filler interactions than that micro-silica-filled composites. The higher values of the glass transition temperatures for the nano- than the micro-filled cross-linked systems are indicated by DMA analysis. The density of the nano-silica-filled CR/CSM rubber blends was lower than that of the micro-silica-filled and was transparent at all filler concentration.

5.4.2 Cure Characteristics

The maximum and minimum torque (M_h and M_l) and delta torque (ΔM) at 160 °C (Table 5.7) increase with the increase of the nano- and micro-silica loading filler of the CR/CSM cross-linking systems. The difference between M_h and M_l could be used as an indirect indication of the cross-linking density of the rubber compound [26]. Both scorch time (t_{s2}) and optimum curing time (t_{c90}) are found to decrease noticeably with increasing silica loading. The nano-silica particle-sized filled rubber blends have higher values than micro-silica particle size for all cure characteristics.

5.4.3 IR Spectral Analysis

ATR spectra carried out on samples of pure CR/CSM rubber blend and of two composites with two different silica particle size contents are displayed in Fig. 5.1 (spectrum 1, 2, and 3). The assignment of the principal bands is presented in Table 5.8. Almost all bands [27] of the pure polymer blend especially those associated with the stretching bands of the CH_2 groups are shifted to lower wave numbers in the filled polymer blend spectra. The bands located at 2943 and 2862 cm^{-1}, respectively, assigned to the asymmetric and symmetric stretch of the methylene group, are found at 2943 and 2860 cm^{-1} for the nano-silica-filled CR/CSM composite and at 2926 and 2858 cm^{-1} for the micro-silica-filled CR/CSM

Table 5.7 The cure characteristics of unfilled and nano- and micro-silica particle-sized filled CR/CSM compounds

Cure characteristics	Sample						
	1	2	3	4	5	6	7
M_h (dNm)	38	34	41	47	37	34	39
M_l (dNm)	8	10	11	13	8	8	8
ΔM (dNm)	30	24	30	34	29	26	31
t_{s2} (min)	2.32	3.25	3.13	2.38	2.14	1.59	1.39
t_{c90} (min)	23.08	19.15	20.21	21.15	21.5	23.07	16.24

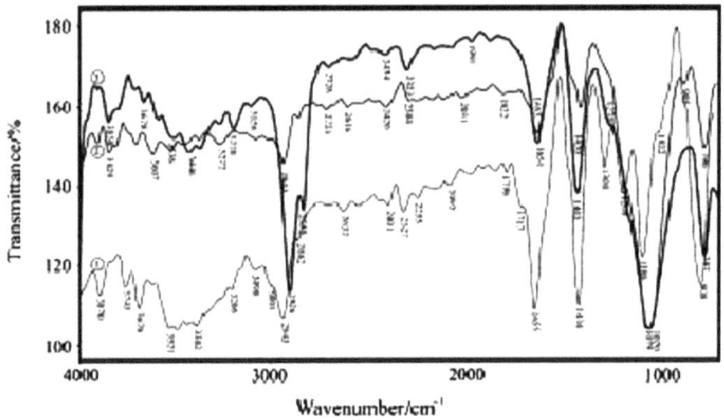

Fig. 5.1 FTIR ATR spectrum of (*1*) unfilled, (*2*) nano-, and (*3*) micro-silica particle-sized filled polychloroprene (CR) and chlorosulfonated polyethylene (CSM) rubber blends CR/CSM (50/50 w/w)

composite sample. The absorption band at 1108 cm^{-1} in the cross-linked CR/CSM rubber blend is assigned to symmetric vibration of SO_2 group.

The peak of –S–Cl which is attributed to the –SO_2Cl group from CSM is absent, and the peak of –C–Cl from CR rubber is absent also. This suggests the cross-linking of the rubber blend macromolecules. Similar observations were made by Ahankaria et al. [28] with reference to the thermally induced self-cross-linking PVC/CR, as well as the PVC/CSM.

The absorption bands at 1655 and 1434 cm^{-1} in the cross-linked CR/CSM rubber blend are assigned, respectively, to symmetric and asymmetric bending vibration of C=C group. The weak absorption bands at 3536 and 3446 cm^{-1} which are attributed to hydrogen bands are strongly attenuated by reaction macromolecules and fillers. New absorption bands centered at 1070 and 1079 cm^{-1} could be assigned to the symmetric stretching modes of SiO–C group from silica filler, and moving symmetric and asymmetric bending vibration of C=C group from 1655–1651 to 1653 cm^{-1} and from 1434–1430 to 1445 cm^{-1} (Fig. 5.1, spectrum 2 and 3) suggests the interaction of the polymers matrix with the silica, also.

5.4.4 Mechanical Properties

Table 5.9 represents the effect of nano- and micro-silica loading on tensile properties (tensile strength, modulus, and elongation at break) and hardness of the rubber vulcanizates. The tensile strength increases with increasing nano- and micro-silica loading in CR/CSM rubber compound. Maximum value of tensile strength, for nano-silica-filled CR/CSM rubber compound, is by samples 4 and 7 with 35 phr nano- and micro-silica (15.8 and 8.5 MPa). The improvement in tensile properties for nano-silica-filled composites seems to be responsible for the

Table 5.8 FTIR-ATR spectroscopic data for unfilled and nano- and micro-silica particle-sized filled polychloroprene (CR) and chlorosulfonated polyethylene (CSM) rubber blends CR/CSM (50/50 w/w)

	Wave number (cm^{-1})	Group	Assignment
CR/CSM (50/50 w/w) rubber blends	2943	CH_3–, –CH_2–, > CH–	ν(–C–H)
	2862	CH_3–, –CH_2–, > CH–	ν(–C–H)
	1655	–CH=CH–	ν(–C=C)
	1434	RHC=CHR	δ(–C–H)
	1108	–SO_2–	$\nu(SO_2)_{sym}$
CR/CSM/nano-silica rubber blend	2943	CH_3–, –CH_2–, > CH–	ν(–C–H)
	2860	CH_3–, –CH_2–, > CH–	ν(–C–H)
	1655	–CH=CH–	ν(–C=C)
	1430	RHC=CHR	δ(–C–H)
	1108	–SO_2–	$\nu(SO_2)_{sym}$
	1070	SiO–C	ν(SiO–C)
CR/CSM/micro-silica rubber blend	2926	CH_3–, –CH_2–, > CH–	ν(–C–H)
	2858	CH_3–, –CH_2–, > CH–	ν(–C–H)
	1651	–CH=CH–	ν(–C=C)
	1445	RHC=CHR	δ(–C–H)
	1108	–SO_2–	$\nu(SO_2)_{sym}$
	1079	SiO–C	ν(SiO–C)

ν—stretching vibrations; δ—bending vibrations in plane

Table 5.9 Some mechanical properties of unfilled and nano- and micro-silica particle-sized filled CR/CSM compounds

Characteristics	Sample						
	1	2	3	4	5	6	7
Tensile strength (MPa)	6.4	13.6	14.6	15.8	6.6	8.3	8.5
Elongation at break (%)	688	572	523	485	600	600	549
Hardness, Shore A	57	65	69	71	55	55	55
Modulus at 300 % elongation (MPa)	2.8	6.5	7	7.1	2.8	3	3

improved dispersion of the nano-silica into the rubber matrix, leading to a higher bound rubber formation [29]. The values of 300 % modulus are increased from 6.5 to 7.1 MPa for nano-silica and 2.8 to 3 MPa for micro-silica-filled compounds, strongly attenuated by reaction macromolecules and fillers, respectively. The hardness values of micro-silica-filled compounds are the same and much lower than nano-silica-filled compounds as shown in Table 5.9.

Values of elongation at break are decreasing with increasing silica loading. The increase of hardness, modulus, and tensile strength and higher values for nano- than micro-silica composites can possibly be explained by the improvement of the degree of adhesion at interfaces between the nano-silica and the CR/CSM rubber blends.

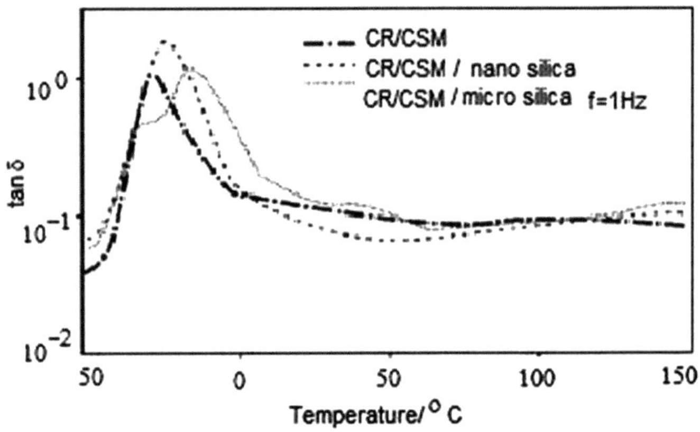

Fig. 5.2 Damping properties (tanδ) as a function of temperature of the unfilled and nano- and micro-silica particle-sized filled CR/CSM (50/50) rubber blend

Table 5.10 The glass transition temperatures (T_g) estimated either by dynamical spectroscopy (T_{gm}), the modulus of elasticity (G), determined from stress strain measurements and cross-linking density (ν), calculated from the storage modulus E' (at 70 °C, 1 Hz) or (V) determined by swelling

Compounds	Sample		
	1	4	7
T_{gm} (1 Hz)/°C	−21	−29	−16
G (MPa)	6.4	15.8	8.5
$E'/10^3$ MPa	5.2	18.1	7.5
V/mol m^{-3}	553	646	173
N_c/mol m^{-3} (from E′)	1115	933	756

5.4.5 Dynamic Mechanical Analysis

Dynamic mechanical analysis (DMA) is based on the differences in mechanical damping properties of unfilled and filled rubber vulcanizates [30]. Figure 5.2 and Table 5.10 show the elastic modulus, E_0, and tan$d$ determined from a dynamic mechanical analyzer under the conditions of 1 Hz and 0.1 % dynamic strain amplitude. The addition of filler to rubber matrix would lower the peak value of tan$d$, designated as (tanδ_{max}), in the plot of tanδ versus temperature. E_0 of blends increases significantly with the addition of silica, particularly at a temperature above the CR/CSM rubber blend glass transition temperature (T_g). This result indicates a strong reinforcement caused by strong interaction between silica and CR/CSM rubber blend via silanol groups and chlorine atoms, respectively (Table 5.10) [31]. Values of glass transition temperature (T_g) determined from the damping peak tanδ are summarized in Table 5.10. In theory, the shift in T_g would be caused by silica–polymer interaction and/or a change in phase compatibility.

Table 5.11 The cross-linking density of unfilled and nano- and micro-silica particle-sized filled CR/CSM compounds

Parameters	sample						
	1	2	3	4	5	6	7
r_p (g cm^{-3})	1.23	1.27	1.3	1.3	1.13	1.16	1.17
R_w	1.07	1.69	1.74	1.07	2.03	1.84	1.07
R_v	1.09	1.22	1.2	1.06	1.21	1.1	1.06
V (mol m^{-3})	553	231	430	646	391	242	173
M_c (gmol^{-1})	1115	2749	1580	1008	1621	2683	3884

5.4.6 Cross-linking Density

The values of R_W and R_V (Table 5.11) decrease with increasing amount of both nano- and micro-silica.

The cross-linking density, V, increased with increasing silica concentration and was the largest for the sample containing nano- and micro-silica of 35 and 10 phr, respectively (samples 4 and 5). The largest chain molecular mass was obtained for micro-silica-filled compounds with amount of 35 phr (sample 7) and the smallest for the sample 4 with the highest amount of nano-silica. The swelling degree decrease can be explained by linkage combination between polymer chains and functional groups on the silica surface and by a possible increase in cross-linking bonds in the polymer matrix because of filler influence on the vulcanization reaction [32]. The cross-linking density (v) of the nano- and micro-silica particle-sized filled CR/CSM rubber blend was calculated from the equilibrium storage modulus in the rubber region over the α-relaxation temperature according to the rubber elasticity theory. The correlations of some data for the modulus of elasticity (G) determined

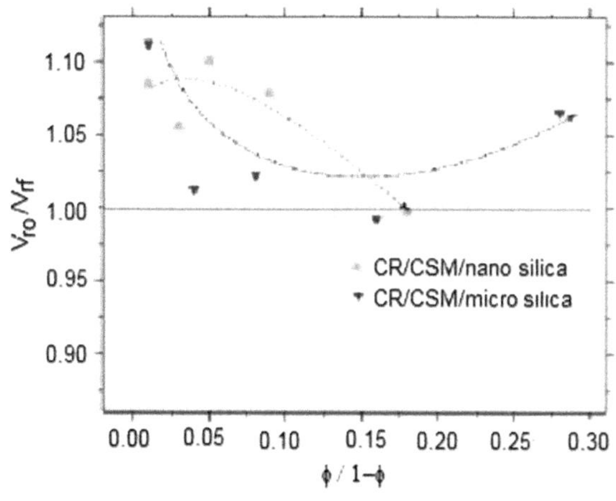

Fig. 5.3 Plot of V_{r0}/V_{rf} versus $\phi(1-\phi)$ according to the Kraus equation for swelling test of nano- and micro-silica particle-sized filled CR/CSM (50/50) rubber blend

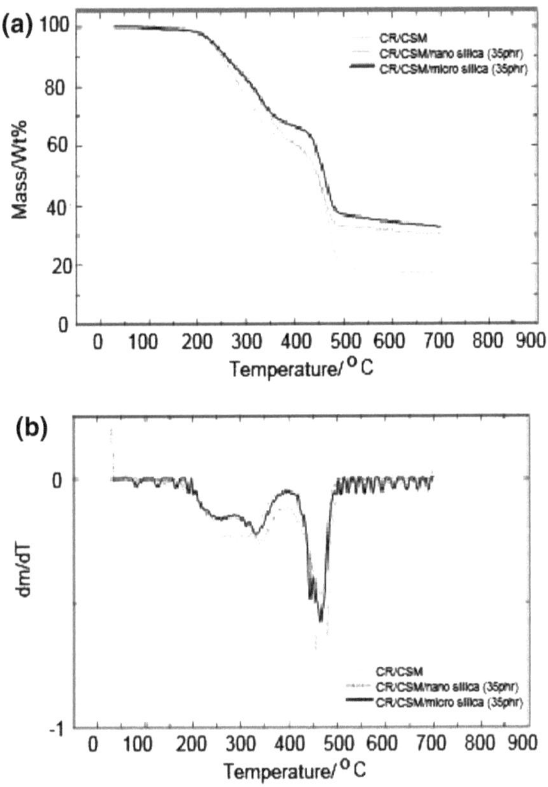

Fig. 5.4 **a** TG curves of unfilled and nano- and micro-silica particle-sized filled CR/CSM (50/50) rubber blend; **b** DTG curves of unfilled and nano- and micro-silica particle-sized filled CR/CSM (50/50) rubber blend

from stress–strain measurement and cross-linking density (ν) calculated from the storage modulus E_0 (at 70 °C, 1 Hz) or (V) determined by swelling are included in Table 5.10. Figure 5.3 shows the plot of V_{r0}/V_{rf} versus $\phi/(1 - \phi)$ according to the Kraus equation [33]. It can be observed exhibit deviations from the proposed linear model, which are greater than 1.

As filler loading increases, the ratio of V_{r0}/V_{rf} decreased. The nano-silica-filled CR/CSM rubber blend has higher V_{r0}/V_{rf} values than micro-silica composites and show better polymer–filler interaction. Thermal stability of CR/CSM compounds filled with nano- and micro-silica particle size. The thermal stability of pure CR/CSM rubber blend and CR/CSM rubber blends filled with nano- and micro-silica particles has been discussed. The thermal decomposition of CR/CSM rubber blend and nano- and micro-silica particle-sized filled CR/CSM rubber blends occurs generally in two main stages (Fig. 5.4a, b; Table 5.12). The first step degradation occurs in the temperature region 250–300 °C, and the second-step degradation is about at 440 °C. The data for the initial decomposition temperature (IDT) (5 % mass loss) [34] and 10 and 30 % mass loss temperatures are given in Table 5.12 and Fig. 5.4a. It can be seen that according to the IDT nano-silica-filled CR/CSM compounds are more stable than unfilled and micro-silica-filled

Table 5.12 Temperature values of unfilled and nano- and micro-silica particle-sized filled CR/CSM compounds for selected mass loss (5, 10, and 30 %)

Compounds	Sample						
	1	2	3	4	5	6	7
DTG peak values (°C)	293.79	326.51	330.35	343.01	334.16	328.87	332.17
	464.64	467.77	468.15	465.14	46.2	459.81	464.51
Mass loss (%)	21.15	26.96	25.44	27.93	24.76	25.67	24.36
	61.58	47.67	58.28	56.85	62.06	50.63	52.66
Total mass loss (%)	84.01	75.73	71.57	69.91	77.52	73.59	67.17
Temperature values for selected mass loss							
$T_{5\%}$ (°C)	225.67	227.81	232.53	233.25	230.5	231.01	232.02
$T_{10\%}$ (°C)	251.61	241.77	262.16	236.3	247.39	254.42	238.64
$T_{30\%}$ (°C)	326.14	353.17	351.45	347.93	338.99	345.12	358.48

$$\text{-[-}CH_2\text{-}CH{=}CH\text{-}CH_2\text{-]x—[-}\overset{\overset{\displaystyle O}{\|}}{\underset{\underset{\displaystyle O}{\|}}{S}}\text{-}CH_2\text{-]y-} \xrightarrow{440\text{-}700^{\circ}C} CH_2\text{-}CH{=}CH\text{-}CH_2 + SO_2 + H_2C{=}CH_2$$

Scheme 5.1 Decomposition mechanism of CR/CSM rubber blend

compounds. In the next step (at 10 % mass loss) with 20 phr of nano-silica and 30 phr of micro-silica particle-sized filled CR/CSM rubber have higher temperature values (262 and 259 °C) than other samples. At 30 % mass loss, the CR/CSM rubber blends filled with 20 phr of nano-silica and 30 phr of micro-silica particle size are the most stable also than other compounds. Figure 5.4a, b shows that unfilled CR/CSM rubber blend compounds are initially more stable than other samples. In the CR/CSM rubber blend compounds filled by nano- and micro-silica types (Table 5.12; Scheme 5.1), the degradation takes place in two stages. Up to 100 °C, there is no change and hence no mass loss. The first-step degradation occurs in the temperature region 250–350 °C and the mass loss in the range 21.15–27.93 and 21.15–28.76 % (depending on the type and amount of silica in CR/CSM rubber blends), most probably indicating the evaporation of oil (naphtalene). The second-step degradation starts at 440 °C and ends at 700 °C. The percent of mass loss at this region is in the range of 47.67–61.58 and 52.66–62.06 % for nano- and micro-silica-filled CR/CSM rubber blend indicating polymer degradation. The large flexible polysulfidic linkages undergo chain scission and convert into monosulfidic and disulfidic linkages. Synthetic rubber decomposes by random chain scission with intramolecular hydrogen transfer. Table 5.12 presents the DTG peak values of CR/CSM rubber compounds filled by nano- and micro-silica particle size. DTG peaks shift to higher temperature values indicating greater thermal stability of samples containing 20 phr nano and micro silica.

Morphology Microscopic Studies

Scanning electron microscopic (SEM) studies of fractured surfaces were observed by scanning electron microscopy (SEM) in order to gain a better insight into the mechanism of failure of SiO_2-reinforced NBR/CSM and CR/CSM cross-linked systems. The SEM microphotographs at 2000 magnification of the surfaces of

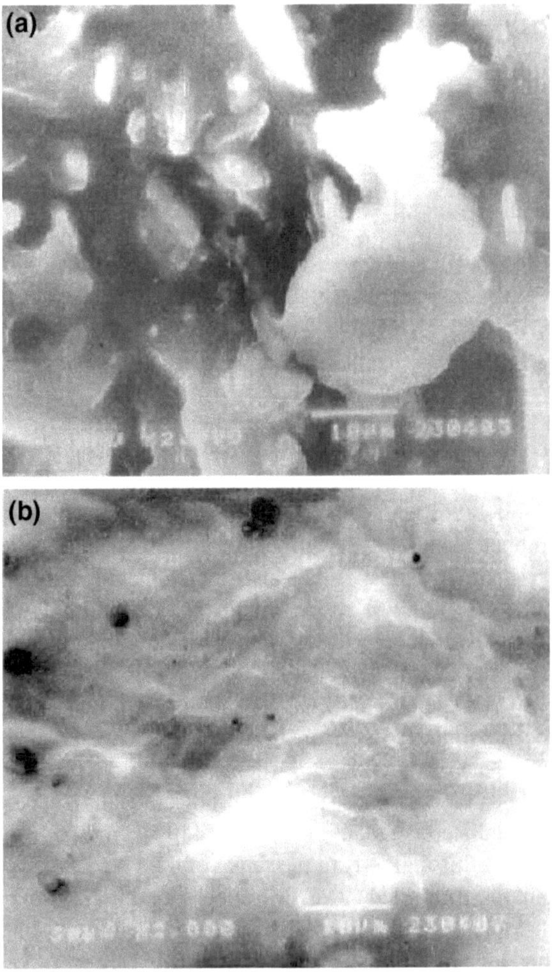

Fig. 5.5 SEM micrograph at 2000× magnification of CR/CSM rubber blends: **a** micro-silica and **b** nano-silica filled

CR/CSM polymer blends are shown in Fig. 5.5. The surface of the CR/CSM polymer blend with nano-SiO_2 is more polished than that with micro-SiO_2. In the microphotographs of micro-SiO2-filled CR/CSM blend a micro-crack of size about 50 μm can be seen.

5.4.7 SBR/CSM/ Carbon Black Rubber Composites

For SBR/CSM blends, the addition of filler leads to a significant enhancement in the physical properties if CSM is the major blend constituent. The blends containing CB N339 show higher hardness, tensile strength, and abrasion resistance when compared to the ones containing CB N772. The experimental data indicate that this can be caused by the improvement in the polymer–filler phase bonding which results from the interfacial interaction of functional groups of the polymer and the active sites of the filler surface leading to a higher apparent cross-linking density. As the concentration of CSM in the rubber blend increases, the cure characteristics, such as optimum cure time, scorch time, and induction time at 160 °C, increase for the blends filled with CB N339. The results for the curing behavior are consistent with the mechanical properties of the SBR/CSM blends.

5.4.8 Mechanical Properties

It was shown that for SBR/CSM blends, the addition of filler leads to a significant enhancement in the physical properties if CSM is the major blend constituent. The blends containing CB N339 show higher hardness, tensile strength, and abrasion resistance when compared to the ones containing CB N772. The experimental data indicate that this can be caused by the improvement in the polymer–filler phase bonding which results from the interfacial interaction of functional groups of the polymer and the active sites of the filler surface leading to a higher apparent cross-linking density. As the concentration of CSM in the rubber blend increases the cure characteristics, such as optimum cure time, scorch time and induction time at 160 °C increases for the blends filled with CB N339. The results for the curing behavior are consistent with the mechanical properties of the SBR/CSM blends. Figure 5.6 shows a comparison of the evolution of the hardness of SBR/CSM blends reinforced by the two types of highly structured carbon black (N339 and N772). It can be seen that the hardness of the blend increases in both systems when the content of CSM rubber increases. However, the increase of the hardness is much more pronounced in the case of the active filler N339 than in the one where N772 was used. Above 50 % of CSM in the blend, the hardness tends to reach a plateau value for N339, whereas it passes through a maximum in the blends filled with N772. The volume fraction of CB affects the mechanical properties of the polymer blends positively [35].

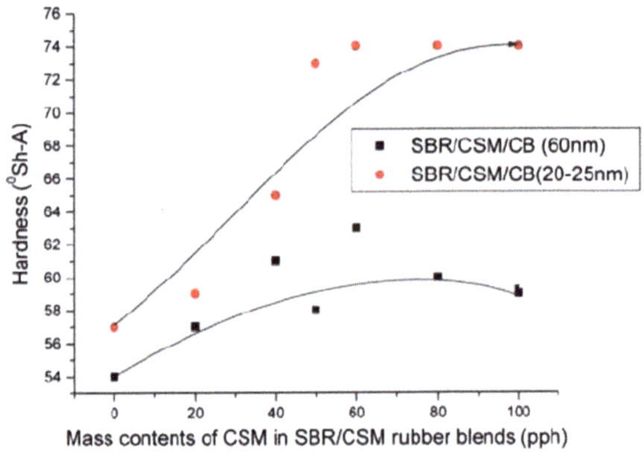

Fig. 5.6 The influence of CSM rubber content on hardness of rubber blends reinforced by highly structured carbon black, with the primary particle size either 20–25 nm or 60 nm

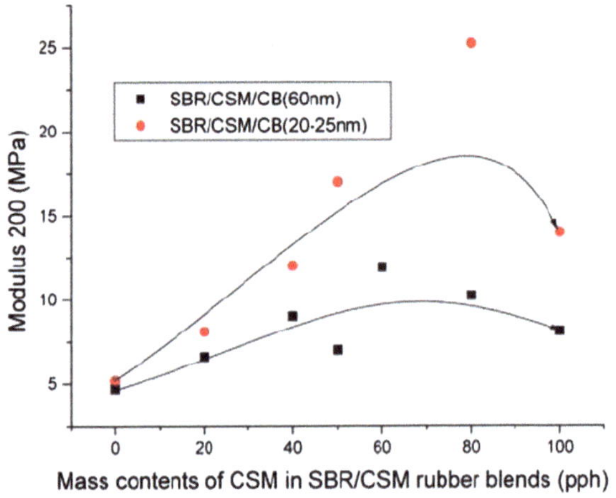

Fig. 5.7 The influence of CSM rubber content on modulus 200 of rubber blends reinforced by highly structured carbon black, with the primary particle size either 20–25 nm or 60 nm

The relationship between the tensile strength, the modulus (M200) at 200 % elongation and the elongation at break (Eb) and the content of CSM in the blends is shown in Figs. 5.7, 5.8, 5.9, and 5.10. All the figures show increasing physical properties with the increasing CSM content in the blend. Again, the reinforcement by CB with smaller primary particle size becomes evident. If the modulus 200 is considered, it becomes evident that the targeted value for SBR/CSM/CB 339 passes a maximum value, whereas for SBR/CSM/N339, the modulus tends to a limiting value (Fig. 5.7).

The reinforcement by CB is shown evidently by the evolution of the tensile strength (Fig. 5.8). For SBR/CSM/CB N339, the values range from 6 to 39 MPa. In comparison with that, the effect of the more coarse black N772 increases the tensile strength from 6 to 18 MPa only. In both cases, the tensile strength increases approximately linearly with the content of CSM in the blend, indicating a preferential distribution of the CB into this polymer phase. If the elongation at break is

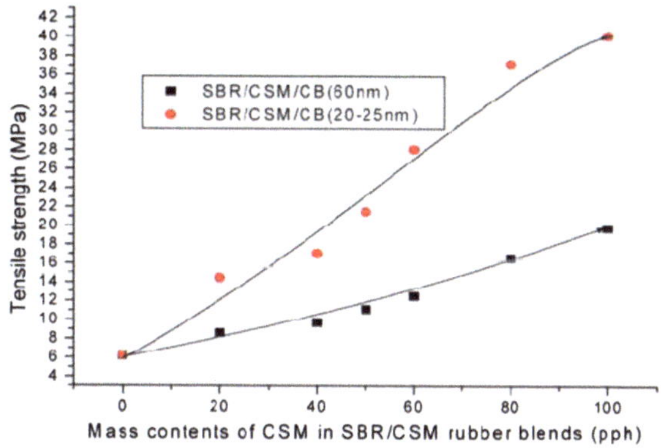

Fig. 5.8 The influence of CSM rubber content on tensile strength of rubber blends reinforced by highly structured carbon black, with the primary particle size either 20–25 nm or 60 nm

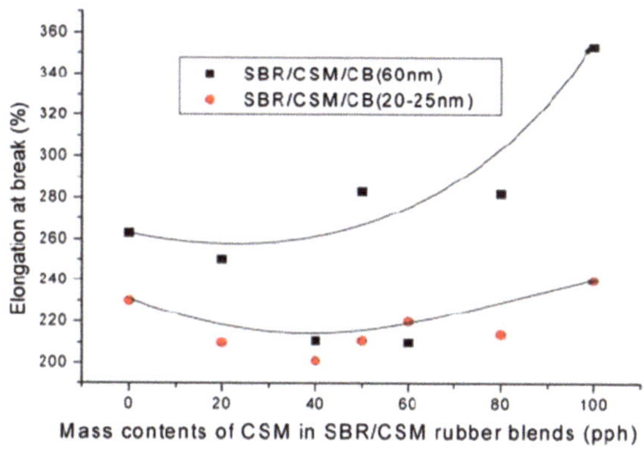

Fig. 5.9 The influence of CSM rubber content on elongation at break of rubber blends reinforced by highly structured carbon black, with the primary particle size either 20–25 nm or 60 nm

considered (Fig. 5.9), it is shown from Fig. 5.8 that the values remain unchanged until 40–50 % of CSM in the blend is reached. After this threshold, the values increase strongly when the CSM content increases. In addition, a slight inversion of the reinforcing effects can be observed. The lower values (about 200 and 240 %) were obtained for systems reinforced by CB N339, and significantly higher values (260 and 350 %) were recorded for all systems reinforced by CB N772. As an overall picture, it seems that the values of tensile properties and elongation at break are synergistic above 50 % CSM leading to higher energies at break. The values gained for the abrasion loss are illustrated in Fig. 5.10. Despite the higher scatter of the experimental data, it can be shown that (i) the abrasion loss is higher for the compounds containing CB N772 than for the ones containing CB N339. However, as a systematic picture in both groups of compounds, the abrasion loss is passing through a maximum at 40–50% CSM underlining again the effect of the reinforcement in the range of high CSM concentrations. More details about the values of the specific mass are tabulated in Table 5.13. It can be seen from this table that the specific mass ranges from 1.082 to 1.312 g/cm^3. The reason for this could be differences in the dispersion and distribution of the fillers in rubber blend.

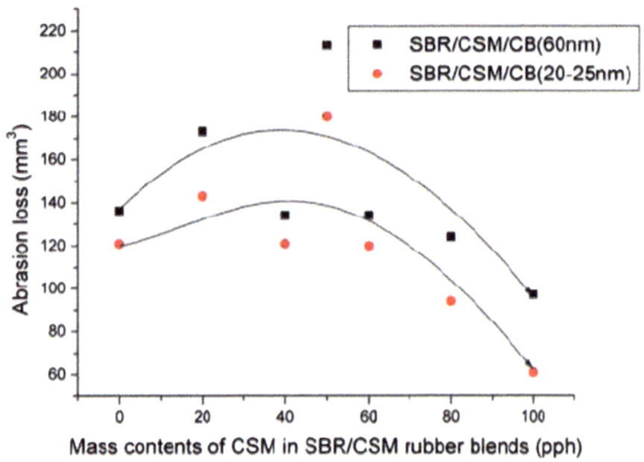

Fig. 5.10 The influence of CSM rubber content on abrasion loss of rubber blends reinforced by highly structured carbon black, with the primary particle size either 20–25 nm or 60 nm

Table 5.13 Curing characteristics and experimental values of specific mass of rubber blends reinforced highly structured carbon black with the primary particle size was either 20–25 nm or 60 nm at 160 °C

Sample name	Maximum torque (M_h) (dNm)	Minimum torque (M_l) (dNm)	Scorch time at 160 °C, t_{s2} (min)	Optimum cure time at 160 ° C, t_{c90}, (min)	Cure rate index (CRI) (min^{-1})	Specific mass (g/cm^3)
1	15	5	2.5	15	0.08	1.08
2	19	4.5	5.3	11	0.17	1.11
3	25	8.5	8	23	0.06	1.14
4	28	5	9.8	21	0.09	1.15
5	29	5	10	30	0.05	1.17
6	29	4	10.5	21	0.09	1.21
7	32	4	4.7	19	0.07	1.25
8	33	4	6.8	12	0.18	1.12
9	38	4	5	16	0.09	1.25
10	38	5	4	14	0.1	1.27
11	38	5	4	19	0.07	1.29
12	39	5	4	17	0.07	1.3
13	39	4	5	16	0.09	1.31
14	32	4	4	19	0.07	1.31

Equilibrium swelling measurements were performed in order to evaluate the apparent cross-linking density in the blend compound, which contains the contribution of the chemical cross-links as well as the one of the strong physical attachments of the polymer to the filler. The values of the cross-links density calculated by using Eq. 5.14 are given in Fig. 5.11. A clear linear increase of the apparent cross-linking density with the CSM content in the blend was observed irrespective of the type of carbon black used. This indicates as the major contributor a strong interaction between the functional groups of CSM and the filler leading to

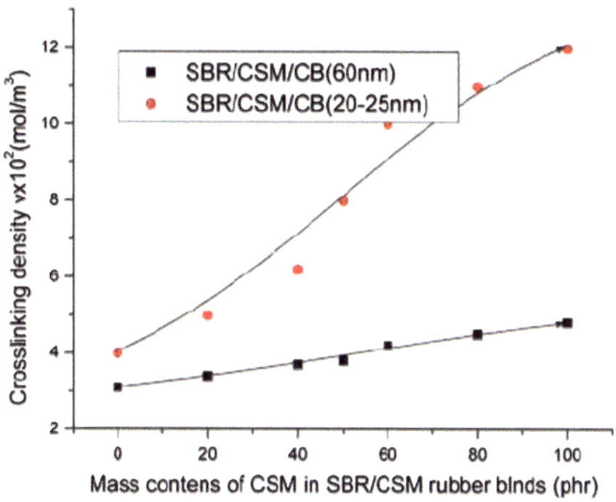

Fig. 5.11 Relation between cross-linking density ν values and CSM mass content of SBR/CSM rubber blends reinforced by highly structured carbon black with the primary particle size either 20–25 nm or 60 nm

strong physical cross-links and eventually also chemical ones which are not affected by the polar solvent used.

$$\nu_{\mathrm{EANC}} = \rho_{\mathrm{p}} N_{\mathrm{A}} / M_{\mathrm{c}} \tag{5.14}$$

$$M_{\mathrm{c}} = -V_1 \rho_{\mathrm{p}} \frac{\left(\varphi_{\mathrm{p}}^{1/3} - \frac{\varphi_{\mathrm{r}}}{2}\right)}{\ln\left(1 - \varphi_{\mathrm{p}}\right) + \varphi_{\mathrm{p}} + x_{\mathrm{p}} + x_1 \varphi_{\mathrm{p}}^2} \tag{5.15}$$

These results are to some extent supported by the observations of the curing characteristics. The maximum torque, the optimum cure time, and the scorch time change significantly with the blend composition. However, concerning the influence of the filler there were no much differences in the scorch and cure time of the blends filled with CB N339 and CB N772. The minimum torque values are for all the compounds in the same range, irrespective of the CSM content in the rubber blends. There was also no difference in the M_l values of the materials filled with CB N339 and the ones filled with CB N772. However, concerning the M_h values, there are two groups of values which show a certain influence by the type of the filler used. For the CB N339, the M_h values increase significantly when the CSM content increases. This behavior is not observed for the group of compounds filled with CB N772. Differences occur between the M_h values of the two series of compounds.

This difference can be caused by a more effective interfacial adhesion between CSM and the active CB. At the micro-level, physical and chemical interactions between the functional groups of CSM and the active sites of the high surface specific area CB can lead to a higher apparent cross-linking density. The formation of a greater number of interfacial cross-links may occur. As the concentration of

CSM in the blend increases, the cure characteristics such as optimum cure time, scorch time, and induction time at 160 °C increase for the blends filled with CB N339. Therefore, the optimum cure time increases. Thus, the results are consistent with the ones from the mechanical behavior.

5.4.9 NR/CSM and NBR/CSM Rubber Blends Reinforced by Carbon Black

The influence of carbon black (with an average particle size of 40 nm) on NR/CSM and NBR/CSM rubber blends reinforced by carbon black was investigated. The mechanical properties of elastomer depend mainly on the carbon black loading. Tensile strength, modulus at 100 % elongation, and hardness were increased with increasing radiation dose, as well as carbon black loading, but the elongation at break values decreases with increasing radiation dose and carbon black content. The FTIR spectra of aged samples confirmed the formation of various oxidation products, i.e., alcohols, ethers and small amounts of lactones, anhydrides, esters, and carboxylic acids during radiation. Significant changes in spectra are induced by chain scission which was registered in the 1620–1450 cm^{-1} region. The formation of shorter polyene sequences and aromatic rings is assumed. The radiation doses lower than 100 kGy influenced only small changes in the rubber macromolecules structure if we consider the development of the bands in carbonyl and conjugated C=C bond regions. On the other hand, 200, 300, and 400 kGy doses caused a serious chemical modification of the prepared rubber blend samples.

Fig. 5.12 Rheometric curves for curing of unfilled and filled rubber blends

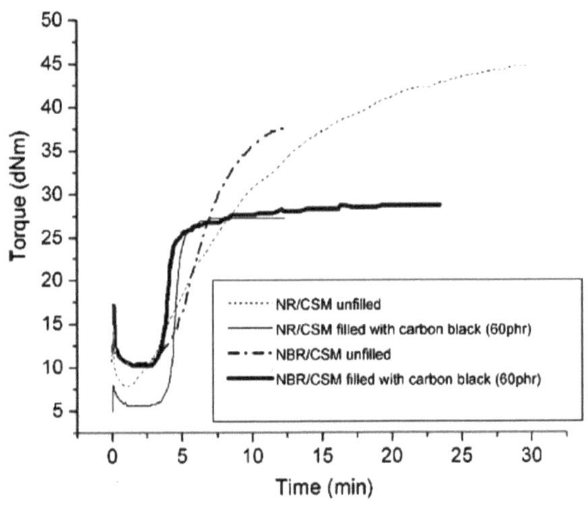

Table 5.14 The obtained data for mechanical properties of cured rubber blends and curing data for compounds with a different content of carbon black

Blend	No	Filler (phr)	Ts (MPa)	Eb (%)	Torque (dNm)		t_{s2} (min)	t_{c90} (min)
					M_{min}	M_{max}		
NR/CSM	1	0	20	573	8.05		2.1	11.3
	2	20	23	529	2.6	25.6	2.3	9.3
	3	40	24	442	3.3	23.6	2.1	8.1
	4	50	25	412	4.1	27.5	1.9	9.7
	5	60	25	343	6.2	32.2	1.8	7.2
	6	80	25	260	6.7	37.3	1.6	8
	7	100	24	220	9.1	40.5	1.6	8.5
NBR/CSM	8	0	14	343	6.8	45.8	8.1	15.2
	9	20	15	381	6.1	48.8	6.3	13.1
	10	40	18	383	6.8	45.7	5.8	13.1
	11	50	20	427	7.1	50.7	4.6	11
	12	60	20	381	7.7	58.6	3.7	5
	13	80	19	373	8.7	52.7	3.3	9.4
	14	100	18	335	9.8	52.7	3.2	9.6

Ts tensile strength; *Eb* elongation at break; t_{s2} scorch time, t_{c90} cure time

5.4.10 Cure Characteristics

The obtained rheometer curves for curing of filled and unfilled rubber blend compounds are given in Fig. 5.12. The cure characteristics of NR/CSM and NBR/CSM rubber blends are summarized in Table 5.14 as the function of the carbon black content together with some other experimental data. The values for curing data depend mainly on the combination of network precursor. The maximum torque, which is a measure of the stiffness of the compound, is directly related to the modulus of the compounds.

The value of the cure time, t_{c90}, is systematically decreased with increasing carbon black content for both rubber blend compounds. The NR/CSM rubber blends have considerably shorter t_{s2} and t_{c90} values than those of carbon black filler of 50 phr or more which resulted again in a noticeable increase in the rate of variation of M100 value with respect to the irradiation dose. These results may be attributed not only to the radiation-induced cross-linking, as well as different interphase linking, but also to the role played by the carbon black filler as a separate phase. In this respect and under a limited extension of 100 % elongation, the carbon black filler would be expected to exist predominantly in its "structure" state, particles aggregated solid and separate form [36]. The effective contribution to the M100 value of vulcanized composites from this solid phase of the filler would then be expected, when taking into consideration that the tensile modulus parameter is an additive in character. The obtained results are in accordance

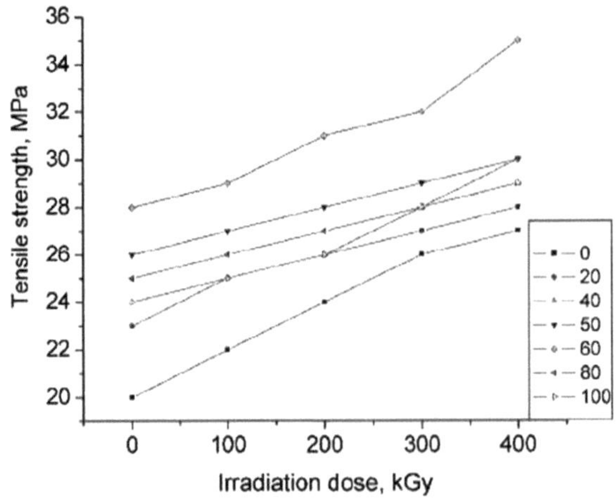

Fig. 5.13 The effect of the radiation dose on the modulus of NR/CSM blend reinforced with a different content of carbon black

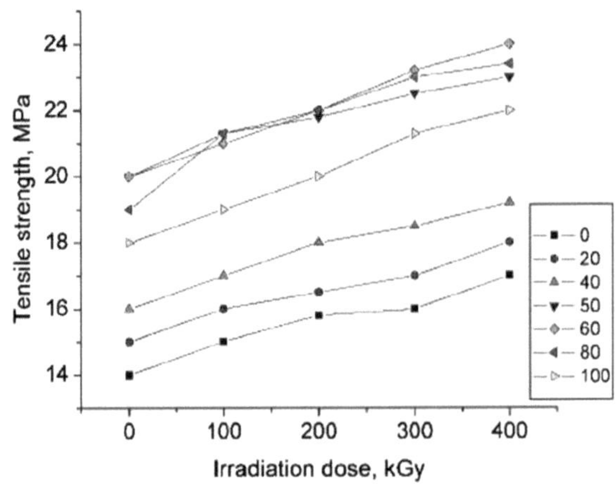

Fig. 5.14 The effect of the radiation dose on the modulus of NBR/CSM blend reinforced with a different content of carbon black

with the comparable results reported before for other rubber composites [37] (Figs. 5.13, 5.14, 5.15, and 5.16).

The variation of the values of elongation at break percent, Eb %, as a function of irradiation dose for unloaded and carbon black-loaded NR/CSM and NBR/CSM blends is depicted in Figs. 5.17 and 5.18, respectively. As expected, the values of Eb for both reinforced blends decrease with increasing irradiation dose. This systematic decrease in Eb values may be attributed to the induced cross-linking by gamma irradiation. On the other hand, the Eb values decreased with the increase in the carbon black filler loading. This reduction is due to stiffening of the matrix by carbon black filler particles [38]. In fact, with any further increase in filler loading, the molecular mobility decreased due to the formation of a physical bonding between the filler particles and the rubber chain. The decrease of Eb values for

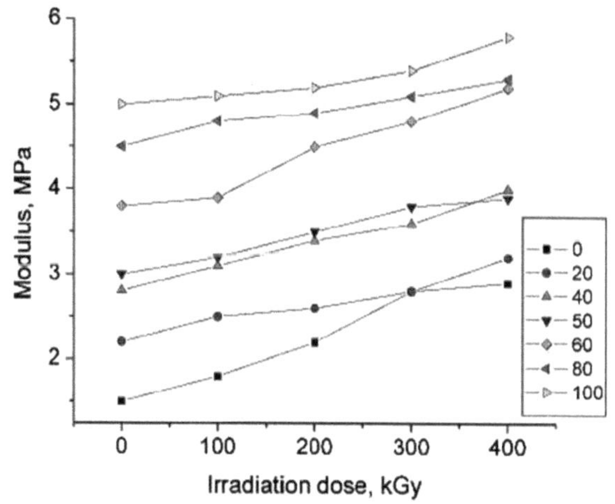

Fig. 5.15 The effect of the radiation dose on the modulus of NR/CSM blend reinforced with a different content of carbon black

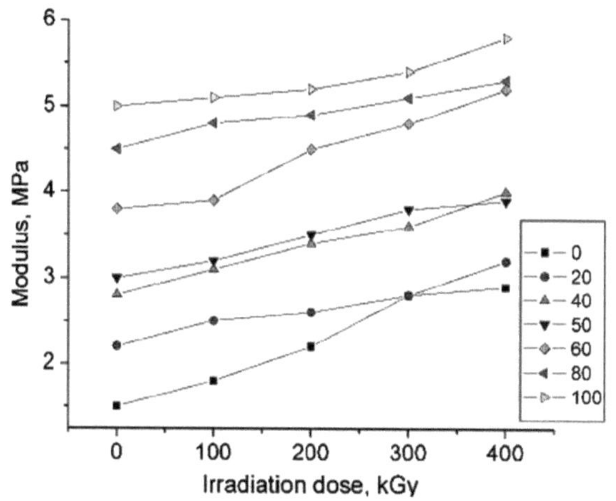

Fig. 5.16 The effect of the radiation dose on the modulus of NBR/CSM blend reinforced with a different content of carbon black

blends loaded with the increasing radiation dose may be explained as an additional bonding between the filler and the rubber due to the large number of free radicals formed under radiation. Therefore, an increased radical recombination is likely to generate strong chemical linkage with an increase of the irradiation dose.

The changes of hardness of loaded and unloaded NR/CSM and NBR/CSM rubber blends as a function of the radiation dose are illustrated in Figs. 5.19 and 5.20, respectively. Also, the data for loaded rubber without irradiation were introduced in the same figure for comparison. It may be observed that the values of hardness of the latter composition have increased effectively with increasing degree of loading with the filler. A similar behavior has been attained for the

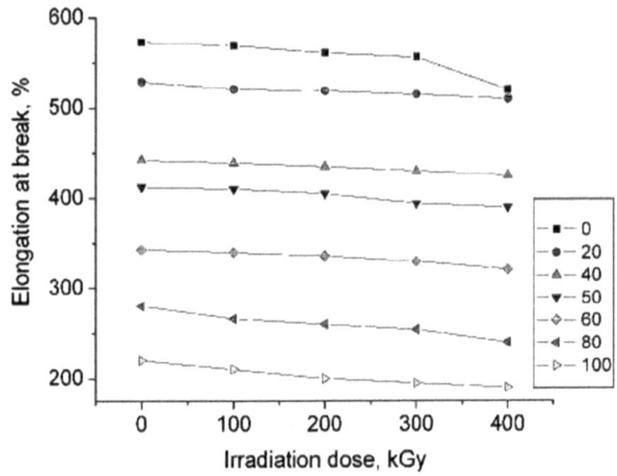

Fig. 5.17 The effect of the radiation dose on the elongation at break of NR/CSM blend reinforced with a different content of carbon black

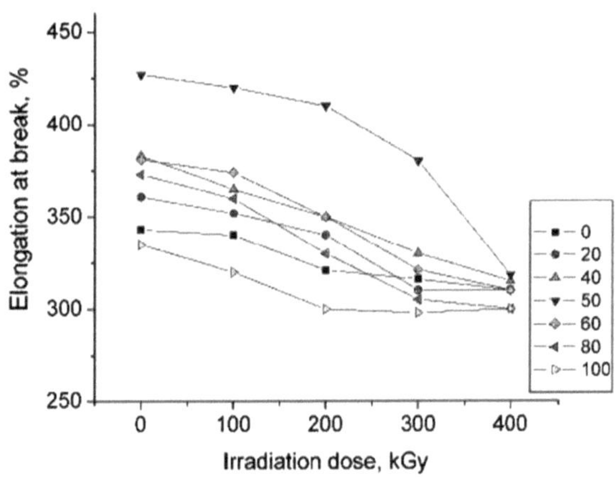

Fig. 5.18 The effect of the radiation dose on the elongation at break of NBR/CSM blend reinforced with a different content of carbon black

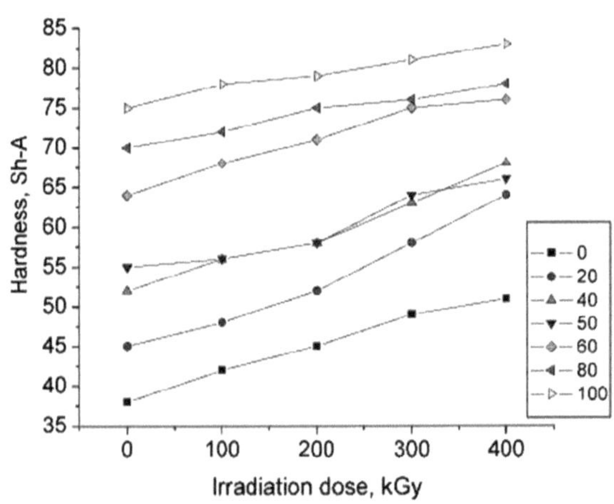

Fig. 5.19 The effect of the radiation dose on the hardness of NR/CSM blend reinforced with a different content of carbon black

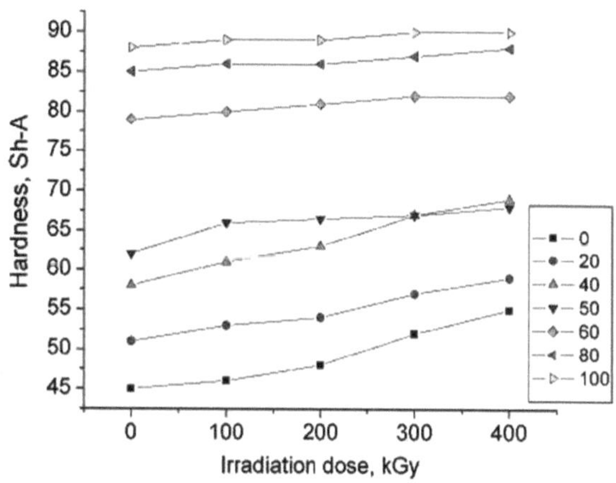

Fig. 5.20 The effect of the radiation dose on the hardness of NBR/CSM blend reinforced with a different content of carbon black

radiation-vulcanized composites when comparing its hardness values at the same irradiation dose. On the other hand, a comparatively limited increase has been attained in the value of hardness for one and the same composition, whether loaded or unloaded, with increasing irradiation dose from 100 to 400 kGy. These data indicate clearly that the contribution to hardness values from the radiation-induced cross-linking that has taken place in the totally amorphous rubber matrix of NBR/CSM blend is a limited one. On the other hand, the main contribution to the hardness value is affiliated with the occurrence of the filler in its aggregated solid and distinct phases, that is, its "structured" form as mentioned before.

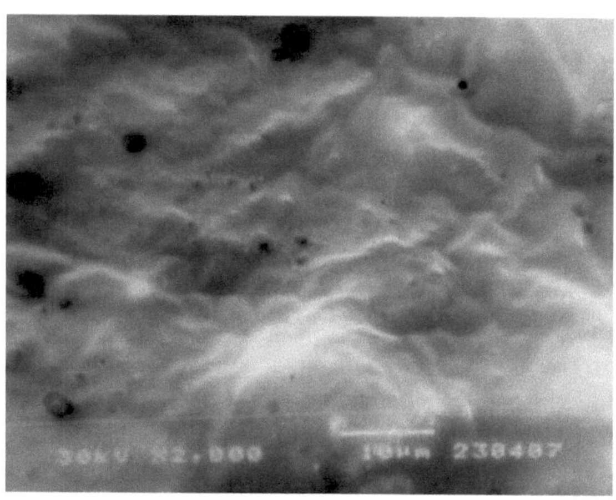

Fig. 5.21 SEM of radiated NR/CSM rubber blend filled with carbon black (60 phr)

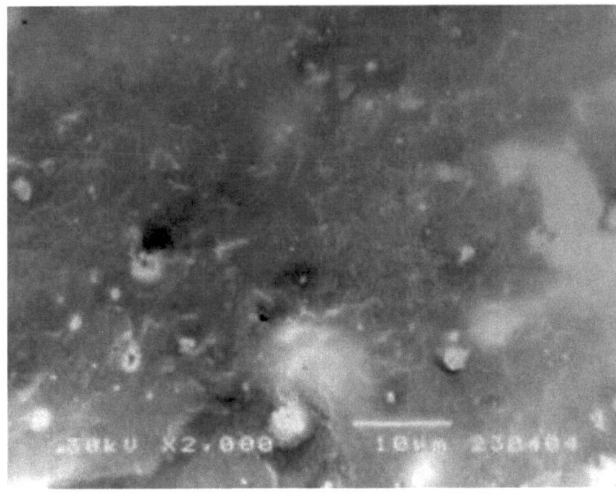

Fig. 5.22 SEM of irradiated NBR/CSM rubber blend filled with carbon black (60 phr)

5.4.11 Electron Scanning Microscopy

SEM micrographs of fractured surfaces for NR/CSM and NBR/CSM rubber blends filled with 60 phr are shown in Figs. 5.21 and 5.22, respectively. The formation of small micro-craters was observed in NBR/CSM rubber blends. A detachment of small micro-craters from the fractured surface is a result of poor polymer–filler adhesion. A comparatively good polymer–filler interaction with carbon black and NR/CSM blend increases the resistance to fracture [39]. Filler distribution in the blend affects the properties of the obtained elastomeric materials.

5.4.12 FTIR Spectra Measurements

Gamma radiation leads to significant changes in the FTIR spectra of carbon black-reinforced elastomers based on NR/CSM (Fig. 5.23a, b) and NBR/CSM (Fig. 5.24a, b) and rubber blend. The broad increase in absorption can be found in C=O, O–H, and C–O stretching vibration domains for both blends. Significant changes appeared in the region of conjugated double bonds, too. Several oxidation products contribute to band at the hydroxyl region, i.e., alcohols (band at 1028–1075 cm^{-1}), ethers and small amounts of lactones, anhydrides, esters, and carboxylic acids. Due to the formation of oxidation products, the bands related to CH_2, CH_3, =C–H, and C=C vibrations decrease. CH_2 deformation band at 1458 cm^{-1}, CH_3 asymmetric deformation at 1372 cm^{-1}, and =C–H wagging at 831 cm^{-1} show a significant increase in the spectra of irradiation reinforced rubber blend.

A decrease in absorption at 1660 cm^{-1} is related to the loss of C=C bonds in 1,4-*cis* units. 1,4-*cis* units also absorb at 831 cm^{-1}. Beside the oxidation of C=C bonds in 1,4-cis units, cis-trans isomerization can take place. The band at 2956 cm^{-1}

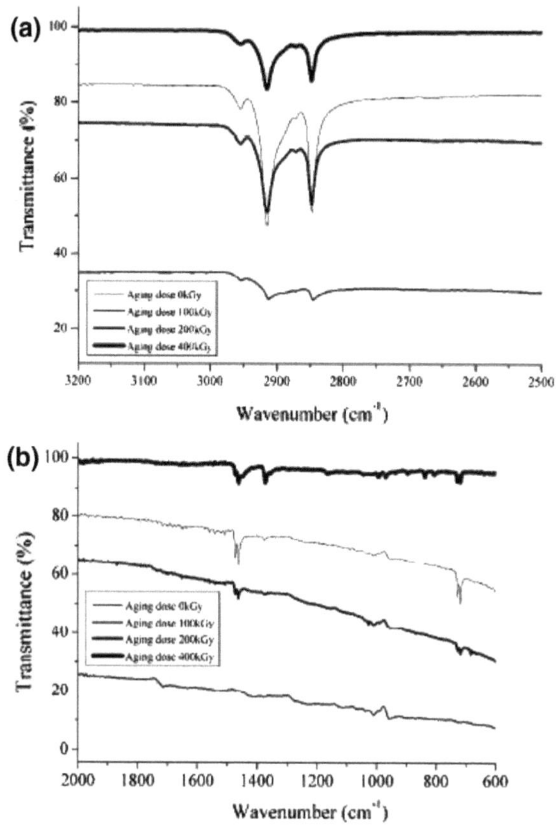

Fig. 5.23 FTIR spectra (**a**, **b**) of elastomers aged under different doses for NR/CSM rubber blend filled with carbon black (60 phr)

related to methyl groups decreases, especially at the highest doses 200, 300, and 400 kGy. Researchers in the field elastomers based on polyisoprene, acrylonitrile, and chlorosulfonated polyethylene rubbers [40] did not report the appearance of new bands related to these processes in the conjugated double-bond region of radiated NR/CSM and NBR/CSM FTIR spectra. In thermally degraded NR/CSM and NBR/CSM rubber blends, there have been identified dipentene, two isomeric forms of dimethylvinylcyclohexene, isoprene, and some aromatic compounds that formed via the Diels–Alder reactions. Coran [41] reported that during thermal degradation of 1,4-*cis*-polyisoprene, the chains with conjugated C=C bonds are also formed. Changes corresponding to similar processes are also present in the region 1620–1450 cm $^{-1}$ of the irradiated polyisoprene and acrylonitrile spectra, where no oxidation product absorbs. Polyenes and polyenals show several bands in the 1500–1650 cm^{-1} range. Benzene C=C stretching vibration bands are located between 1600 and 1450 cm^{-1}; hydrogen bond can cause the reduction of the stretching vibration frequency. The subtraction spectra reveal the occurrence of the broad band with a significant increase in absorption at 1470 cm^{-1}. This band can be assigned to aromates and shorter polyene structures. Absorption at 1470 cm^{-1}

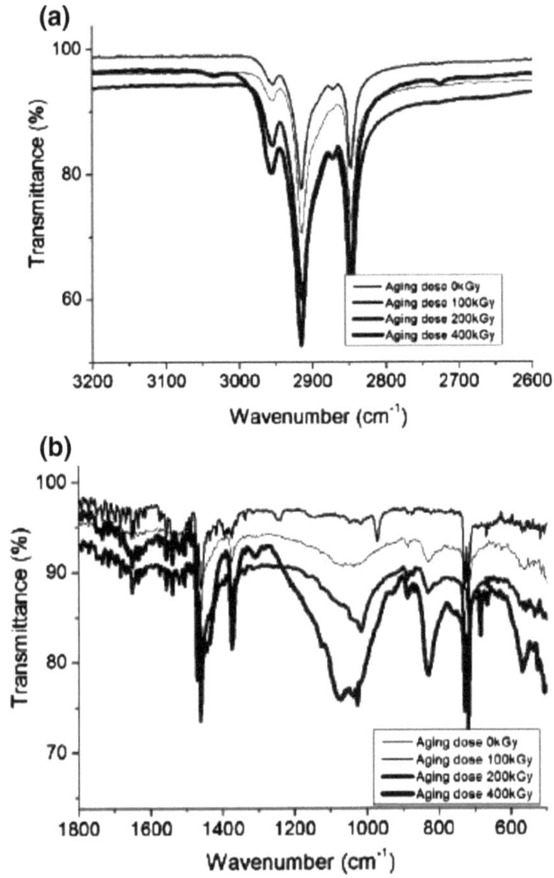

Fig. 5.24 FTIR spectra (**a**, **b**) of elastomers aged under different doses for NBR/CSM rubber blend filled with carbon black (60 phr)

remains the same (the maximum is slightly shifted to higher wave numbers) for the highest doses, and absorption at 1461 cm–1 was lower in the case of the sample irradiated with the highest dose (Figs. 5.23 and 5.24). It implies that formed species are a subject of consecutive reactions. It can be concluded that radiation induces serious changes in NR/CSM and NBR/CSM rubber blends' backbone chain scissions and various consecutive reactions take place. These changes are noticeable in the 1470–1450 cm^{-1} region that indicates the occurrence of the polyene sequences and aromatic rings. Contrary to the chain scission reactions [42], the formation of cross-links is difficult to detect by using IR spectroscopy due to the overlap of many bands in the region of C–C stretching vibrations. Doses exceeding 200, 300, and 400 kGy caused serious modification to the polymer structures.

The influence of carbon black (with an average particle size of 40 nm) on gamma radiation resistance of rubber blends was investigated. The mechanical properties of elastomer depend mainly on the carbon black loading. Tensile strength, modulus at 100 % elongation, and hardness were increased with increasing radiation dose, as well as carbon black loading, but the elongation at break values decreases with increasing radiation dose and carbon black content. The FTIR spectra of aged samples confirmed the formation of various oxidation products, i.e., alcohols, ethers and small amounts of lactones, anhydrides, esters, and carboxylic acids during radiation. Significant changes in spectra are induced by chain scission which was registered in the 1620–1450 cm^{-1} region. The formation of shorter polyene sequences and aromatic rings is assumed. The radiation doses lower than 100 kGy influenced only small changes in the rubber macromolecular structure if we consider the development of the bands in carbonyl and conjugated C=C bond regions. On the other hand, 200, 300, and 400 kGy doses caused a serious chemical modification to the prepared rubber blend samples.

5.4.13 CSM/IIR and CSM/CIIR Rubber Blend

The incorporation of CSM in CSM/IIR and CSM/CIIR blends increases the scorch time, t_{s2}, and cure time, t_{c90}. At a similar CSM content, the CSM/CIIR blend exhibits shorter scorch time t_{s2} and cure time t_{c90} than the CSM/IIR blend. The incorporation of CSM in the CSM/CIIR blend also increases the maximum torque (M_h) and torque difference ($M_h - M_l$). However, for the CSM/IIR blend, the maximum torque and torque difference decrease with the incorporation of CSM, an observation which is attributed to the maldistribution of cross-link between the two rubber phases. M100 and M300 increase with increasing CSM content in both blends. The former increase is attributed to the enhancing effect of CSM, and the latter increase is associated with strain-induced crystallization of the respective rubbers. Tensile strength and elongation at break of the blends decrease gradually with CSM content, an observation which is ascribed to the decreasing effect of strain-induced crystallization of IIR and CIIR as CSM content is increased in the respective blends. The cross-linking density is found to slightly change with an increase in the concentration of the CSM rubber in CSM/CIIR and CSM/IIR rubber blends. Higher cross-linking density is obtained for CSM/CIIR rubber blends. The micrograph of CSM/CIIR exhibits that the composite's surface is rough with many tear lines and branching indicates the higher tensile strength and a weak interaction between CSM and CIIR rubbers.

Figures 5.25 and 5.26 show the effect of CSM content on the scorch time t_{s2} and cure time, t_{c90}, of the CSM/IIR and CSM/CIIR rubber blends. It can be seen that the t_{s2} and t_{c90} of these blends increase with increasing CSM content. CIIR and CSM are polar rubber. As the content of CSM in the blends is increased, the curative agent, i.e., sulfur, becomes more soluble in CSM and CIIR compared to IIR. Consequently, the curing rate of the blend increases with increasing content of CSM. Researches [43–45] reported that for blends of two rubbers with similar polarity, distribution of cross-links can be equal through phases. It can be seen also in Figs. 5.25 and 5.26 that at a similar CSM content, t_{s2} and t_{c90} of CSM/CIIR blend are shorter than those of CSM/IIR blend. The effect of CSM content on the maximum torque

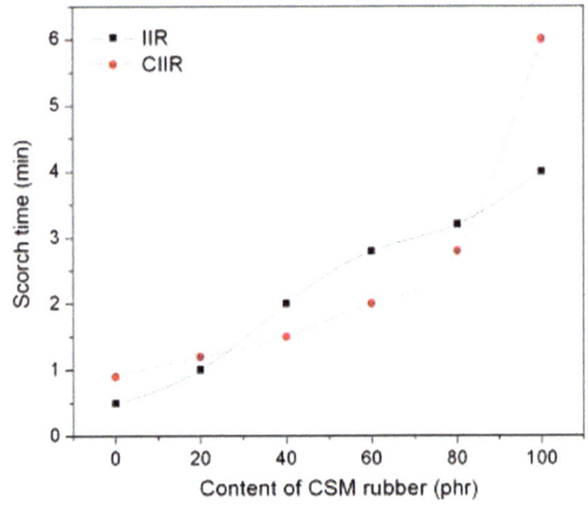

Fig. 5.25 Variation of scorch time t_{s2} with CSM content at 160 °C for CSM/IIR and CSM/CIIR blends

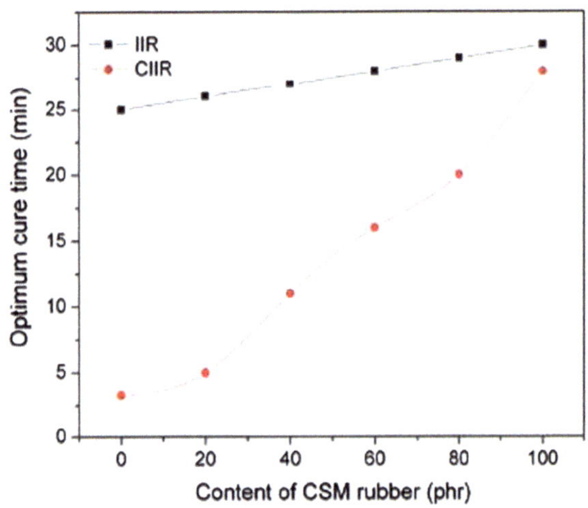

Fig. 5.26 Variation of cure time t_{c90} with CSM content at 160 °C for CSM/IIR and CSM/CIIR blends

(M_h) is shown in Fig. 5.27. It can be seen that the M_h increases with increasing CSM content in the CSM/CIIR blend. Figure 5.27 also shows that above 40 phr of CSM, M_h of CSM/CIIR blends has higher values than CSM/IIR blend as a result of the miscibility presence between CSM and CIIR rubber. The relationship between the torque difference, $\Delta M = M_h - M_l$ (max. torque − min. torque), and content of CSM is shown in Fig. 5.24. It is known that the torque difference indirectly relates to the cross-linking density of the blends [46]. This figure clearly shows that CSM/CIIR blend is a more compatible blend than CSM/IIR blend. The decreasing torque difference for CSM/IIR indicates that a distribution of cross-link occurred which results in an over-cross-linked phase and a poorly cross-linked one.

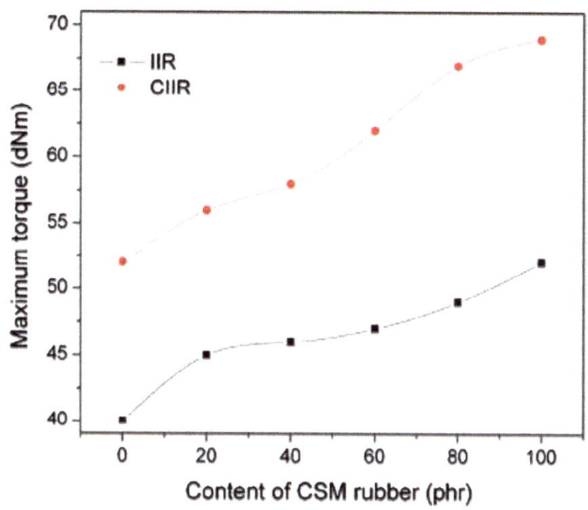

Fig. 5.27 Maximum torque M_h versus CSM content at 160 °C for CSM/IIR and CSM/CIIR rubber blend

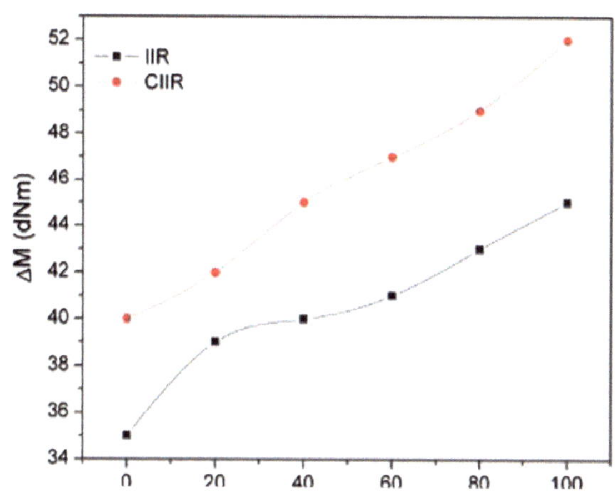

Fig. 5.28 Torque difference $(M_h - M_l)$ versus CSM content at 160 °C for CSM/IIR and CSM/CIIR blends

Again at a similar CSM content, CSM/CIIR blend exhibits higher $M_h - M_l$ than CSM/IIR blend. Markovic [47] reported that with normal sulfur-vulcanizing systems, the efficiency of CIIR vulcanization seemed to be improved. The isolated double bonds in CIIR inhibit the formation of intramolecular sulfide links, thus resulting in the increase of the cross-linking rate (Fig. 5.28).

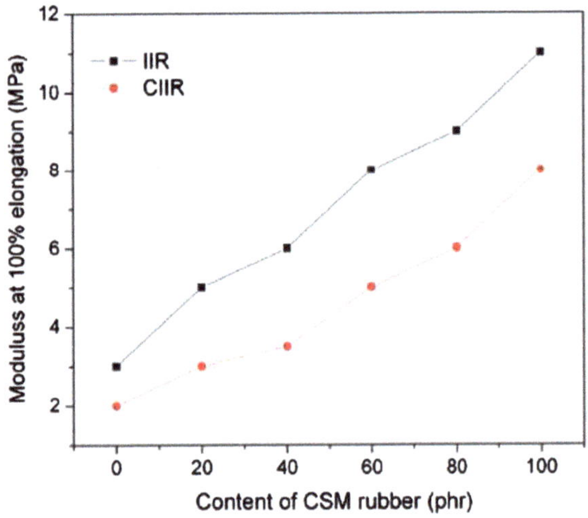

Fig. 5.29 Modulus at 100 % elongation versus CSM content at 160 °C for CSM/IIR and CSM/CIIR rubber blend

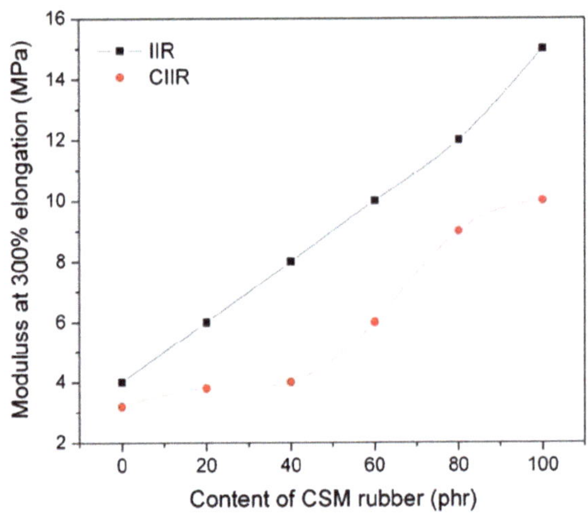

Fig. 5.30 Variation of M300 with CSM content in the blends

5.4.14 *Mechanical Properties*

Figure 5.29 shows the variation of M100 with CSM content in the two CSM/IIR and CSM/CIIR rubber blends. It is obvious that as the percent of CSM increases, M100 also increases correspondingly.

This means that the incorporation of CSM in the blend would enhance cross-linking as M100 is a measure of degree of cross-linking. CSM/CIIR blend gives a higher M100 value than the CSM/IIR blends. This observation may be attributed to greater enhancement effect of CSM on CIIR. In fact, the cross-linking density is increased for both blends. As in the case of CSM/IIR blend, the lower

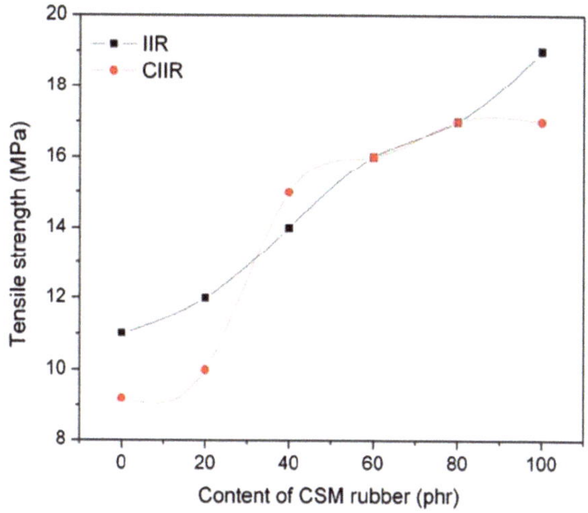

Fig. 5.31 Variation of tensile strength with CSM content in the blends

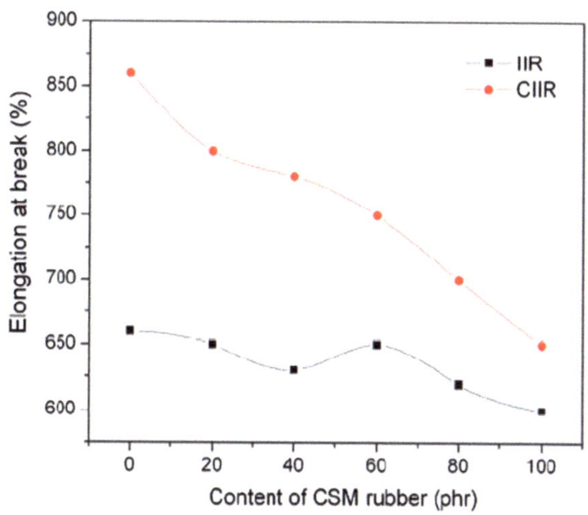

Fig. 5.32 Variation of elongation at break with CSM content in the blends

M100 is attributed to the non-enhancement effect of CSM which is a polar polymer compared to IIR, a nonpolar rubber. Thus, the CSM/IIR blend has a lower cross-linking density than CSM/CIIR blends. The variation of M300 with CSM content is presented in Fig. 5.30. In this case, CSM/IIR blend consistently shows a higher M300 than CIIR blends.

The increase of M300 values indicates that CSM reinforces IIR and CIIR blends: The degree of reinforcement is greatest in IIR, followed by CIIR as the degree of crystallization decreases with increasing % of halogenated. Figure 5.11 shows the dependence of tensile strength of the two rubber blends on CSM content. Again, IIR blend consistently indicates a higher tensile strength than the CIIR blend as observed for M300. However, in this case, tensile strength decreases gradually with increasing CSM content for all the blends studied. This observation is associated with decreasing effect of strain-induced crystallization as CSM content is increased. As discussed earlier in M300, IIR exhibits the greatest strain-induced crystallization, followed by CIIR. This sequence of decreasing crystallization is reflected by the decreasing value of tensile strength for the respective single-component rubbers [48]. In the case of CSM, it has the lower tensile strength due to the non-crystallizable nature of the rubber. Figure 5.31 also shows that for CSM content greater than 60 phr, greater drop in tensile strength for the blends is observed. This finding may be ascribed to the dominant effect of CSM at higher loading of CSM.

The effect of CSM on elongation at break for CSM/IIR and CSM/CIIR rubber blends is shown in Fig. 5.32. The effect, in fact, is very similar to that of tensile strength; that is, elongation at break decreases gradually with increasing CSM content. Owing to the higher strength of IIR, its blend gives a much higher elongation at break compared to CIIR blend. For the single-component CSM, its elongation at break is the lowest among the rubbers studied due to the non-crystallizable property of CSM.

5.4.15 Swelling Properties

Some calculated data for the "relative" cross-linking density values, v, obtained from the swelling measurements for rubber blends are given in Table 5.15. The cross-linking density is found to slightly change with an increase in the concentration of the CSM rubber in CSM/CIIR and CSM/IIR rubber blends. Higher cross-linking density is obtained for CSM/CIIR rubber blends; this is because of the restriction of swelling due to the similar (polar) nature of rubbers and formation of

Table 5.15 The data of the cross-linking density reciprocal swelling ratio ($1/Q$) volume fraction of CSM/CIIR and CSM/IIR rubber blends in swollen gel (Φ_p), molecular weight between cross-link (M_c) and Flory Huggins parameter (χ) from swelling measurements

Sample	$v \times 10^3$ (mol dm^{-3})	Qr	$1/Q$	M_c (gmol^{-1})	χ
CSM/IIR/CIIR (100/0/0)	710	0.32	0.48	750	0.42
CSM/IIR/CIIR (80/20/0)	795	0.33	0.41	762	0.42
CSM/IIR/CIIR (60/40/0)	800	0.33	0.47	778	0.41
CSM/IIR/CIIR (40/60/0)	810	0.34	0.47	780	0.41
CSM/IIR/CIIR (20/80/0)	830	0.34	0.47	790	0.4
CSM/IIR/CIIR (0/100/0)	832	0.34	0.47	790	0.41
CSM/IIR/CIIR (80/20/0)	812	0.33	0.47	791	0.42
CSM/IIR/CIIR (60/40/0)	824	0.33	0.47	770	0.42
CSM/IIR/CIIR (40/60/0)	831	0.34	0.47	814	0.41
CSM/IIR/CIIR (20/8/0)	847	0.34	0.47	831	0.41
CSM/IIR/CIIR (100/0/0)	850	0.34	0.47	821	0.41

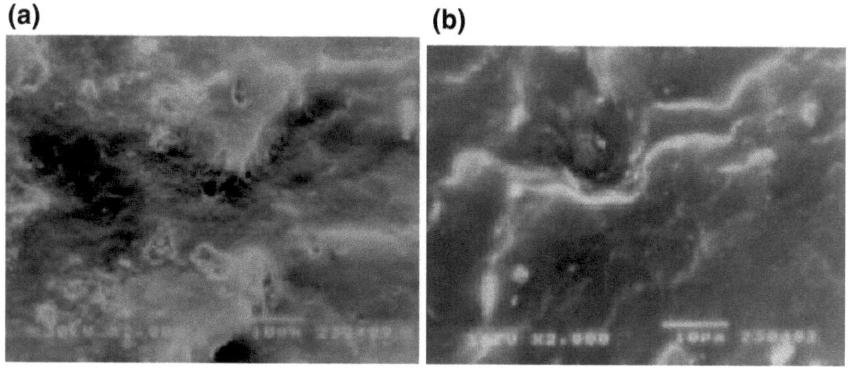

Fig. 5.33 SEM micrograph of CSM/IIR (60/40) (**a**) and CSM/CNR (60/40) rubber blends (**b**)

new cross-links according to free sulfur and residues of curatives react with rubbers, which causes an increase in Φ_p and increases the cross-linking density. Thus, the observed tensile strength variation can be exactly correlated with the variation of v from swelling studies. In swelling studies of these samples, cross-links are flexible to penetrate, but as the loading of CSM increases, they restrict the swelling. The swelling ratio, $1/Q$, calculated using Eq. (5.1) is found to increase (Table 5.15).

5.4.16 *Morphological Study*

In Fig. 5.33a, b, the SEM photomicrograph of CSM/IIR and CSM/CIIR (60/40) rubber blends at 2000× magnification is shown. The micrograph of CSM/CIIR (Fig. 5.33b) exhibits that the composite's surface is rough with many tear lines and branching.

This type of failure indicates the higher tensile strength and a weak interaction between CSM and CIIR rubbers.

5.5 Recent Advances on Chlorosulfonated Rubber Nano-blends

Thermostability and surface morphology of nano- and micro-filled NBR/CSM and CR/CSM rubber blends. Cross-linked NBR/CSM and CR/CSM rubber systems with a mass ratio of 50:50 (Table 5.5) have good weather resistance, color stability, high ozone resistance, and high extension values.

Table 5.16 Glass transition temperature (T_g) for 10 and 30 % weight loss and other characteristic temperatures of unfilled and nano- and micro-particle-sized SiO_2-filled NBR/CSM and CR/CSM cross-linked systems

Compounds	T_{max} (°C)	$T_{10\%}$ (°C)	$T_{30\%}$ (°C)	T_{end} (°C)	M_{rest} (%)	T_g (°C)
NBR	434	403	427	707	3.37	−22
CSM	348	319	354	655	25.4	−37
NBR:CSM (50:50)	449	312	429	472	3.8	−20
NBR:CSM + SiO_2 (15 nm)	452	303	424	486	33.1	−27
NBR:CSM + SiO_2 (28 μm)	450	296	425	485	38.6	−28
CR	282	283	300	590	28.9	−39
CR:CSM (50:50)	466	252	326	537	16.9	−38
CR:CSM + SiO_2 (15 nm)	463	257	349	605	30.3	−40
CR:CSM + SiO_2 (28 μm)	465	259	361	565	32.6	−45

5.5.1 *Thermal Stability*

The thermal stability analysis was based on the temperature for certain weight losses, namely 10 and 30%, and the rate of weight loss changes as a function of temperature (first derivative of the dynamic TG curve) i.e., the initial and overall shapes. The first corresponds to the partial degradation of the sample, and the second is the result of complete thermal decomposition. The samples could be characterized by the position of the decomposition peaks and their size, the difference being the result of the different influence of their structure and the differences in the degradation processes. The data for the 10 and 30 % weight loss temperatures are given in Table 5.16.

From the values of the temperatures at 10 % weight loss and T end, it can be seen that the nanoparticle-sized SiO_2-filled NBR/CSM cross-linked systems (303, 486 °C) are more stable than the micro-particle-sized SiO_2-filled NBR/CSM cross-linked systems (296, 485 °C). In the CR/CSM cross-linked systems, the temperature values at 10 % weight loss show little difference in the thermostability between the micro-particle-sized SiO_2 and the nanoparticle-sized SiO_2-filled cross-linked systems. In the next step (30 % weight loss), the micro-filled cross-linked system had more stability than the nano-filled cross-linked system. Among the most important factors that can influence the initial TG behavior are the strength of the bonds in the main CR/CSM and NBR/CSM polymer chains and the bonds with SiO_2. The single glass transition temperature for the NBR/CSM and CR/CSM cross-linked systems indicates that NBR, CSM, and CR are compatible. The higher values of the glass transition temperatures for the nano- than for the micro-filled cross-linked systems indicate that nano-filled cross-linking systems have greater number of SiO–C links than micro-filled cross-linking systems and more stability.

5.6 Industrial Values and Applications

In recent years, economic, technological, and other regulatory pressures have gradually narrowed the further development of new chemical varieties of polymers [49, 50]. A blend can offer a set of properties that may give it the potential of entering application areas not possible with either of the polymer comprising the blends. Among the authors investigating rubber blends, Baker [51] reported that replacement of polychloroprene by a 20/80 natural rubber/neoprene GRT blend could be accessible for many applications where neoprene GRT is used; Patra and Das [52] studied the flame retardancy and heat shrinkability of polyolefin/elastomer blends. The blending together of natural rubber (NR) and chlorosulfonated polyethylene rubber (CSM) is intended to produce a vulcanizate with the best properties from each component, i.e., the good strength properties of NR and the good weather resistance, color stability, and high extension values of CSM.

5.6.1 CSM/SMR 20 CV and CSM/Pergut S 40 Rubber Blends

The Mooney scorch time, t_5, of CSM/SMR 20 CV and CSM/Pergut S 40 blends decreases with increasing SMR20 CV and Pergut S 40 content in the blends. The reduction is more pronounced at lower temperatures, i.e., below 130 °C. For CSM/SMR 20 CV blend, the cure index shows a maximum at 20 phr of SMR 20 CV in the blend at lower vulcanization temperature, an observation which is attributed to the incompatibility between CSM and SMR 20 CV and the faster cure rates of the latter. In the case of CSM/Pergut S 40 blends, the cure rate virtually decreases with increasing Pergut S 40 content, as a result of the better compatibility between CSM and Pergut S 40, both of which are polar in nature. For temperature greater than 140 °C, the cure index for both blend systems exhibits less dependence on the blend ratio due to the availability of thermal energy to overcome the

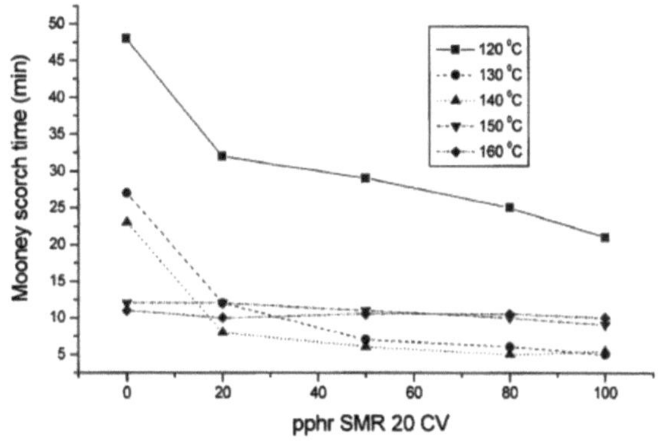

Fig. 5.34 Variation of the Mooney scorch time with blend ratio of SMR 20 CV (phr) in the CSM/SMR 20 CV blend for various vulcanization temperatures

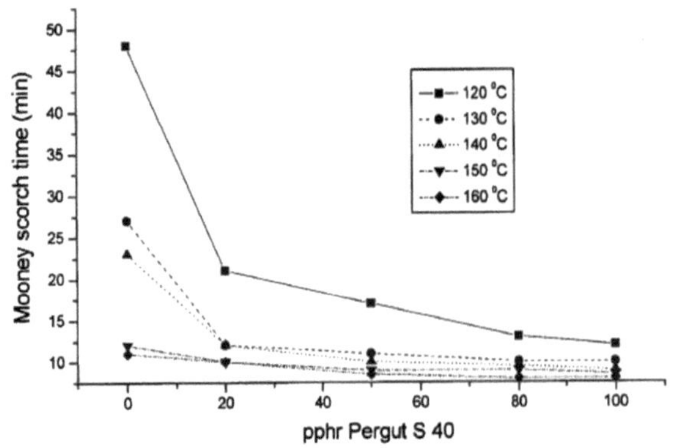

Fig. 5.35 Variation of the Mooney scorch time with blend ratio of Pergut S40 (phr) in the CSM/Pergut S40 (phr) blend for various vulcanization temperatures

activation energy of vulcanization. Based on the morphological investigations, the fracture surfaces of CSM/SMR 20 CV (80:20) rubber blends are less homogeneous than those of CSM/Pergut S 40 (80:20).

Scorch Time

The variation of the Mooney scorch time, t_5, of the CSM/SMR 20 CV and CSM/Pergut S 40 blends with a blend ratio of SMR 20 CV and Pergut S 40 is shown in Figs. 5.34 and 5.35, respectively.

It can be seen that the t_5 of the blends decreases with increasing SMR 20 CV and Pergut S 40 content. CSM is a polar rubber. As the content of CSM in the blends decreases, the curing agents, i.e., sulfur, magnesium oxide, and tetramethylthiuram disulfide, become more soluble in the SMR 20 CV and Pergut S 40 than in the CSM. Consequently, the cure rate of the blend increases with decreasing content of CSM. According to Lewan [53], for blends with two rubbers differing in polarity, such as SMR 20 CV or Pergut S 40 with CMS, a distribution of cross-links can arise through

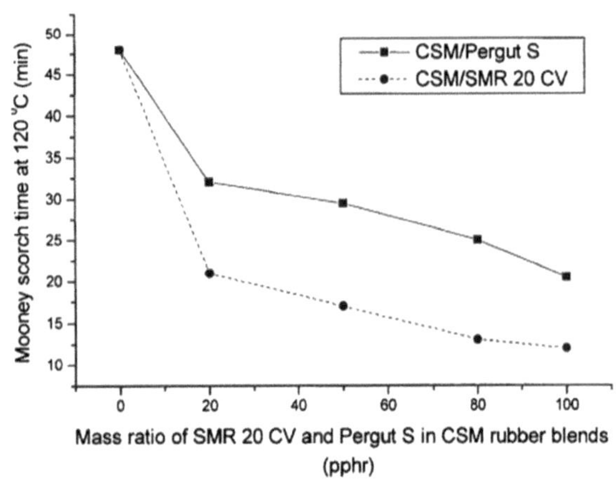

Fig. 5.36 Variation of the Mooney scorch time at 120 °C with the ratio of Pergut S40 and SMR CV60 (phr) in the CSM/SMR 20 CV and CSM/Pergut S40 rubber blend

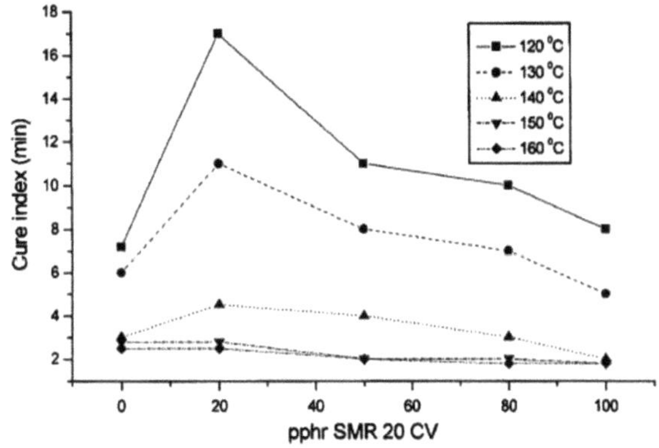

Fig. 5.37 Variation of the cure index blend ratio of SMR 20 CV (phr) in the CSM/SMR 20 CV rubber blend for various vulcanization temperatures

preferential solubility of the curing agents and vulcanization intermediates. Both figures also indicate that the reduction of t_5 with a blend ratio of SMR 20 CV and Pergut S 40 is more pronounced at temperatures lower than 130 °C. At higher temperatures, i.e., higher than 140 °C, t_5 does not show a significant dependence on the blend ratio because enough thermal energy is available to overcome the activation energy of vulcanization. A comparison of t_5 for CSM/SMR 20 CV and CSM/Pergut S 40 blends at 120 °C is shown in Fig. 5.36. It can be seen that with a similar blend ratio, the t_5 of the CSM/SMR 20 CV blend is shorter than that of the CSM/Pergut S 40 rubber blend. According to Poh and Wong [54], more activated precursors to cross-links are formed which accelerate the vulcanization process.

Cure Index

The dependence of cure index of the CSM/SMR 20 CV blend on the blend ratio of SMR 20 CV for the various temperatures investigated in this study is shown in Fig. 5.37. For temperatures lower than 140 °C, the cure index is significantly dependent on the blend ratio of SMR 20 CV. It passes through a maximum at 20 phr of SMR 20 CV in the blend. This observation is attributed to the incompatibility between CSM (a polar rubber) and SMR 20 CV (a nonpolar rubber), the respective solubility parameters of which are 9.5 and 8.1.11 [55]. Also, owing to the higher unsaturation in SMR 20 CV, it cures faster than CSM, resulting in uneven vulcanization of the blend. For instance, at 120 °C, the cure index of CSM and that of SMR 20 CV are 7.2 and 8 min, respectively. As the SMR 20 CV content in the blend is increased from 0 to 20 phr, more sulfur reacts with the SMR20 CV rubber than with CSM. This means that less sulfur and magnesium oxide are available for the vulcanization of CSM, which is the dominant rubber component in the blend. However, beyond 20 phr of SMR 20 CV, the role of SMR 20 CV becomes more significant and results in a drop of the cure index of the blend. For the temperatures greater than 140 °C, the cure index shows less dependence on the blend ratio of SMR 20 CV, a phenomenon that is associated with the increase in cure rates for

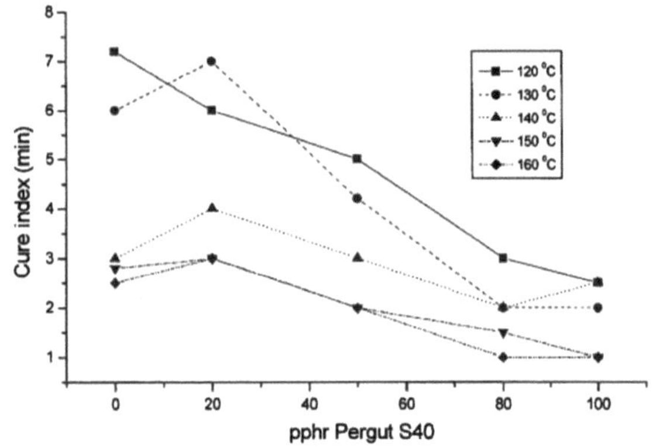

Fig. 5.38 Variation of the cure index blend ratio of Pergut S (phr) in the CSM/Pergut S rubber blend for various vulcanization temperatures

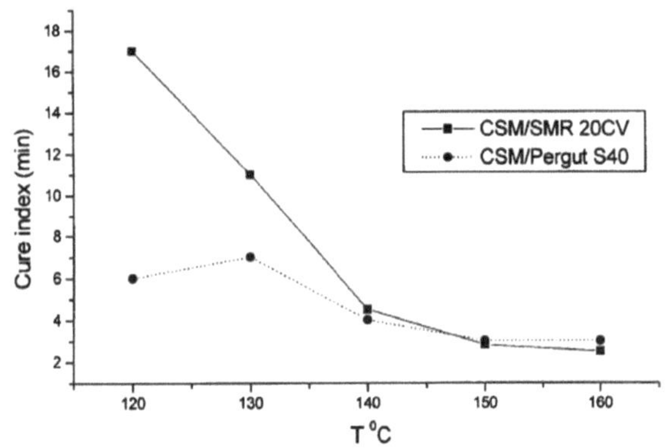

Fig. 5.39 Temperature dependence of the cure index at 20 phr of Pergut S40 and SMR CV60 (phr) in the CSM/SMR 20 CV and CSM/Pergut S40 rubber blend

both rubbers resulting from an ample supply of thermal energy to overcome the activation of vulcanization for both rubbers, i.e., even vulcanization of the blend is obtained. In fact, above 150 °C, the cure index is almost independent of the blend ratio of SMR 20 CV.

The effect of blend ratio of Pergut S 40 on the cure index of CSM/Pergut S 40 blend is shown in Fig. 5.38. Generally, a gradual drop of cure index is observed as the phr of Pergut S 40 is increased, although a slight maximum is displayed in some cases. This finding is attributed to the better compatibility between CSM and Pergut S 40, both of which have a polar nature. The respective solubility parameters are 9.5 and 9.2.9 [56] which are closer than the corresponding values for CSM and SMR 20 CV, as discussed earlier. As in the case of the CSM/SMR 20 CV system (Fig. 5.38), the cure index of the blend was less dependent on the blend ratio of Pergut S 40 at higher vulcanization temperature, i.e., greater than 140 °C. Again, this is attributed to the availability of thermal energy to overcome the activation energy of vulcanization.

In order to compare the cure index of both blends, the cure index was plotted against the cure temperature for blends, containing either 20 phr SMR 20 CV or

Pergut S 40 (Fig. 5.39). The cure index decreases with increasing temperature for both blends. The cross-linking of CSM takes place according to a completely different mechanism from those involved in the cross-linking of natural rubber i.e., the activation energy is different.

5.7 Conclusion

Blending of two or more different types of rubber is considered a useful technique for developing materials with properties superior to those of individual constituents or ones which demonstrate useful property combinations. The approach is important especially from a technical point of view. Elastomeric materials based on the chlorosulfonated polyethylene (CSM) rubber nano-blend have a quite good crack resistance, wet grip, and weather resistance. On the other hand, it shows a good resistance to interactions with alcohols as well as strong acids and bases. This type of material also exhibits excellent ultraviolet light and oxygen stability. In addition to applications for glovebox gloves, this rubber has additional uses such as in lining and sheath materials, coatings, and adhesives. CR/CSM rubber blend study illustrates the intrinsic potential of nano-silica particle size as reinforcing filler in elastomeric materials. The nano-silica-filled CR/CSM rubber blends have higher values for cure characteristics than micro-silica particle-sized filled compounds. Nano-silica type activated the vulcanizing process through the promotion of hydrogen sulfide formation and the rupture of S–N linkage when heated with sulfonamides in rubber either in the presence or in the absence of other compounding ingredients. The strong peak from SiO–C at 1070 and 1079 cm^{-1} in the IR spectrum of nano- and micro-silica-filled CR/CSM rubber blend suggested the presence of interaction with silica which leads to increased stability of the filled CR/CSM polymer blends. According to the Kraus equation, it can be observed that the nano-silica-filled CR/CSM rubber blend has higher V_{r0}/V_{rf} values and better rubber filler interactions than micro-silica-filled composites. The higher values of the glass transition temperatures for the nano- than the micro-filled cross-linked systems are indicated by DMA analysis. The density of the nano-silica-filled CR/CSM rubber blends was lower than that of the micro-silica-filled and was transparent at all filler concentration.

For SBR/CSM blends, the addition of filler leads to a significant enhancement in the physical properties if CSM is the major blend constituent. The blends containing CB N339 show higher hardness, tensile strength, and abrasion resistance when compared to the ones containing CB N772. The experimental data indicate that this can be caused by the improvement in the polymer–filler phase bonding which results from the interfacial interaction of functional groups of the polymer and the active sites of the filler surface leading to a higher apparent cross-linking density. As the concentration of CSM in the rubber blend increases, the cure characteristics, such as optimum cure time, scorch time, and induction time at

160 °C, increase for the blends filled with CB N339. The results for the curing behavior are consistent with the mechanical properties of the SBR/CSM blends.

The influence of carbon black (with an average particle size of 40 nm) on gamma radiation resistance of rubber blends was investigated. The mechanical properties of elastomer depend mainly on the carbon black loading. Tensile strength, modulus at 100 % elongation, and hardness were increased with increasing radiation dose, as well as carbon black loading, but the elongation at break values decrease with increasing radiation dose and carbon black content. The FTIR spectra of aged samples confirmed the formation of various oxidation products, i.e., alcohols, ethers and small amounts of lactones, anhydrides, esters, and carboxylic acids during radiation. Significant changes in spectra are induced by chain scission which was registered in the 1620–1450 cm^{-1} region. The formation of shorter polyene sequences and aromatic rings is assumed. The radiation doses lower than 100 kGy influenced only small changes in the rubber macromolecules structure if we consider the development of the bands in carbonyl and conjugated C=C bond regions. On the other hand, 200, 300, and 400 kGy doses caused a serious chemical modification of the prepared rubber blend samples.

The incorporation of CSM in CSM/IIR and CSM/CIIR blends increases the scorch, t_{s2}, and cure time, t_{c90}. At a similar CSM content, the CSM/CIIR blend exhibits shorter scorch time t_{s2} and cure time t_{c90} than the CSM/IIR blend. The incorporation of CSM in the CSM/CIIR blend also increases the maximum torque (M_h) and torque difference ($M_h - M_l$). However, for the CSM/IIR blend, the maximum torque and torque difference decrease with the incorporation of CSM, an observation which is attributed to the maldistribution of cross-link between the two rubber phases. M100 and M300 increase with increasing CSM content in both blends. The former increase is attributed to the enhancing effect of CSM, and the latter increase is associated with strain-induced crystallization of the respective rubbers. Tensile strength and elongation at break of the blends decrease gradually with CSM content, an observation which is ascribed to the decreasing effect of strain-induced crystallization of IIR and CIIR as CSM content is increased in the respective blends. The cross-linking density is found to slightly change with an increase in the concentration of the CSM rubber in CSM/CIIR and CSM/IIR rubber blends. Higher cross-linking density is obtained for CSM/CIIR rubber blends. The micrograph of CSM/CIIR exhibits that the composite's surface is rough with many tear lines and branching indicates the higher tensile strength and a weak interaction between CSM and CIIR rubbers.

Nano-silica filler interacts more with NBR/CSM and CR/CSM polymer matrix than micro-silica, which is the result of the greater contact area of the nano-filler, greater adsorption, and lower mobility of the polymer chains or the filler surface. The higher values of the glass transition temperatures for the nano- than the micro-filled cross-linked systems indicate that the nano-filled cross-linked systems have a larger number of SiO–C links than micro-filled cross-linked systems and hence increased stability. According to morphological investigation, can be concluded: (i) The surface of NBR/CSM polymer blends is more polished than that of CR/CSM polymer blends and without crack; (ii) the presence in both NBR/CSM and CR/CSM rubber blends with micro-particle size of SiO_2 of rubber balls and some amount of fibrosity in the matrix, voids and crazes, as well as micro-cracks and crystallinity contribute toward failure of the macromolecular network structure. The increased number of cracks and voids in the CR/CSM polymer blends contribute to the earlier loss of physical and mechanical properties, compared to the NBR/CSM polymer blends.

The Mooney scorch time, t_5, of CSM/SMR 20 CV and CSM/Pergut S 40 blends decreases with increasing SMR20 CV and Pergut S 40 content in the blends. The reduction is more pronounced at lower temperatures, i.e., below 130 °C. For CSM/SMR 20 CV blend, the cure index shows a maximum at 20 phr of SMR 20 CV in the blend at lower vulcanization temperature, an observation which is attributed to the incompatibility between CSM and SMR 20 CV and the faster cure rates of the latter. In the case of CSM/Pergut S 40 blends, the cure rate virtually decreases with increasing Pergut S 40 content, as a result of the better compatibility between CSM and Pergut S 40, both of which are polar in nature. For temperature greater than 140 °C, the cure index for both blend systems exhibits less dependence on the blend ratio due to the availability of thermal energy to overcome the activation energy of vulcanization. Based on the morphological investigations, the fracture surfaces of CSM/SMR 20 CV (80:20) rubber blends are less homogeneous than those of CSM/Pergut S 40 (80:20).

Acknowledgments The financial support for this study was granted by the Ministry of Science and Technological Development of the Republic of Serbia (Project nos. 45022, 45020, and 172056).

References

1. Marković G, Samaržija-Jovanović S, Jovanović V, Marinović-Cincović M, Budinski-Simendić J (2009) Chem Ind Chem Eng Q 15(4):291–298
2. Omran AM, Youssef AM, Ahmed MM, Abdel-Bary EM (2010) Kautch Gummi Kunst 6:197–202
3. Wu S (1985) Polymer 26(12):1855–1863
4. Marković G, Samaržija-Jovanović S, Jovanović V, Marinović-Cincović M (2010) J Therm Anal Calorim 100:881–888
5. Essawy H, El-Nashar D (2004) Polym Testing 23(7):803–807
6. Marković G, Radovanović B, Marinović-Cincović M, Budinski-Simendić J (2006) Kautch Gummi Kunst 5:251–255
7. Voet A (1980) J Polym Sci 15:327–373
8. Flory PJ (1946) Ind Eng Chem 38(4):417–436
9. Roychoudhury A, De PP (1997) J Appl Polym Sci 63(13):1761–1768
10. Zhang L, Wang Y, Wang Y, Sui Y, Yu D (2000) J Appl Polym Sci 78(11):1873–1878
11. Mostafa A, Abouel-Kasem A, Bayoumi MR, El-Sebaie MG (2009) Mater Des 30:1561–1568
12. Liu Q, Zhang Y, Xu H (2008) Appl Clay Sci 42:232–237
13. Das A, Costa FR, Udo W, Heinrich G (2008) Eur Polym J. 44:3456–3465
14. Odlyhal M, Cohen NS, Foster GM, Aliev A, Verdonck E, Grandy D (2003) J Therm Anal Calorim 71:939–950

15. Furtado CRG, Leblanc JL, Nunes RCR (1999) Eur Polym J 35:1319
16. Yatsuyanagi F, Suzuki N, Ito M, Kaidou HE (2001) Polymer 42:9523
17. Akovali G, Ulkem I (1999) Polymer 40:7417
18. Mann DG, Evans KM (2007) Molecular genetics and the neglected art of diatomics. In: Brodie J, Lewis J (eds) Unravelling the algae—the past, present and future of algal systematics. CRC Press, Boca Raton (Florida)
19. Williams DM, Kociolek JP (2007) Eur J Phycol 42:313–319
20. Markovic GS, Veljkovic O, Marinovic-Cincovic MT, Jovanovic V, Samarzija-Jovanovic S, Budinski-Simendic JK (2013) Compos Part B Eng 45 br. 1, str. 178–184
21. Stojčeva-Radovanović B, Marković GS, Marinović-Cincović MT, Babić D , Nedeljković JM (2002) Hemijska industrija 56(10):415–421. doi:10.2298/HEMIND0210415S
22. Markovic GS, Marinovic-Cincovic MT, Jovanovic V, Samarzija-Jovanovic S, Budinski-Simendic J (2014) Chapter: Non-linear. Non-linear viscoelasticity of rubber composites and nanocomposites, pp 193–271
23. Da H. M. da Costa, R. C. R. Nunes, L. L. Y. Visconte, C. R. G. Furtado (2001) Kautschuk Gummi Kunststoffe 54: 5 242-249
24. Markovic GS, Marinovic-Cincovic MT, Jovanovic V, Samarzija-Jovanovic S, Budinski-Simendic J, Natural rubber materials: blends and IPNs (Chap. 1). Composites and nanocomposites, vol 1, pp 1–27
25. Markovic GS, Marinovic-Cincovic MT, Jovanovic V, Samarzija-Jovanovic S, Budinski-Simendic J, Composites and nanocomposites, pp 181–204
26. Janos M, Bela P (2008) J Ind Eng Chem 14:535–563
27. Das A, Stockelhuber KW, Jurk R, Saphiannikova M, Fritzsche J, Lorenz H et al (2008) Polymer 49:5276–5283
28. Ahankaria SS, Kamal KK (2008) Mater Sci Eng A 491:454–460
29. Fu SY, Xi-Qiao F, Bernd L, Yiu-Wing M (2008) Compos B 39:933–961
30. Leblanc JL (2002) Prog Polym Sci 27:627–687
31. Marković G, Marinović-Cincović M, Valentova H, Ilavsky M, Radovanović B, Budinski-Simendić (2005) Mater Sci Forum 494:475–481
32. Janowska G, Rybinski P (2004) J Therm Anal Calorim 78:839–847
33. Grauic N (1967) Encyclopedia of polymer science and technology, vol 4. Wiley, New York, p 159
34. Tarrıo-Saavedra J, Lopez-Beceiro J, Naya S, Artiaga R (2008) Polym Degrad Stab 93:2133–2137
35. Tanrattanakul V, Udomkichdecha W (2001) J Appl Polym Sci 82:650
36. López-Manchado MA, Valentín JL, Carretero J, Barroso F, Arroyo M (2007) Eur Polym J 43:4143–4150
37. Rattanasom N, Saowapark T, Deeprasertkul C (2007) Polym Testing 26:369–377
38. Chakraborthy SK, Setua DK, De SK (1982) Rubber Chem Technol 55:1286
39. Setua DK, De SK (1984) J Mater Sci 19:983–988
40. Coran AY, Hamed P, Goettler LA (1976) Rubber Chem Technol 49:1167–1181
41. Chakraborthy SK, Setua DK, De SK (1982) Rubber Chem Technol 55:1286
42. Arumugam N, Tamareselvy K, Venkata RK (1989) J Appl Polym Sci 37:2645–2659
43. Ding R, Leonov AI (1966) J Appl Polym Sci 61(3):455–463
44. Poh BT, Wong KW (1998) J Appl Polym Sci 69(7):1301–1305
45. de León R, Morales G, Acuña P, Soriano F (2010) Polym Eng Sci 50(2):373–383
46. Choi SS (2000) J Appl Polym Sci 75(11):1378–1384
47. Marković G, Marinović-Cincović M, Radovanović B, Budinski-Simendić J (2007) Chem Ind Chem Eng Q 13(4):186–191
48. Radovanović B, Marković G, Radovanović A (2008) Polym Compos 29:692–697
49. Ray S, Das CK (2001) Polym Plast Technol Eng 40

50. Zanzing DJ (1993) Rubber Chem Technol 66:456
51. Baker CSL (1974) Rubber Chem Technol 5:5
52. Patra PK, Das CK (1998) J Appl Polym Sci 68:597
53. Lewan MV (1998) Blends of natural rubber. Champman & Hall, London, p 52
54. Poh BT, Wong KW (1998) J Appl Polym Sci 69:1301
55. Bridson JA (1978) Rubber chemistry. Applied Science Publishers, London, p 83
56. Lewan MV (1998) Blends of natural rubber. Champman & Hall, London, p 52

Chapter 6
Polybutadiene Rubber-Based Nanoblends (PBR-nB): Preparation, Characterization and Applications

Dharmesh Hansora and Satyendra Mishra

Abstract This chapter deals with various preparation methods, characterization, properties, and applications of polybutadiene rubber (PBR)-based nanoblends (PBR-nBs). Various classes of PBR-based nanocomposites and PBR-nBs are addressed by giving importance to the interfacial compatibility of different phases. PBR-nBs such as immiscible PBR blends, PBR/thermoplastic, and PBR/thermoset-based nanostructured blends prepared by interphase modification and compatibilization using micro- and nanofillers have been briefly discussed. In addition, various characteristic properties such as thermal, mechanical, morphological, rheological, and electrical as well as permeability and scattering information of PBR-nBs have been studied. Finally, the applications, new challenges, and opportunities of these PBR-nBs are also discussed.

Keywords Polybutadiene rubber · Nanoblends · Nanocomposites · Preparations · Properties · Applications

6.1 Introduction

6.1.1 History

Almost 55,000 years ago, one of the greatest achievements of mankind was the invention of the wheel, which is now used on cars, planes, and farm equipment. The first air-filled, pneumatic tire was patented in 1888, which has now replaced hard rubber tires to provide a smooth and high speed for riders. An improvement in mileage of car tires was mainly due to the brilliance of rubber scientists and technologists who used chemical ingredients such as processing aids, accelerators,

D. Hansora · S. Mishra (✉)
Department of Plastics Technology, University Institute of Chemical Technology, North Maharashtra University, Jalgaon 425001, Maharashtra, India
e-mail: profsm@rediffmail.com

D. Hansora
e-mail: 2568dharmesh@gmail.com

G. Marković and Visakh P.M. (eds.), *Rubber Nano Blends*,
Springer Series on Polymer and Composite Materials,
DOI 10.1007/978-3-319-48720-5_6

activators, fillers, and anti-degradants for better processing and mechanical properties. To improve the mechanical properties (hardness, tear strength, tensile strength (TS), and elongation at break) of a rubber, micro-/nanofillers (colloids of carbon blacks, synthetic silica, organoclays, and metal oxides) with surface areas from 150 to 400 m^2/g are generally added. In addition to this, anti-degradants and processing aids have been used to protect the rubber against environmental aging and also for processing properties. Curing chemicals can produce cross-linking points between the rubber chains at temperature between 140 and 240 °C. Use of cross-linking fillers such as precipitated silica pretreated with organosilanes and curing chemical was observed in various rubbers [natural rubber (NR), styrene-butadiene rubber (SBR), Polybutadiene rubber (PBR), and nitrile-butadiene rubber (NBR)] without compromising their mechanical properties. Accelerators and activators are added to cross-link the rubber and optimize the chemical bonding between the rubber and filler [1].

6.1.2 Blends of Elastomers, Polymers, and Rubbers

A blending of two or more elastomers is carried out (i) to improve the physical and mechanical properties of the first elastomer, (ii) to obtain good processing characteristics of rubber compound, and (iii) to decrease the compound cost. Elastomers can be compounded as well as blended with thermoplastics and thermosets also. Basic types of rubber elastomer nanoblends are illustrated in Fig. 6.1. A rubber–polymer blend such as rubber-thermoplast can be obtained by mixing rubber either with soft thermoplastic elastomers or with toughened thermoplastic, e.g., hPBR/PP blend with nanoscale morphology [2]. Rubber-thermoset nanoblends can be toughened by rubber particles which are either the use of core-shell (rubbery core and an outer shell of a glassy polymer) rubber particles or initially miscible reactive rubbers. The rubbery core is generally based on PBR. The outer glassy polymer shell is required to prevent coalescence of rubbery particles during synthesis and enhances a good interface with the matrix. The outer glassy polymer shell is usually based on styrene/acrylonitrile copolymers [3, 4].

6.1.3 Present Scenario on PBR-Based Nanostructured Composites and Blends

The uncross-linked, blended rubber polymers or pre-polymers must be initially miscible to form an interpenetrating polymer–rubber network (IPN) at the molecular level. IPN can be formed by simultaneous and sequential polymerization or cross-linking to give a co-continuous network of chemically separated polymers. Both polymers could be rubbers, or one rubber-other a polymer to enhance the

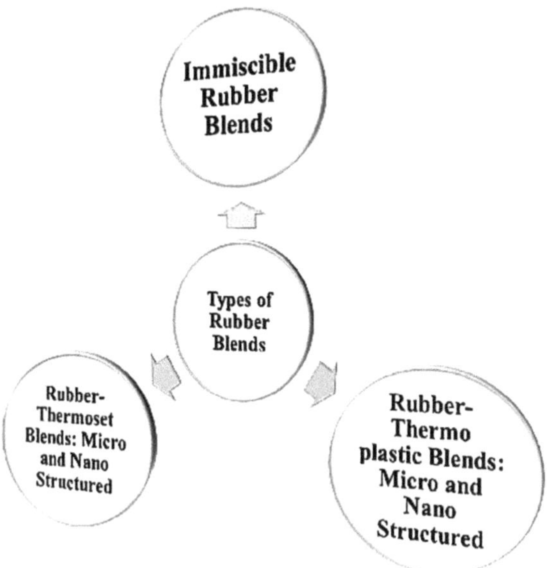

Fig. 6.1 Types of rubber blends

toughness of resulting blend. Important PBR-based nanoblends (PBR-nBs) and nanocomposites are reported in Table 6.1.

Kaynak et al. [4] investigated that rubber can be toughened by using epoxy resin by incorporating various modifiers, such as hydroxyl terminated PBR (HTPBR), a silane coupling agent, recycled scrap tire rubber particles, and a liquid elastomer. A novel route has been tried by Visakh et al. [4] to synthesize and couple PBR with epoxidized soybean oil (ESO). Earlier studies on filler distribution in the blend of PBR/NR showed that more filler was preferentially absorbed in the low viscosity PBR [4]. Also rubber blend of PBR/NBR (50/50) was prepared by using carbon black as conducting filler [5]. Chough et al. [6] studied the relationship between vulcanization reactivity and chemical structure for a ternary rubber blend of PBR/NR/SBR which can be synthesized using 98 % *cis*-1,4 PBR, 98 % *cis*-1,4-polyisoprene NR, SBR (23.5 % bound styrene, 68 % *trans*-1,4, 17 % *cis*-1,4, and 15 % 1,2-butadiene), and commercial additives (carbon black, sulfur, zinc oxide, stearic acid, *N*-1,3-dimethylbutyl-2,*P*-phenyl-phenylenediamine, and TBBS). Go and Ha [7] studied the morphological behavior of PBR and ethylene-propylene-diene (EPDM) rubber blend and reported that rubber blend is incompatible with each other, but the addition of a mixture of aliphatic and aromatic hydrocarbon (AARH) effectively enhances the compatibility. El Sabbagh et al. [8] showed that the poor compatibility of NR/EPDM rubber blend could be enhanced by the addition of the compatibilizers such as PBR, chlorinated rubber, and polyvinyl chloride (PVC). As the brittleness of PS limits its applications, PBR is added to improve the toughness of PS which is commercially known as high-impact polystyrene (HIPS) [9]. Because of nanometer sizes of filler dispersed in rubber nanoblends of PBR/PS/nanoclay [9] and PBR/NR/nanoclay [10], markedly

Table 6.1 PBR-based rubber nanoblends and nanocomposites

S. no.	Name of PBR blend	Additional components and nanofillers	Reference
1	PBR/SBR	TBBS, TMTD, ZnO, Nanosilica	[1]
2	hPBR/PP	–	[2]
3	PBR/ER	Epoxy resin, soybean oil	[4]
4	PBR/NR	Carbon black	[5]
5	PBR/NR/SBR	Carbon black, ZnO	[6]
6	PBR/EPDM	ZnO, TBBS, sulfur, hydrocarbons	[7]
7	PBR/NR/EPDM	–	[8]
8	PBR/PS	Vinyl nanoclay	[9]
9	PBR/NR	TBBS, ZnO, stearic acid, nanoclay	[10]
10	PBR	Nano-$CaCO_3$, fly ash	[11]
11	PS/PBR/PS	Nanosilicate, epoxy resin	[12]
12	PBR/SBR	CNTs, alcoholic-CNT	[13]
13	PBR/NR/RR	NanoSiO_2, ZnO, TNTD, sulfur	[14]
14	PBR/NBR	–	[15]
15	PBR/PS	NaOH, nanoclay	[20]
16	PS/PBR/PS	–	[22]
17	PBR	Nano-$CaCO_3$	[23]
18	PBR/SBR	Nanoclay	[25]
19	PBR/PBR/PBR	–	[26]
20	PBR	OMMT	[27]
21	PBR	OMMT	[28]
22	PBR	Organo-MMT	[29]
23	PBR/NR	OMMT, LDH	[30]
24	PBR	OMMT	[31]
25	PBR, NBR, SBR	MMT, Na-MMT	[32]
26	PBR	–	[33]
27	PBR/BIMS	Carbon black, nanosilica	[34]
28	PBR/NR	OMMT	[35]
29	PBR/SBR	Graphite nanoplatelets	[36]

improvement in their properties was observed as compared to that of pure polymers or their traditional composites.

Mishra et al. [11] observed significant improvement in physical, mechanical, thermal, and flame retarding properties of nano-$CaCO_3$-filled PBR as compared to commercial $CaCO_3$ and fly ash-filled PBR. PBR/PS block copolymers can be epoxidized by using meta-chloroperoxy benzoic acid [12]. Then, the blends were prepared with epoxy resin, boehmite clay, and layered silicate nanofiller in the presence of methylenedianiline as a hardener. The conducting filler CNTs can also be incorporated into (50/50) rubber blend solution of PBR/SBR [13]. Silica was incorporated into a ternary blend system of PBR/NR/reclaim rubber (RR) by the sol–gel technique at different temperatures. The physicochemical properties of vulcanized PBR/NR/RR nanoblend indicated that the reinforcing efficiency was found to be increased with increment in RR content. These vulcanized PBR/NR/RR nanoblend prepared at 50 °C had superior mechanical properties than that of other vulcanizates prepared in 30 and 70 °C [14]. Corish et al. [15] observed better heat stability and abrasion resistance of PBR/NBR blend with various advantages. Mishra et al. [16, 17] observed decrement in activation energy with increment in ethylene vinyl acetate (EVA) content in EVA/EPDM (80/20) blend with maximum thermal stability. The breakdown of EVA/EPDM rubber blend in the mixer was minimum in the temperature range of 110–120 and 150–160 °C [18]. Maximum curing rate, energy, and total torque were observed in the case of butyl rubber/PP blend because of fast saturation of double bond present in butyl rubber [19]. Prabhakar et al. [20] showed that the thermal stability of PS can be improved by the addition of Kaolinite nanoclay and PBR as a compatibilizer. The effect of high-density polyethylene (HDPE)/EPDM blend was observed on the cross-linking characteristics and dynamic elastic properties, which reveal that cross-link density (a measure of TS) first increases and then decreases with increment in EPDM content. The dynamic elastic property (a measure of impact strength) also showed a peculiar trend. Both studies showed that less than 40 parts loading of EPDM gave the best mechanical properties [21]. Triblock copolymer (PS/PBR/PS) with ultra-thin nanosections had lower surface energy than PS block. In bulk, PS and PBR blocks are micro-phase separated, leading to morphology with curved nanostructures of 20–30 nm in width [22]. Mishra et al. [23] reported a comparative study of PBR filled with various loadings of fly ash and nano-$CaCO_3$ and showed that the nano-$CaCO_3$ can improve the TS more than 50 % than fly ash and reduce flammability up to a 75 %. An improvement of 100 % was observed in tear strength, while the toughness and hardness also were increased significantly [23]. Drastic improvement was observed in the mechanical properties of nano-$CaCO_3$-filled rubber containing linseed oil [24].

Tire components comprise a PBR/SBR blend containing exfoliated clay nanoplatelets, which has been already patented by the Goodyear Tire and Rubber Company in USA [25]. A crystallizable blend of PBR can also be compatibilized by the addition of 5–10 % amorphous diblock copolymers of (1,2 PBR/1,4 PBR) [26]. PBR/Organomontmorillonite (OMMT) nanocomposites [27, 28] were prepared by the solution intercalation method. These nanocomposites also showed excellent tensile properties with <12 wt% content OMMT due to their nanoscale dispersion and stronger interactions between OMMT and PBR chains [27]. The introduction of a small amount of OMMT showed greater improvement in the thermal stability and swelling behavior due to better barrier properties of OMMT layers. In addition to this cure time, scorch time, and their difference for PBR/OMMT, composites were also found reduced as compared to that of PBR [29].

PBR/NR nanoblends have been developed using with pristine and organically modified two layered nanofillers, i.e., montmorillonite (MMT) and layered double hydroxide (LDH) [30]. Faster curing was observed in PBR/NR blends containing OMMT as compared to unmodified one due to the functional groups present in the interlayer regions of the clays [31]. Various butadiene rubber-based nanoblends

[32] were developed using 19, 34, and 50 % of acrylonitrile content in nitrile rubber, SBR (23% styrene content), and PBR with stearyl amine-modified Na-MMT clay. Modeling and simulation of a PBR process, reaction mechanism, and reactor design were successfully carried out to study low molecular weight PBR products [33]. Zhang et al. [34] studied the effect of carbon black and silica fillers on the interfacial properties of elastomer blends of (PBR, terpolymer [brominated poly(isobutylene-*co*-*p*-methylstyrene) (BIMS)]). Rubber blends of PBR/NR/OMMT were prepared by direct mechanical blending [35]. Expanded graphite (EG) and isocyanate-modified graphite nanoplatelets (i-MG)-filled PBR/SBR blends can substitute NR in tire-tread application [36].

6.2 Preparation Methods

An elastomer-based blend can be prepared by three different techniques (Fig. 6.2), namely latex, solution, and dry blending methods [4, 37]. Latex blending technology has potential for fine dispersion of the components in the latex using surfactants, e.g., latex blending of PBR/NR and NR/SBR/carbon black. Solution blending technology gives coarse particles because of the low viscosity solution which promotes rapid coalescence, e.g., solution blending of PBR/NR and NR/SBR blend. Important methods, including latex blending, two-roll mixing, and extrusion blending for preparation of different PBR-nBs using various ingredients (e.g., nanofillers), are discussed here.

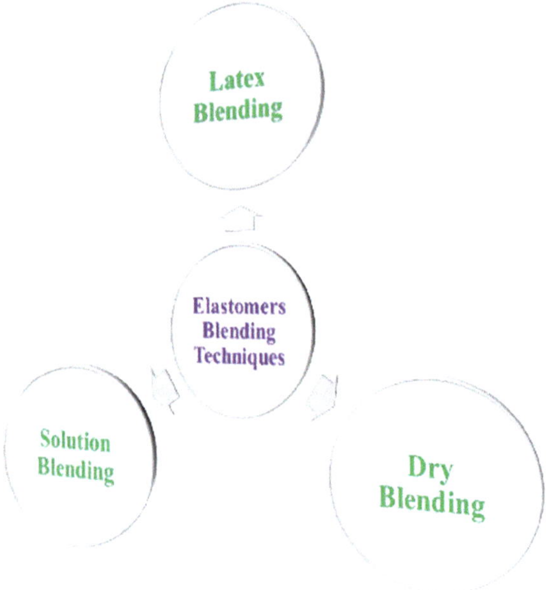

Fig. 6.2 Techniques for blending of elastomers

6.2.1 PBR/SBR/Nanosilica

Saeed et al. [1] prepared rubber blend of PBR/SBR by mixing 1,4-*cis* 96% PBR and SBR along with filler and ingredients such as *N*-tert-butyl-2-benzothiazole sulfonamide-TBBS (a safe-processing delayed action non-sulfur-donor accelerator), Tetramethyl thiuram disulfide (TMTD) (a fast-curing sulfur-donor accelerator), zinc oxide (ZnO) (an activator), *N*-(1,3-dimethylbutyl) *N*-phenyl-phenylenediamine (PPDA) (an anti-degradant to protect the rubber against environmental aging), and heavy paraffinic distillate solvent extract containing aromatic oils (a processing oil to reduce the rubber viscosity). (i) The PBR/SBR/ZnO rubber blends were prepared at ambient temperature in a small banbury mixer with counter-rotors rotating at 45 rpm. Ingredients such as TBBS and ZnO were added for 4 min; after the filler and PBR were mixed together for 10 min, and mixing was continued for another 6 min before the removal of compounds from the mixer. (ii) To make a blend of rubber without ZnO, SBR was mixed with the processing oil for 2 min and then PBR was blended for 1 min. Subsequently, silica was mixed for 3 min, and then after, TMTD and anti-degradant were mixed for another 6 min prior to remove from the mixer. At this time, mixing temperature of the both rubber blends was observed around 53 °C. Chemicals were added before the curing, the rotors were stopped, and the rubber was cooled down to 50 °C to avoid scorching during the mixing. Finally, both rubber blends were removed after complete mixing and milled to obtain 68-mm-thick sheets which were stored for at least 24 h before their analysis [1].

6.2.2 hPBR/PP

Svoboda et al. [2] synthesized rubber blends of hPBR (hydrogenated PBR) and isotactic PP by mixing them at 200 °C for 7 min at 100 rpm in a mini-mix molder in total three weight ratios of hPBR/PP chosen as 75/25, 50/50, and 25/75. The melt-mixed blends were extruded and then compression molded to a 0.2-mm-thick sheet at 200 °C which was then quenched in dry ice–methanol at −73 °C, in ice water at 0 °C followed by in the water at 26 °C [2].

6.2.3 Recycled PBR/(HDPE-LDPE-TiO_2)/PET Blend

Rubber researchers pay much attention to recycling of rubber products such as used tires. One important step is the processing or reduction of whole tires into useful parts. The best way to recycle the rubber products would be devulcanizing and then reuse them in the rubber industry. Visakh et al. [4] reported a blending of the recycled rubber so it can be shaped into useful articles at a reasonable cost. The main focus should be to find solutions for recycling those polymeric wastes that are traditionally difficult to use in re-processing, such as HDPE, LDPE, and PET. Therefore, recycled rubber/PET composites represent a tempting alternative. Tire rubber (e.g., PBR) is a low cost second raw material that can easily act as matrix, but the amount of PET able to be mixed with rubber is limited by the low compatibility between these two components [4]. To obtain rubber blend various recycled materials are used, which include (i) PET, obtained from soft drink bottles, (ii) rubber waste from tires and bicycle tubes, (iii) HDPE from yogurt containers, and (iv) LDPE from packaging. The mechanical properties and the environmental behavior of these recycled materials can be tailored by dispersing nanoparticles of titanium, calcium and zinc oxide, and fly ash which give efficient hybrid inorganic–organic composites. Optimizing the composition of recycled rubber–polymer matrix was also reported, e.g., rubber-PET, rubber-HDPE, and recycled rubber/PET/HDPE (85/10/5). To obtain recycled rubber waste polymer–metal oxides PET composites based on recycled materials, follow the steps that have been outlined by Viskah et al. [4].

6.2.4 PBR/MA-g-EPDM

Go et al. [7] prepared 98 % *cis* content PBR blend with maleic anhydride (MA)-treated EPDM and ethylidene norbonene (ENB) as a termonomer. MA-g-EPDM was prepared by reacting EPDM with 2.0 phr of MA in the presence of 0.2 phr dibenzothiazyl disulfide in a banbury mixer with a speed of 40 rpm at 230 °C for duration of 3 min. Bis (3-triethoxysilyl propyl) tetrasulfide (TESPT) was used as a coupling agent. A binary blend of PBR/EPDM can be mixed easily in a Brabender using different rubber ratios (75/25, 50/50, and 25/75 by weight) for 30 min at 100 °C. Mixing of the carbon black with PBR/EPDM blend was done by a two-stage procedure. In the first stage (without curatives), masterbatch blends of MA-g-EPDM and PBR were mixed with 40 phr of carbon black for 6 min at 155 °C, and rotor speed was kept at 40 rpm. In the second stage for vulcanization and evaluation, masterbatch stocks were prepared from each of the PBR/EPDM and the PBR/MA-g-EPDM blends by the addition of 2 phr sulfur, 5 phr ZnO, 1 phr sulfenamide, and 10 phr process oil, on a two-roll mill at 100 °C for 5 min followed by curing in a thermofluid press. PBR/EPDM blends containing hydrocarbon resin (AAHC) or TESPT, designated as PBR/EPDM/AAHC or PBR/EPDM/TESPT ternary blends, which were also prepared using different ingredients [7].

6.2.5 PBR/PS/Vinyl Nanoclay

Subana et al. [9] studied the combined effect of nanoclay and PBR/PS blend (also known as HIPS) on commercially available blend, which was manufactured by in situ method. This rubber nanoblend was obtained by adding 1, 2, and 3 wt% of vinyl nanoclay to HIPS prepared by melt mixing. These HIPS nanoblends were prepared by a typical internal mixer at temperature of 180 °C with rotational speed of 50 rpm for duration of 8 min. The specimens were prepared by the injection molding at the temperature of 200 °C for evaluations [9].

6.2.6 PBR/NR/Nanoclay

Alipour et al. [10] prepared nanoblend of PBR/NR (100/0, 50/50, and 75/25) in open two-roll mill with rotor speed of 80 rpm for duration of 18 min at room temperature. For preparation of this nanoblend, 5 wt% nanoclay was pre-dried at temperature of 80 °C for 24 h, which was then added to blend of PBR/NR along with curing agents and compounding ingredients [(zinc oxide (3 phr), stearic acid (2.5 phr), aminox (2 phr), PVI, *n*-cyclohexylthiophthalimide (2 phr), sulfur (2.5), TBBS (0.6)].

6.2.7 PBR/PS/Epoxy Resin and PBR/PS/Nanosilica

Di-block, tri-block and star-block copolymers (BCP) blends of PS/PBR/PS (containing 50 % PBR) and PBR/PS (50–70 % PBR) were epoxidized by using meta-chloroperoxy benzoic acid (MCPBA) [12]. Then, the PBR/PS blend with epoxy resin and along with their nanocomposites were prepared using boehmite and layered silicate as nanofiller in the presence of methylenedianiline as a hardener. The reaction mixture was prepared by dissolving 1 g of the BCP in 10 mL of dichloromethane and then allowed to mix under inert atmosphere in the round bottom flask equipped with a magnetic stirrer and thermometer. Calculated amount of MCPBA was added to the polymer reaction mixture under constant stirring at the 0 °C for 90 min. The completion of the reaction and the excess of the MCPBA were confirmed using potassium iodide/starch paper. After the completion of the reaction, the mixture was filtered and separated out from the residue. Then filtrate was extracted using aqueous $NaHCO_3$ followed by drying with Na_2SO_4 and filtered out. The organic portion of the solution was separated and the epoxidized BCP was recovered by separating dichloromethane in a rotary evaporator. The residue was dried till constant weight [12]. The neat epoxy system (diglycidyl ether of bisphenol-A/methylene diamine) and the blends with epoxidized PBR/PS were prepared by taking 30 wt% of epoxidized BCP and 70 wt% of epoxy resin dissolved in toluene. The resultant solution was heated up to 110 °C for complete evaporation of the solvent and the addition of curing agent (i.e., methylene dianiline) was followed under constant stirring and heating to make homogeneous mixture. The content was poured into the Teflon mold. For the preparation of nanocomposites, the same procedure was repeated by applying 3 wt% of boehmite and layered silicate separately to the mixture in the required proportion. Finally, blends and nanocomposites were pre-cured at the 80 °C in the vacuum oven for 12 h, which was kept for post-curing at 140 °C for additional 12 h [12].

6.2.8 PBR/SBR/MWCNT

Das et al. [13] prepared a high-performance solution of PBR/SBR (50/50) blend which was incorporated using modified and unmodified multi-walled carbon nanotubes (MWCNTs). Because of the outstanding properties of CNTs, its good dispersion in a rubber blend was carried out by a novel mixing approach for achieving good dispersion of CNT-alcohol suspension, which was then mixed with the rubber blend at elevated temperature. These rubber nanoblends exhibited significantly enhanced physical properties at low concentration and high aspect ratio of CNT, which also enabled the formation of a conductive percolating network in the blend at 2 wt% concentrations. Finally, the Payne effect, an indication of filler–filler interactions, was observed at low concentrations of CNT in the rubber matrix [13].

6.2.9 PBR/NR/RR/Nanosilica

Debapriya et al. [14] prepared a tire-tread formulation of ternary blend of PBR/NR/RR using different proportions of RR (20–60 wt%) with (30/70) rubber blend PBR/NR incorporated within nanosilica ion by the sol–gel technique at different temperatures. All formulations contained no RR as well as varied proportions (20–60 wt%) of RR, additives such as ZnO (5 phr), stearic acid (2 phr), sulfur (0.5 phr) and TMTD (2.16 phr), which were mixed in proportion based on total 100-g rubber irrespective of the proportion of RR. The compounded rubber stocks were then placed in a mold and pressed between the platens of a compression-molding machine, which were then cured at the 160 °C by applying pressure of 5000 psi for the respective optimum cure times [14]. PBR/NR/RR blends contain nanoSiO_2, which were prepared by the sol–gel technique [14]. The vulcanizates of these ternary blends were prepared by soaking the cured sheets in TEOS for 48 h at various temperatures (30, 50 and 70 °C). After soaking, the swollen sheets were immersed in a 10 % aqueous n-butyl amine for 24 h at the respective temperature with the swelling. The catalyst solution was then removed, and the sheets were dried at the 50 °C for 72 h for drying and further drying in a vacuum oven at the 50 °C to obtain constant weight [14].

6.2.10 PBR/PS/Nanoclay

Prabhakar et al. [20] prepared nanoblend of PBR/PS using in situ polymerization of PBR and PS by adding Kaolinite nanoclay. For that he washed styrene using 5 % NaOH and took to a three-neck round bottom flask with water circulation through the condenser followed by the addition of inhibitors to prevent polymerization during storage. A 25 mL of styrene and equal amount of 5 % NaOH were shaken well and allowed to settle in a separating funnel so that NaOH layer can be separated out. This procedure was repeated 3 times with DM water and also with NaOH solutions followed by a final rinse with DM water. As the moisture present in styrene could affects polymerization, it could be removed by adding pellets of molecular sieves into the styrene for 15 min so that it can absorb the water drops if any. The pellets can be removed by decantation. The experiment was done in a three-neck flask connected to a condenser and kept under heating to the 90 °C in water bath with continuous stirring so vapors of styrene can be escaped out. It was covered using a piece of cotton. Nitrogen gas was introduced to create an inert atmosphere above the monomer surface to avoid any external factors affecting the polymerization. Weighed amount of rubber was added when the temperature became steady at 80 °C, and 30 min time was given for dissolving the rubber. When the solution becomes homogeneous, benzoyl peroxide (BPO) was added as the BPO becomes active at 80 °C. Stirring was continued for about 25 min or more to obtain viscous and the turbid solution from clear solution. At this time, nanoclay was added and allowed to disperse 30 min or more and whole liquid was transferred to a Teflon mold with curing time of one week. It was desired to vary the composition from 0.5 to 6 wt% of both PBR and nanoclay [20].

6.2.11 PBR/NR/Nanofiller

Bottazzo et al. [30] prepared rubber blends (43/57) of PBR/NR/MMT and PBR/NR/LDH in two-roll mixing mills. The Cloisite was taken as unmodified MMTs and organically modified with dimethyl dihydrogenated tallow quaternary ammonium and methyl tallow bis-2-hydroxyethyl quaternary ammonium, respectively. Similarly, LDHs were taken as unmodified clay and modified with hydrogenated fatty acid. The 3 phr of nanoclays was used as optimum value. The compounding of the blends was done by a laboratory size open two-roll mill by mixing PBR and NR for 5 min. And then, nanoclays, ZnO, stearic acid, plasticizer, sulfur, and accelerators were added in sequence. After completion of mixing, the compounded composites were vulcanized at the 152 °C in an electrically heated hydraulic press for 30 min.

6.2.12 PBR/BIMS/Nanosilica/Carbon Black

Zhang et al. [34] prepared two elastomer PBR and brominated poly(isobutylene-co-*p*-methylstyrene) (BIMS), *p*-methylstyrene (PMS), and *p*-bromomethylstyrene (BrPMS). Consequently, in standard rubber tires, hard sidewall and the soft inner liner can be made of carbon-reinforced BIMS blended with butyl and NR. Both polymers are elastomeric in nature with a low T_g (−50 °C). Carbon black (1.85 g/cm^3) particles having size of 290 and 500 Å were used, while silica particles (2.2 g/cm^3) were used of particle size of 90 Å and specific surface area of 295 m^2/g. Automotive BIMS was taken as elastomer material because these materials possess low permeability to gases, which can control the mechanical properties by varying the level of BIMS and PMS as well as filler content and cross-link density [34].

6.2.13 PBR/NR/OMMT

Gu et al. [35] prepared PBR/NR/OMMT blends by direct mechanical blending. The bentonite clay (50 g) was dispersed in 5000 mL of distilled water at 65 °C and purified by sedimentation and washing. The dried bentonite clay was dispersed under vigorous stirring. A solution of 30 g octadecylamine, 200 mL distilled water, and 6 mL concentrated H_2SO_4 was prepared at 65 °C and slowly added to the bentonite dispersion. The reaction was stirred for 4 h. The sediment was then collected by filtration and washed ten times with hot water, followed by vacuum drying at 90 °C, grounding and sieving to 30 μm. The compounding of PBR/NR with fillers and vulcanization ingredients was done in a laboratory size open two-roll mixing mill. Firstly, PBR (70 phr) and NR (30 phr) were mixed for 5 min followed by the addition of filler, and other ingredients such as OMMT, ZnO (5 phr), stearic acid (1 phr), accelerator mercaptobenzothiazole (0.8 phr), antioxidant (1.5 phr), and sulfur (2.5 phr) were added in sequence for total compounding cycle of 15 min. Finally, the compounded blends were compression-molded at 145 °C under 15 MPa for the optimum cure time (T_{90}) [35].

6.2.14 PBR/SBR/Graphite Nanoplatelets

Malas et al. [36] used EG- and isocyanate-modified graphite (i-MG) nanoplatelets to prepare nanoblend of PBR/SBR. First, EG was prepared from natural graphite (NG) by a well-known acid process combining with thermal exfoliation. To incorporate more number of polar groups in the surface of EG, it was then further oxidized by H_2SO_4/HNO_3 (3:1) at the 90 °C for 12 h with vigorous stirring. A homogeneous suspension of oxidized EG (2 g) in 100 mL DMF was prepared in a 100 mL three-neck flask equipped with a magnetic stirring bar under nitrogen atmosphere. During continuous stirring, 2,4-Toluene di-isocyanate (2 g) was then added under a nitrogen atmosphere at the 80 °C for 24 h. The reaction mixture was next poured into 200 mL of acetone to coagulate the product. The resultant product i-MG was then filtered and washed completely with additional acetone. Lastly, the resulting product was dried at the 100 °C under vacuum for 48 h. The EG and i-MG containing PBR/SBR nanoblends in the presence of carbon black (CB) were fabricated by melt blending. EG/CB and i-MG/CB with other curatives were incorporated into the PBR, SBR, and PBR/SBR (1/1) blends in separate batches by melt blending in a laboratory scale open two-roll mixing mill (the speed ratio of the rotor was 1:1.4) at room temperature. The amount of nanofillers (wt%) with respect to rubber was kept constant at ~3. The compositions of the PBR/SBR rubber blends are kept 0/100, 50/50, and 100/0 with 3 wt% of both EG and i-MG. Vulcanization of the rubber blends was carried out on a compression-molding machine at the 160 °C for an optimum cure time of 30 min [36].

6.3 Tensile Testing

To improve the rubber's mechanical properties such as hardness, tear strength, TS, and elongation at break, fillers with surface areas from 150 to 400 m^2/g are used. Fillers such as colloidal carbon blacks, synthetic silicas, organoclay, and metal oxides have been added [1]. The compatibility and miscibility of a blend are parameters that can improve the tensile properties of blends. It is important to control the phase morphology and better interfacial adhesion between the blend constituent via chemical or process approach by using compatibilizing agents or by dynamic cum chemical compatibilization [4].

The main aim of the studying mechanical properties of different rubber blends is to identify the potential use in development of passenger car tire-tread formation. For determining the tensile properties of PBR/SBR (25/75) at different strain amplitudes, cyclic fatigue life and tan δ of the rubber blends were cured at 170 °C and cut into sheet size of 23 cm × 23 cm × 2.7 mm thick prior to mechanical analysis. Tensile properties of PBR/SBR rubber nanoblends cured with ZnO and TBBS were observed better (26 MPa) than those cured with only TMTD (14 MPa); i.e., TS was higher by 46 % [1]. Tensile stress–strain curves of hPBR/PP blends were measured at room temperature with universal tensile testing machine (UTM) with a crosshead speed of 2 mm/min. After the pre-set strain was attained, the crosshead returned at the same speed as stretching. After the strain reached zero, the sample was released from the clamps and residual strain was measured at certain intervals [2].

Figure 6.3a, b shows tensile stress–strain behavior of the hPBR/PP blend (in 50/50 and 75/25 ratios) [2]. In Fig. 6.3a, it can be clearly observed that the blends were stretched to 50 % elongation and then the movement of the clamp was reversed at the same speed, while inserted small picture shows curves for neat PP. The behavior of PP showed clear yield with plastic deformation at about 8 %

elongation, even though the quenching temperature was reported to be low at −73 °C. The residual strain of neat PP was high (~43 %), which suggested plastic blend deformation, while miscible hPBR/PP blend had no yield point and less residual strain as compared to neat PP [2]. Residual strain decreases with the resting time increases as shown in Fig. 6.3b. Steep recovery was observed during the first hour of resting, and then the recovery speed was found to be decreased. After 24 h, the recovery of neat PP was quite small from 175 to 170 % during 24 h, after stretching of 200 %.

The influence of hPBR content on the residual strain was also discussed, which shows the lowest residual strain values were observed for blends containing 50 and 75 % hPBR after stretching 50 % elongation. Residual strain was plotted as a

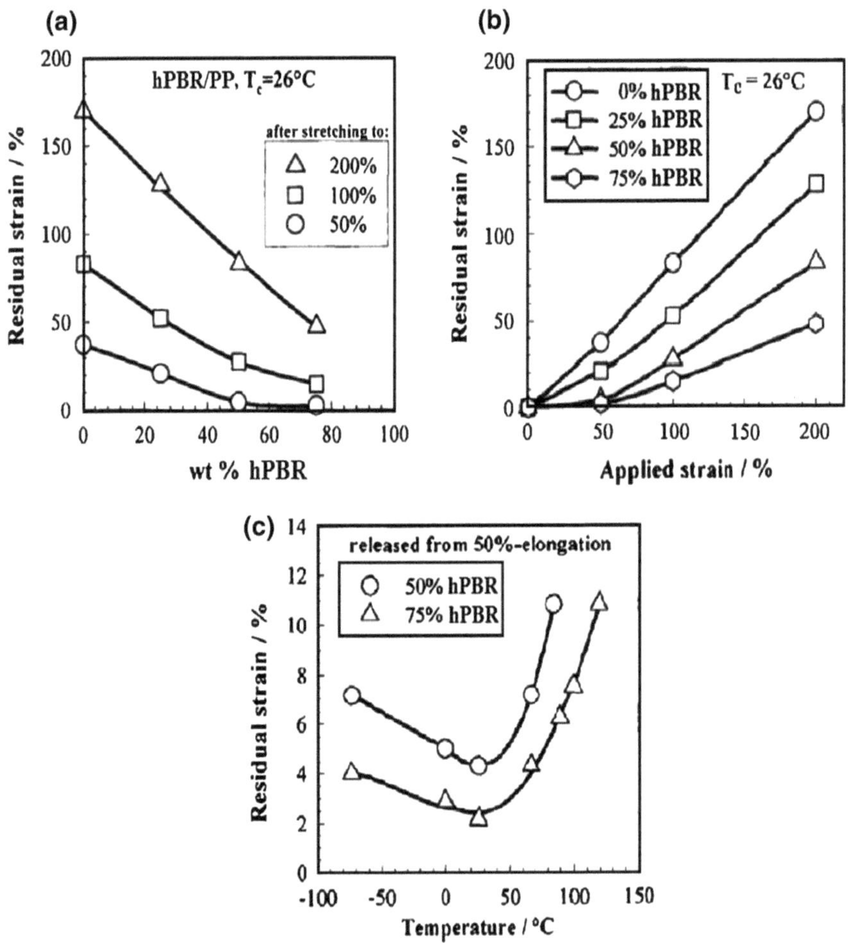

Fig. 6.3 **a** Tensile stress versus strain curves and **b** residual strain versus resting time curves of hPBR/PP rubber blend **c** Residual strain vs Applied strain at different wt% of hPBR. (by permission of [2])

function of applied strain for blends with different blend compositions as shown in Fig. 6.3c, which shows optimum crystallization conditions get best elastic properties of two blends stretched to 50 % elongation. Crystallization at high temperature is very harmful for elasticity; i.e., when the blend was crystallized at −73 °C, elasticity was observed a little bit worse, but blend still remained elastic. Elastomeric behavior can be improved by annealing the blend at higher temperature after quenching, when the blend crystallizes at high temperature, large crosshatch lamellae may be formed which showed blend had very poor elastic properties [2].

Visakh et al. [4] reported greater tensile and tear strengths of PBR/NR blends, when the mixing was sufficient to reduce the domain size below 1 m; no further improvement in properties was observed with further mixing. Clarke et al. [38] obtained greater tensile and tear strengths in blends of NR and PBR when mixing was sufficient to reduce the domain size below 1 μm; no further improvement in properties was observed with further mixing (Fig. 6.4) [4, 38].

Saban et al. [9] reported tensile modulus (TM) of commercial and direct blends of PBR/PS, which was found to be increased with the addition of vinyl nanoclay from 0 to 3 wt%. Values of TM were reported to be higher (1749 MPa for commercial blend and 1590 MPa for direct blend) for 1 wt% vinyl nanoclay content which was due to the presence of compatibilizer in PS grafted onto the PBR backbone chain, but TM was found to be decreased in the case of 3 wt% vinyl nanoclay [9]. Das et al. [13] observed stress–strain curves of PBR/SBR blend loaded with different CNTs (in phr) and reported that stress was increased with increasing CNT loadings up to 5 phr. Also, alcoholic-CNT mixture showed higher stress as compared to a dry CNT mixture [13].

TS is mainly associated with the stress distribution within rubber and the effective increase in the crack path. Strong silica–rubber interaction is required to maximize micro-dispersion, which results in the highest possible level of TS for ternary PBR/NR/RR-nanosilica-based blend [14]. Debpriya et al. [14] found a pronounced increase in the TS by increasing silica content in RR using the sol–gel technique. However, increasing trend was observed for TS of all ternary rubber nanoblends, but the maximum TS was obtained at the 50 °C because of adequate cross-link density, uniform dispersion of nanosilica in RR, and superior reinforcing efficiency due to chemical bond formation between RR and nanosilica. Although the dispersion of nanosilica in all ternary rubber nanoblends was homogeneous, ternary blends have highest cross-link density at the 70 °C as compared to other ternary nanoblends, which made the vulcanizates hard and brittle [14].

The rubber nanoblends filled with in situ dispersed, partially exfoliated and intercalated clay particles had higher TS of 17.6 MPa for a blend of PBR/sSBR/eSBR (70/0/30), while it was lower near about 16 MPa for a blend of PBR/sSBR/eSBR (30/70/0) [25]. Tensile properties of PBR/NR/MMT and PBR/NR/LDH nanoblends were measured at room temperature with a UTM with a crosshead speed of 500 mm/min. The TS of PBR/NR/LDH blend was found to be decreased between 3.8 and 5.2 MPa as compared to that of pure PBR/NR blend which was 6.5 MPa, while the opposite trend was observed for PBR/NR/OMMTs with TS value of 5.6–7.5 MPa due to the dispersions of clays in the elastomeric matrix and also intensity of the

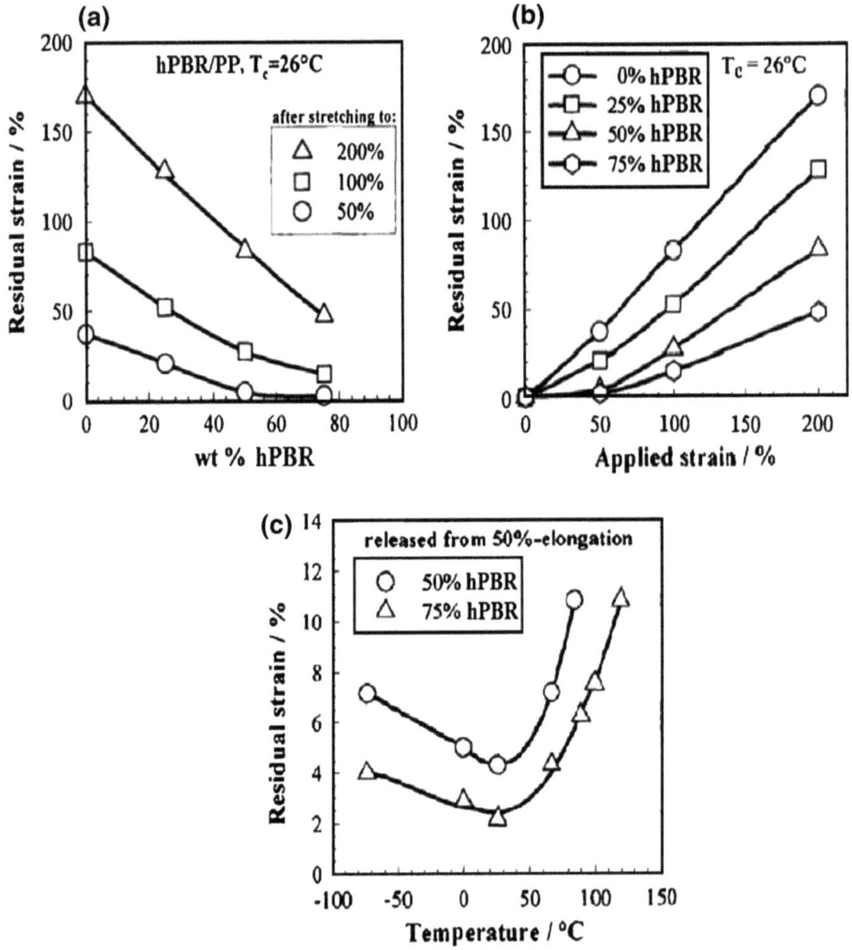

Fig. 6.4 **a** residual strain versus hPBR content (wt%) in blend at different stretching, **b** residual strain versus applied strain at different hPBR content (wt%) and **c** residual strain versus crystallization temperature at different hPBR content (wt%) (by permission of [2])

interactions between the fillers and the rubber chains [30]. PBR/NR nanoblends loaded with 6 wt% OMMT had the TS value of 19.6 MPa, which was 90.3 % higher than that of pure PBR/NR hybrid blends. This higher value was due to the homogeneous dispersion of OMMT particles. The TS of PBR/NR/OMMT nanoblends found to be increased with increasing OMMT content up to 6 wt%, but decreased slightly for nanoblend with 8 wt% OMMT content. The enhancement in TS was due to the strong interaction between the clay mineral layers and the rubber chains, while low TS of 6 wt% OMMT-loaded PBR/NR blend can be attributed to the aggregation of the particles at high OMMT content [35]. Mechanical properties of the PBR/SBR-based nanoblends were recorded by UTM under ambient conditions at 25 ± 2 °C according

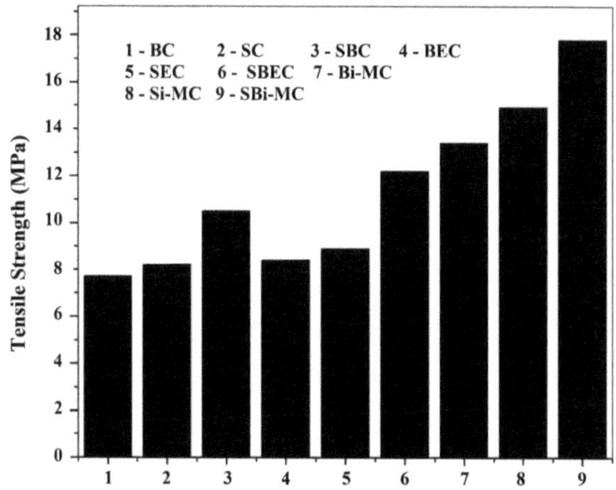

Fig. 6.5 Tensile strength of the PBR/SBR rubber nanoblends (by permission of [36])

to ASTM: D412 standard [36]. The effects of EG and i-MG addition with CB on the tensile properties of PBR, SBR, and PBR/SBR blends are represented in Fig. 6.5. Incorporation of 3 wt% of EG and i-MG with 30 wt% of CB into PBR/SBR blends increased the TS of blend and it was higher as compared to 40 wt% CB-loaded PBR/SBR blends. It was evident that the TS of the various blends such as PBR/EG/CB (BEC), SBR/EG/CB (SEC), PBR/SBR/EG/CB (SBEC), PBR/i-MG/CB (Bi–MC), SBR/i-MG/CB (Si–MC), and PBR/SBR/i-MG/CB (SBi–MC) were approximately 9, 8, 16, 74, 82, and 70 % higher than that of their respective controls (PBR/CB, SBR/CB, and PBR/SBR/CB blends) [36].

6.4 Differential Scanning Calorimeter Studies

Thermal analysis such as differential scanning calorimetry (DSC) is frequently used to determine the micro-phase structure of rubber blends. A modulated temperature DSC can be used to measure the glass transition temperature (T_g) of the pure PBR, SBR, and their PBR/SBR nanoblends filled with nanofiller [1]. T_g of PBR/SBR rubber nanoblend cured with ZnO and TBBS was observed at −79 °C, which was similar to the (−80 °C) for PBR/SBR blends cured with only TMTD. For pure PBR and SBR, T_gs were reported as −105 and −50 °C, respectively [1]. Using a DSC, the calorimetric runs of PBR/NR/SBR/ZnO blends were observed at different scan rates (5, 10, 20, and 40 °C/min) under a nitrogen atmosphere. At a scan rate of 5 °C/min, peak temperature lied in the range of 208–215 °C with an enthalpy range of 14–20 J/g, while at scan rates of 10, 20, and 40 °C/min, peak temperatures were recorded in the ranges of 193–202, 182–190, and 170–178 °C with enthalpy ranges of 12–17, 8.6–14.3, and 8–13 J/g, respectively. Activation energies were observed in the range of 97–104 kJ/mol [6]. The calorimetric T_g of PBR/NR/RR/nanosilica blend was

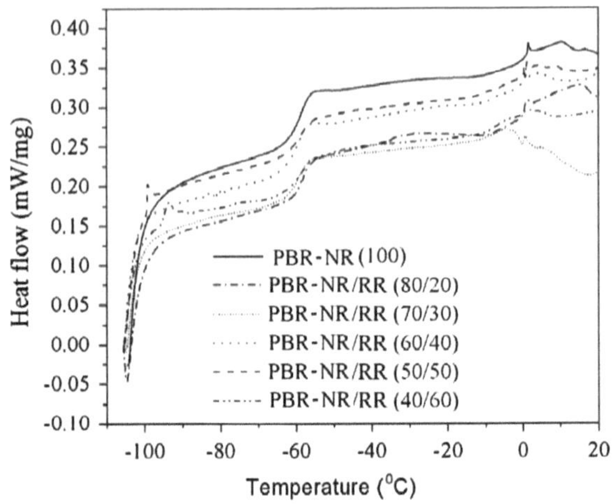

Fig. 6.6 DSC thermograms in the region of the glass transition for the PBR/NR/RR blends vulcanized at 50 °C (by permission of [14])

observed in the range of −100 to 25 °C [14]. However, the reinforcing efficiency, mechanical properties, and thermal behavior of PBR/NR/RR/nanosilica blend vulcanized at 50 °C are found to be superior as compared to that blend vulcanized at 30 and 70 °C. Therefore, a DSC study was reported out only for PBR/NR/RR/nanosilica blends vulcanized at 50 °C, which had a single T_g of about −55 °C. All NR/PBR/RR/nanosilica blends vulcanized at 50 °C are shown in Fig. 6.6 [14].

The silica nanoparticles slightly influenced T_g of NR/PBR/RR blends, which was observed 23 °C lower as compared to that of neat rubber independent of silica content [14], while the heat capacity increment (ΔC_p) was found to be decreased with increasing silica content. As ΔC_p is a measure of the amount of polymer that participates in the glass transition, this decrement is often studied in terms of an immobilized layer of polymer around the filler nanoparticles or crystallites, respectively. The fraction of immobilized polymer can then be calculated by following Eq. (6.1):

$$\chi_{im} = 1 - \left(\frac{\Delta C_p}{\Delta C_p^0 (1 - W_{Sio_2})} \right) \tag{6.1}$$

In Eq. (6.1), ΔC_p^o is the heat capacity increment of the neat rubber and W_{SiO_2} is the weight fraction of filler (e.g., silica). In PBR/NR/RR blend vulcanized at 50 °C, the χ_{im} was found to be increased from 269 % with increasing silica and RR content. With increasing nanosilica content, the polymer filler interaction or reinforcing efficiency of the vulcanizates were found to be increased; hence, fraction of immobilized polymer chains is also increased [14]. DSC was carried out of PBR/PS nanoblends containing 0, 3.25, and 6 wt% of PBR and 0, 3.25, and 3.25 wt% of nanoclay [20]. Prabhakar et al. [20] studied the DSC thermograph of PBR/PS blend containing 3.25 wt% of both PBR and nanoclay, and he recorded T_g at 144 °C indicated from increasing in heat flow. As it reached its melting point, where it melted completely at 196 °C, the heat absorption reduced and continued till its degradation starts. T_g of 0.5 wt% PBR/PS blend was found to be increased marginally from 135 to 138 °C as clay content increased from 0, 3.25, and 6 wt%. Similarly, 3.25 wt% PBR/PS blend also showed a significant increment in T_g from 138 to 144 °C with increasing clay content from 0, 3.25 and 6 wt%. But 6 wt% PBR/PS blend showed decrement in T_g at 149, 143, and 146 °C with increasing clay content from 0, 3.25, and 6 wt% [20]. Design equation obtained by regression analysis is given by Eq. (6.2), in which 'A' shows clay composition in weight percentage, 'B' shows rubber composition in weight percentage, considering A and B with lesser effect than its power terms.

$$T_g = (128.341) + (3.31585\,A) + (6.50043) - \left(0.346537\,A^2\right) - \left(0.743232\,B^2\right) - (0.0661157\,AB) \tag{6.2}$$

Figure 6.7 shows the contour plot showing T_g at different rubber and clay compositions. The different ranges are shown in different colors. From the graph, we can understand that the cay composition increases T_g [20]. A blend of 0.5 wt% PBR/PS showed initial degradation at 373, 364, and 344 °C and final degradation at 468, 478, and 454 °C with increasing clay content from 0, 3.25, and 6 wt%, respectively. Similarly, 3.25 wt% PBR/PS blend showed initial degradation at 370,

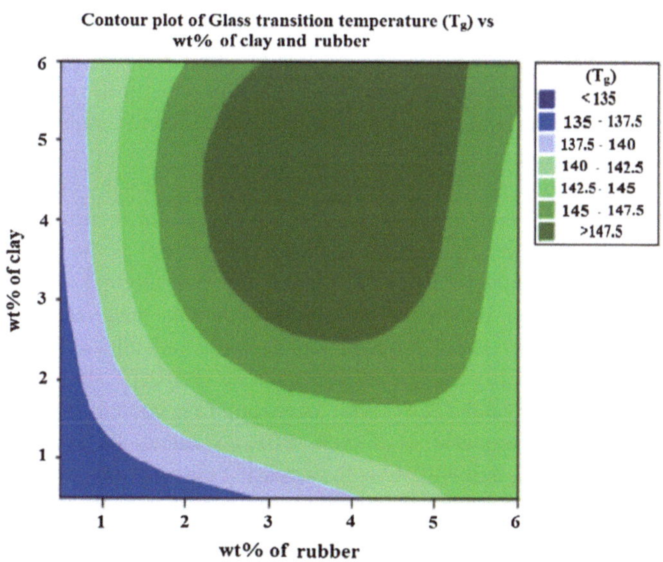

Fig. 6.7 Contour plot of glass transition temperature (GT) versus composition of clay and PBR (by permission of [20])

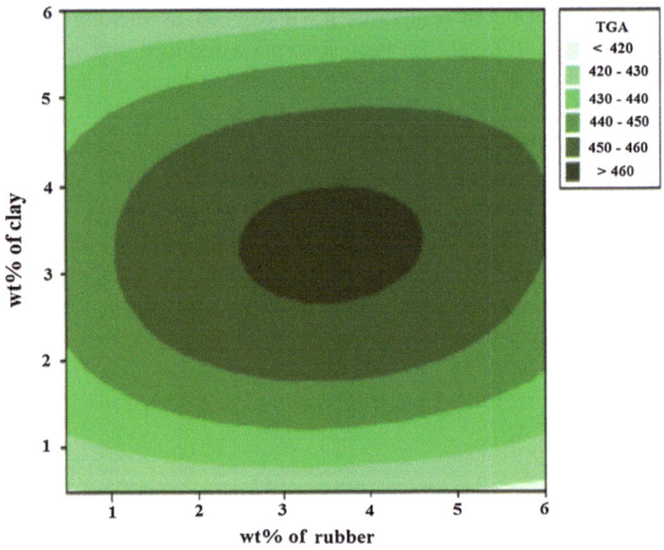

Fig. 6.8 Contour plot of maximum degradation temperature versus composition of clay and PBR (by permission of [20])

357, and 370 °C and final degradation at 489, 489, and 434 °C with increasing clay content from 0, 3.25, and 6 wt%, respectively. But, 6 wt% PBR/PS blend showed initial degradation at 378, 356, and 365 °C and final degradation at 502, 500, and 477 °C with increasing clay content from 0, 3.25, and 6 wt%, respectively. 3.25 wt % of the PBR/PS blend containing 3.25 wt% of clay showed initial degradation at 355–357 °C and final degradation at 488–489 °C [20].

Figure 6.8 shows the contour plot showing the maximum degradation temperature at different rubber and clay compositions. The different ranges are illustrated in various colors, also desired to have a higher degradation temperature. The inner circle shows the region of maximum degradation, in which clay and rubber composition ranges in optimum values from 2.5 to 4.5 wt% for thermal stability. The effect of nanoclay was very low at lower compositions, but as the percentage was increased the fillers dispersed successfully and imparted thermal resistance to the material. Also a further increment in the clay contents made the formation of agglomerates which ultimately reduced the strength and affected negatively on composite properties [20]. A blend of 0.5 wt% PBR/PS showed maximum degradation at 427, 426, and 427 °C with increasing clay content from 0, 3.25, and 6 wt%, respectively. Similarly, 3.25 wt% PBR/PS blend showed a maximum degradation at 460, 435, and 440 °C with increasing clay content from 0, 3.25, and 6 wt%, respectively. But 6 wt% PBR/PS showed a maximum degradation at 467, 471, and 434 °C with increasing clay content from 0, 3.25, and 6 wt%, respectively. A blend of 3.25 wt% of PBR/PS containing 3.25 wt% of clay showed maximum degradation at the 434–436 °C [20]. The vulcanization behavior of the uncured PBR/NR/MMT and PBR/NR/LDH nanoblends has been studied by DSC.

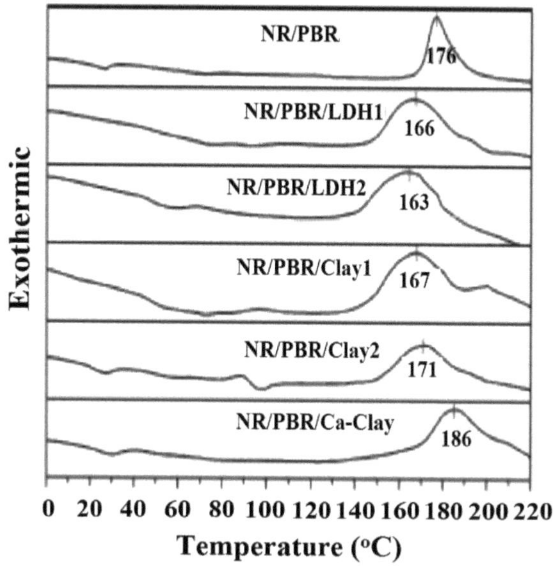

Fig. 6.9 DSC thermograms of uncured PBR/NR/MMT and PBR/NR/LDH rubber nanoblends (by permission of [30])

In DSC thermograms, the curing reaction, being a thermally activated process, originates an exothermic peak [30].

Figure 6.9 shows that the vulcanization temperatures were different depending on the nanoclay used in the PBR/NR blends. DSC analysis corroborates the results obtained by moving die rheometer; i.e., while vulcanization of the PBR/NR/nanoclay blend was taking place at the highest temperature, the others rubber nanoblends showed a lower vulcanization temperature than that of the pure PBR/NR blend. Also, it is possible to observe that the exothermic peak for the PBR/NR/clay mixtures is much broader than that of pure PBR/NR ones. This peak is probably due to overlapping of two exothermic peaks, also rubber chains close to the clay that cross-link at different temperatures with respect to the chains apart, resulting in a broader exothermic effect starting at 155–175 °C. All of the PBR/NR blends were submitted to two identical heating cycles; vulcanization process was completed during the first cycle and no thermal variation was observed during the second one. From the DSC and rheometric results, it can be said that the organic modifications of the clay can bring opposite effects on the vulcanization process, while the presence of cationic surfactant in MMT accelerates the curing reaction, but anionic one in LDH slows it down [30].

Gu et al. [35] observed onset temperature of curing for PBR/NR blend at 141 ° C, PBR/NR/OMMT (4 wt%) blend at 139 °C, and PBR/NR/OMMT (8 wt%) blend at 136 °C. Two vulcanization exothermic peaks of PBR/NR were contributed to the different affinity of NR and PBR on the curing agents. The smaller peak at lower temperature was due to the vulcanization process of PBR, while higher temperature was due to the NR vulcanization. However, the onset curing temperature of the NR phase in PBR/NR/OMMT blend decreased significantly, and only one vulcanization exothermic peaks was observed, which indicated that the addition of OMMT led to the synchronous vulcanization of PBR and NR [35]. The octadecyl ammonium ions in the MMT may accelerate the vulcanization of NR. The torque of NR in the absence of the curing accelerator was found to be increased slowly with time, while that of NR using an organic modifier as curing accelerator significantly was found to be increased rapidly with time. The accelerating effect of octadecyl ammonium ions on vulcanization process was mainly due to the heterolytic cleavage of elemental sulfur by the amine, resulting in ionic cross-linking [35].

6.5 Dynamical Mechanical Analysis

The ratio between loss modulus and elastic modulus is defined as Tan δ, which represents the viscous component of modulus and includes all the energy dissipation processes during dynamic strain. The Tan δ can be measured with DMA instruments using test pieces of approximately 35 mm long × 10 mm wide × 2.7 mm thick with different (1, 10, and 100 Hz) frequencies [1]. The energy loss in car tires during dynamic strain can affect their service performance, such as rolling resistance, which is related to the movement of the whole tire corresponding to deformation at a frequency of 10–100 Hz and a low Tan δ value at a temperature of 50–80 °C was required to meet high-performance tires, to reduce rolling resistance and save energy and fuel. A high Tan δ value (high hysteresis) at −50 to −30 °C is required to obtain high skid resistance and ice- and wet-grip tire.

Figure 6.10 shows the temperature dependencies of the storage moduli, the loss moduli, and the loss factors Tan δ of the CNT/PBR/SBR nanoblends. With an increase in temperature, the storage modulus of all CNT/PBR/SBR nanoblends was reported to be decreased, which is associated with the glass transition phenomenon

Fig. 6.10 DSC thermograms of uncured (*a*) PBR/NR (*b*) 4 wt% PBR/NR/OMMT (*c*) 8 wt% PBR/NR/OMMT (by permission of [35])

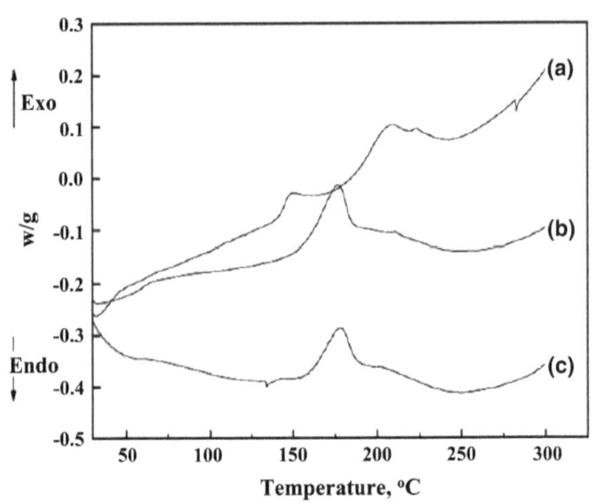

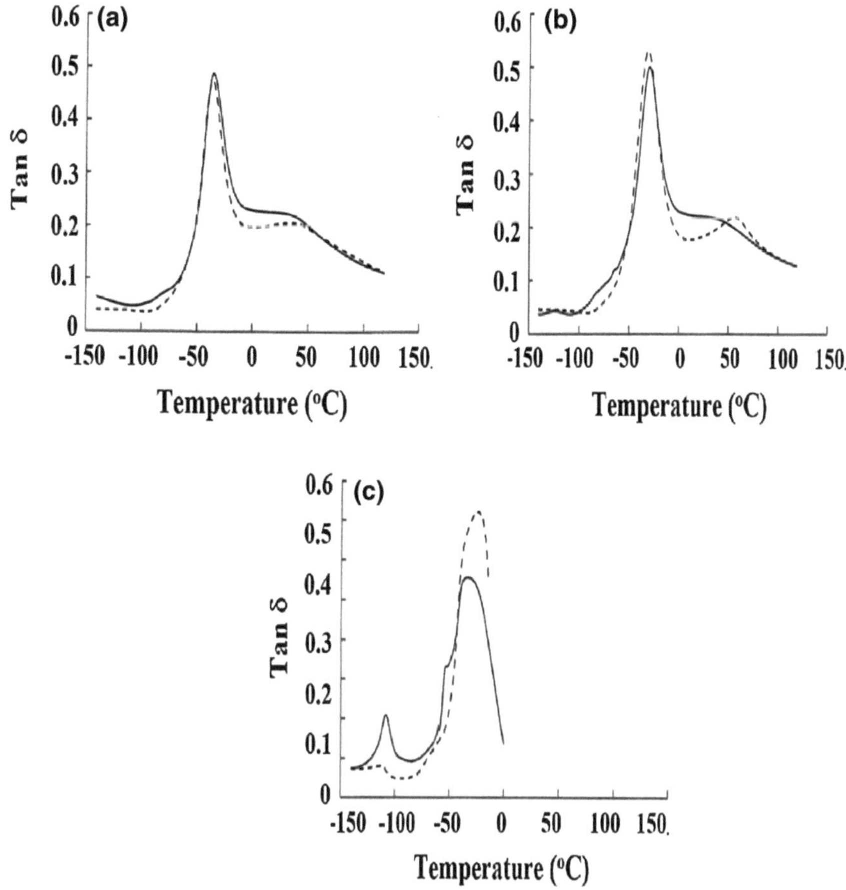

Fig. 6.11 Tan δ versus temperature curves of PBR/SBR rubber nanoblends at different frequencies of **a** 1, **b** 10, **c** 100 Hz; (*solid line*) for cured with ZnO and TBBS, (*dash line*) for cured with only TMTD (by permission of [1])

of the elastomer chains. As shown in Fig. 6.10, the glass transition temperature at the maxima of the Tan δ plot does not change with the tube content in the rubber matrix. However, the peak height reduces considerably with higher CNT content. This behavior also indicates the strong reinforcement efficiency with only low content of CNT [35].

Figure 6.11a–c shows Tan δ as a function of the test temperature at different frequencies, respectively [1]. At frequency of 1 Hz, Tan δ peaks of PBR/SBR rubber nanoblend were cured with (i) ZnO and TBBS,; (ii) only TMTD were reported 0.47 at −36.6 °C and 0.49 at −35.5 °C, respectively, while at 10 Hz, Tan δ peaks were found to be slightly higher to 0.53 at −31.7 °C and 0.50 at −30.7 °C, respectively. Similarly, at frequency of 100 Hz, Tan δ peaks were observed to be increased up to 0.72 at −24.0 °C and 0.55 at −32.4 °C. Clearly, PBR/SBR rubber nanoblend cured with ZnO and TBBS possessed a better skid resistance and ice- and wet-grip at 10 and 100 Hz because it had higher Tan δ values at low temperatures. However, at above 50 °C, which is important to reduce the rolling resistance and save energy, the Tan δ of both rubber blends (i.e., cured with ZnO and TBBS, and cured only with TMTD) was fairly similar at a frequency of 1 Hz. Notably, an increment in the test frequency had a significant effect on the Tan δ peak of these rubber blends. Hence, it can be said that PBR/SBR rubber nanoblend had much higher abrasion resistance index and a lower heat build-up, which could replace the passenger car tire treads that are currently in use [1].

A continuous phase of higher T_g gave rise to dramatic increment in the modulus of PBR/NR blend which can be observed [4]. Optimized performance was observed in a (50/50) blend of PBR/NR to be used for radial truck tire-tread stock with a wide distribution black with 75 % location in the NR phase. This black gave equivalent modulus and 15 % lower modulus as compared to different loadings of carbon black [5]. Go et al. [7] studied Tan δ versus temperature curves of PBR/EPDM blends with different rubber ratios and reported that the location of the peak in the Tan δ versus temperature of the respective curve corresponded to the T_g of the rubber blend. As expected, both PBR and EPDM were incompatible and also showed two separate T_gs. The plasticizing function of AAHC for the PBR/EPDM blend can be observed in the form of Tan δ behavior. It was also reported that when AAHC or TESPT was added to the PBR/EPDM (50/50) blend, the maximum point of Tan δ, i.e., T_g of EPDM, shifted downward to that of PBR, and this was because of AAHC acted as a lubricant and TESPT acted as a plasticizer as well as a coupling agent to the buildup of silica-to-rubber bonds. However, (50/50) blend of PBR/MA-g-EPDM did not indicate clear shift of T_g [7]. Das et al. [13] reported that the incorporation of MWCNTs into PBR/SBR blends can affect the glass transition behavior by reducing the height of the Tan δ peak considerably. Above the T_g, the storage modulus was found to be increased after the incorporation of a small amount of MWCNTs into PBR/SBR nanoblends.

In Fig. 6.12, it can be clearly observed that when BIMS was added to the PBR/NR blend, loss factor (Tan δ) was found to be increased at low temperature at −30 to 10 °C but decreased above 30 °C [37]. The loss factor of the PBR/NR blend was measured in the temperature range from −120 to 80 °C. DMA indicated that the peak loss factor was lower than that of PBR/NR and shifted to higher temperatures, especially for the NR phase.

Two Tan δ peaks are clearly shown in Fig. 6.12: One is from the small glass transition peak of PBR at lower and another is of NR at higher temperature. Both of the NR and BR phases of OMMT/NR/PBR (4 wt%) showed lower Tan δ peak values than the NR/BR hybrids, while Tan δ of NR phase shifted to higher temperature. These observations indicated a decrease in mobility of the rubber molecules due to the strong interaction between the rubber matrix and OMMT. The Tan δ value of OMMT/NR/PBR (4 wt%) at 0 °C was slightly lower than that of the pure NR/PBR, which indicated that the nanocomposite had better wet skid resistance properties [35].

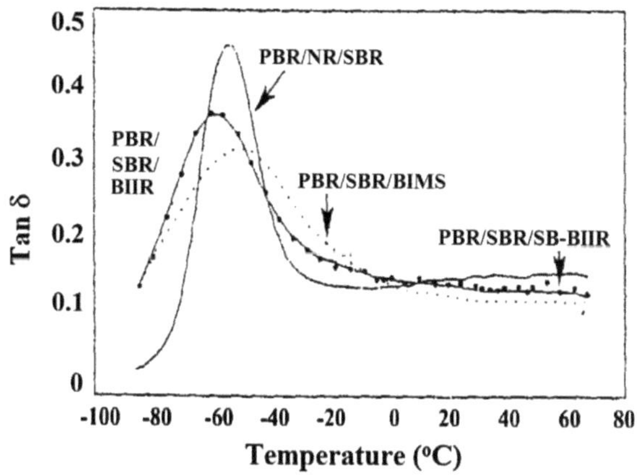

Fig. 6.12 Tan δ of PBR/IR/SBR/BIMS blends as a function of temperature (by permission of [37])

Malas et al. [36] investigated the effects of EG and isocyanate-modified graphite (i-MG) on the dynamic mechanical properties of PBR, SBR, and PBR/SBR blends by DMA cum thermal analysis. The temperature-dependent storage moduli (E_0) and loss factor of the rubber blends are presented in Fig. 6.13a–d. Both the elastic and viscous behaviors of the rubber blend affected the resulting strain due to the application of an oscillating force. The storage modulus can be regarded as the elastic modulus of the rubber blends and loss tangent was interconnected to the energy drenched due to energy dissipation as heat [36].

It is clearly shown in Fig. 6.13a, b that PBR/SBR blends containing EG and i-MG in the presence of CB showed a drastic increment in the storage modulus in a wide range of temperature as compared to the PBR-based nanocomposites. But, in comparison with SBR-based nanocomposites, storage modulus of PBR/SBR blends was found to be increased at very low temperature region and then decreased from a low-to-high temperature region which was because of very low transition temperature of PBR. Homogeneous mixing of PBR with SBR along with good dispersion of nanofillers increased its stiffness and also storage modulus of PBR/SBR blends. Figure 6.13c, d displays the temperature-dependent Tan δ curves of the various PBR/SBR blends, which indicate that greater Tan δ value of the PBR/SBR blends in the range of −20 to 10 °C. This can be taken into account for superior antiskid properties of the PBR/SBR blends under wet conditions while lower Tan δ of PBR/SBR blends in the range of 50–60 °C indicates the low rolling resistance of the rubber stuffs. It is shown in Fig. 6.13 that the Tan δ values of the EG- and i-MG-filled PBR/SBR blends in the region of −20 to 10 °C were higher than that of PBR-based nanocomposites. So, after the homogeneous mixing of PBR and SBR in the presence of different nanofillers, the antiskid property under wet conditions of the PBR vulcanizates was found to be improved. At the same time, rolling resistance of the PBR/SBR blends was found to be lowered (because of lower Tan δ of the PBR/SBR blends in the temperature range of 50–60 °C) as compared to that of SBR vulcanizates. EG/i-MG containing PBR/SBR blends in the presence of CB showed superior storage modulus, antiskid properties, and lower rolling resistance

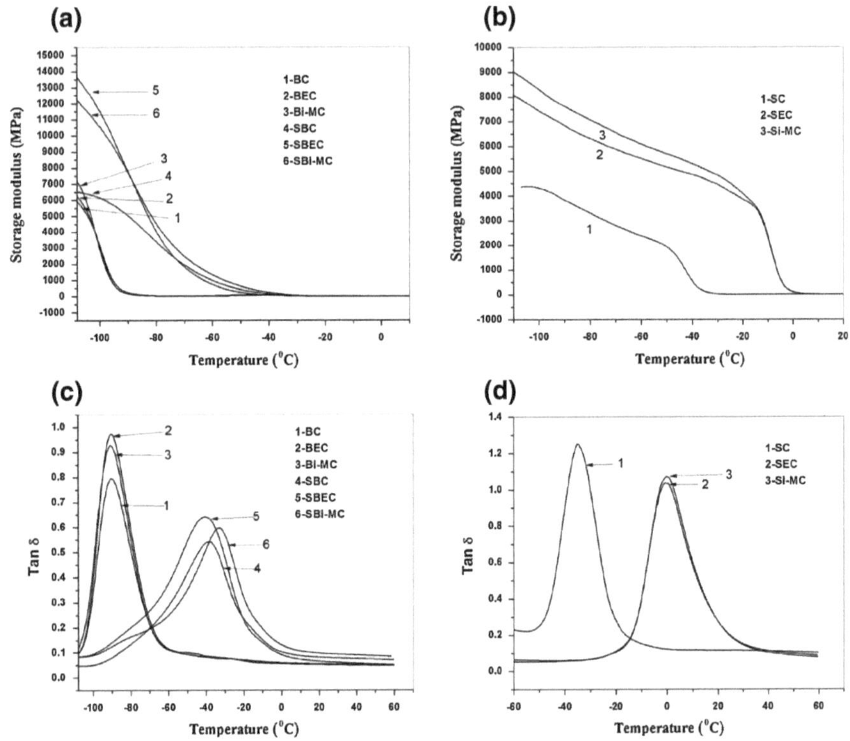

Fig. 6.13 **a–b** Storage modulus versus temperature curves and **c–d** Tan δ versus temperature curves of the PBR/SBR blends (by permission of [36])

as compared to the only CB-loaded PBR/SBR blends, which was due to the better interactions and interfacial adhesion between i-MG sheets and the rubber matrices [36].

6.6 Thermo-Gravimetric Analysis Studies

Rubber scientists and technologists investigated the thermal stability of rubber nanoblends by thermo-gravimetric analysis (TGA) and compared them with that of neat rubber matrix. Debpriya et al. [14] observed the thermal stability of PBR/NR/RR ternary blend (vulcanized at 30, 50 and 70 °C) at different sol–gel reaction temperatures, which was determined through TGA by using nitrogen as a purge at heating rate 10 °C/min.

Figure 6.14a–c represents the TGA profiles of the PBR/NR/RR ternary blend vulcanized at 30, 50, and 70 °C and their DTG curves are also presented in the inset. The DTG curves for ternary blend vulcanized at 30 and 70 °C show two-stage degradation, but this ternary blend vulcanized at 50 °C shows only one-stage degradation distinctly and in the second stage, it is very less prominent. In the case of ternary blend vulcanized at 30 °C, its first decomposition of NR matrix occurred between 280 and 408 °C, while the second decomposition was attributed to the PBR phase degradation in the range of 408–510 °C. However, ternary blend vulcanized at the 70 °C shows the first decomposition between 297 and 412 °C and the second decomposition between 407 and 515 °C. But ternary blend vulcanized at the 50 °C shows only first decomposition between 290 and 499 °C. The TGA of RR indicates three-stage degradation: first decomposition occurs between 160 and 336 °C, second stage between 336 and 507 °C, and third stage between 507 and 671 °C. Incorporation of nanosilica by sol–gel technique increased the temperature of first degradation peak of NR and RR due to homogeneous dispersion of nanosilica. Various parameters such as the temperature interval of all degradation stages observed from DTG curves, temperature of all stages, maximum rate of degradation, sample weight loss at the different temperature, and char residue values were also reported. It is also reported that first degradation of PBR/NR/RR blends vulcanized at 30 and 50 °C was found to be increased with increasing RR as well as nanosilica intake, but the second degradation appeared only for ternary blends vulcanized at 30 °C which did not show any appreciable change with increasing RR content. Here, ternary blends vulcanized at 30 °C show two characteristic degradation peaks in DTG curve between 280 and 510 °C. The % weight loss of ternary blends vulcanized at 30 °C was found to be decreased from 38.3 to 30.4 and 36.3 to 29.7 % after the first and second degradations, respectively, and it also showed an increment in thermal stability of the ternary blends with increasing the RR content. Moreover, an increment in the char residue values with RR content was due to the increasing nanosilica content in the ternary blends. Ternary blends vulcanized at 50 °C showed one characteristic degradation peak in DTG curve between 290 and 500 °C and % weight loss with a decreasing trend as similar to that of ternary blends vulcanized at 30 °C. In Fig. 6.14c of ternary blends vulcanized at 70 °C, it is clear that thermal stability increases with increasing RR content and shows two-stage degradation. First degradation started from 295 and 308 °C and completed between 407 and 413 °C with weight loss of 33.1–30.4 %, while the second degradation started from 407 and 413 °C and completed between 494 and 515 °C with weight loss of 34.4–26.9 %. This indicated that the thermal stability of ternary blend was found to be increased with increasing RR content. In Fig. 6.14a, c, it is shown that two-stage thermal degradation occur for ternary blends vulcanized at 30 and 70 °C probably indicating two types of solid state reactions. For all different ternary blends vulcanized at 30 and 70 °C, char residue was found to be increased as silica intake increased along with increasing RR content [14]. Prabhakar et al. [20] studied TGA and showed changes in weight of a 3.25 wt% PBR/PS blend containing 3.25 wt%. He reported that when the temperature reached 380 °C, the degradation started and reached maximum at a temperature of 440 °C [20].

The thermal degradation of pure PBR/NR blend takes place in two steps, reported by Bottazzo et al. [30]. He reported that the first decomposition of the NR

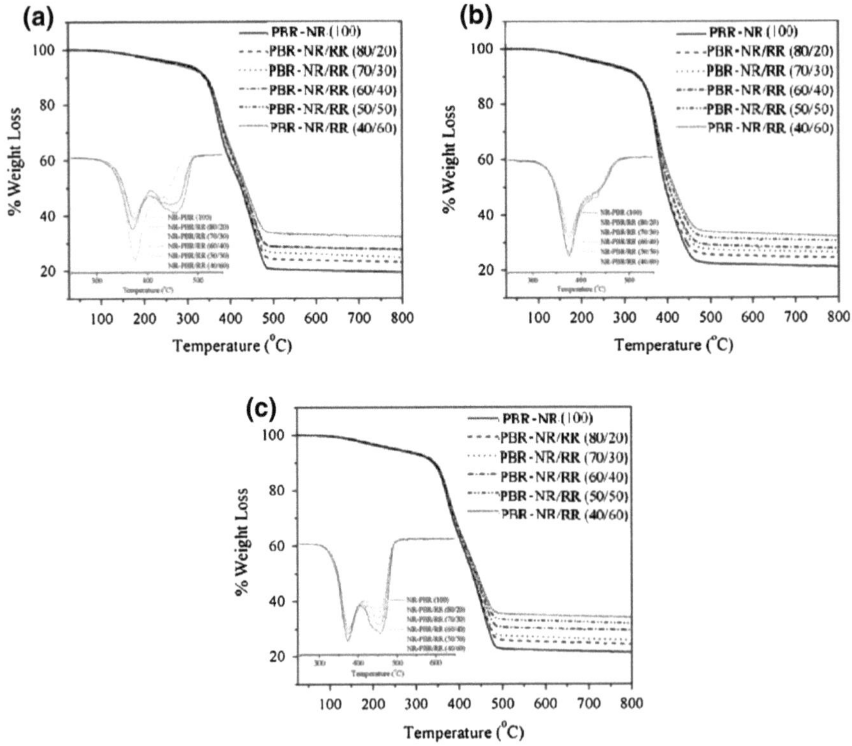

Fig. 6.14 TGA profiles of the PBR/NR/RR ternary blend vulcanized at **a** 30, **b** 50, and **c** 70 °C (by permission of [14])

matrix occurs between 280 and 410 °C, while the second step decomposition is attributed to the PBR phase degradation and it is observed in the range of 410–520 °C. The addition of nanoclay lowers peak of first decomposition temperature due to a poor dispersion of the nanofillers (MMTs and LDHs) in the PBR/NR blend, but OMMTs increase the temperature of 4.4 and 5.2 °C in the second thermal degradation due to its good compatibility with the PBR blend. The increase of thermal stability of rubber nanoblend is generally attributed to the barrier effect that hinders the diffusion of small molecules generated during thermal decompositions [30].

Two decomposition peaks can be observed in the thermal degradation curves, reported by Gu et al. [35]. He observed the first decomposition due to the thermal degradation of NR chains starting at 250 °C and reaching a maximum at around 367 °C, while the second decomposition at 440 °C due to degradation of PBR matrix. Both rubber nanoblends (4 and 8 wt%) PBR/NR/OMMT showed basically the same onset and first peak decomposition temperature as compared to pure PBR/NR. This can be attributed to the decreased diffusion of low molecular gases

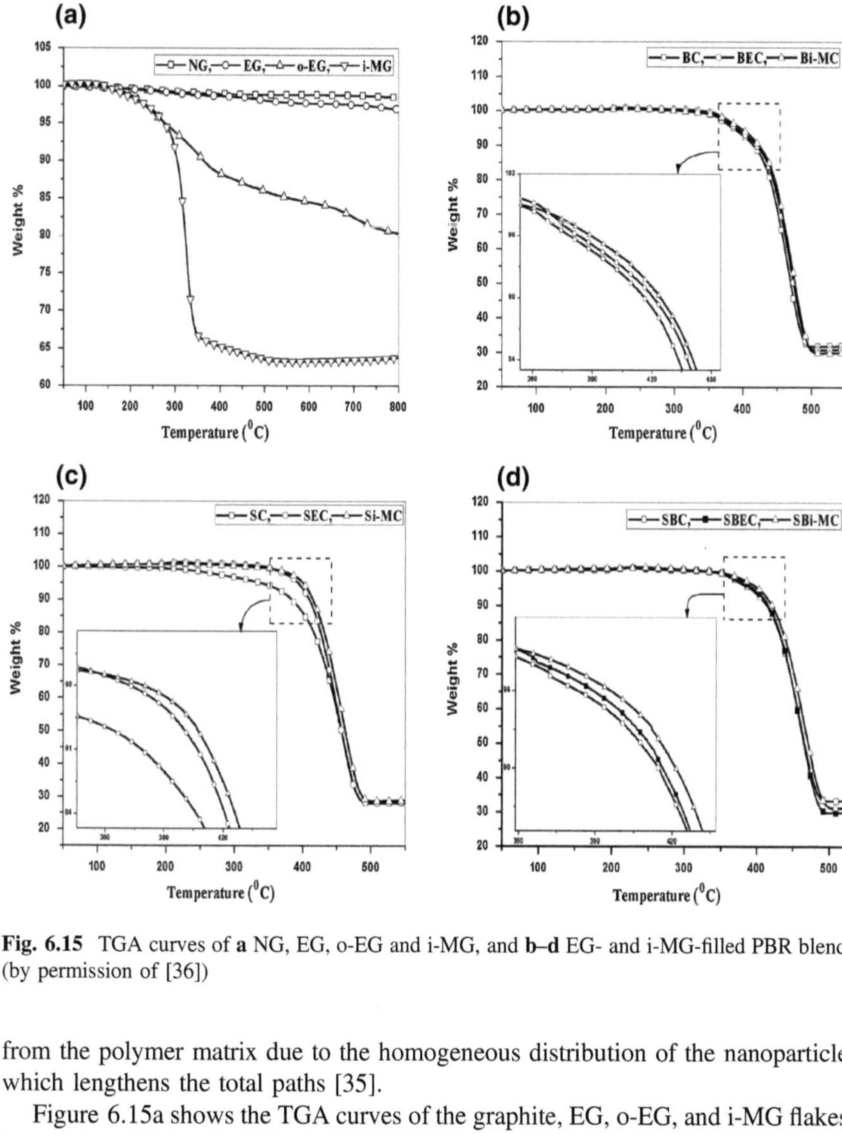

Fig. 6.15 TGA curves of **a** NG, EG, o-EG and i-MG, and **b–d** EG- and i-MG-filled PBR blends (by permission of [36])

from the polymer matrix due to the homogeneous distribution of the nanoparticle, which lengthens the total paths [35].

Figure 6.15a shows the TGA curves of the graphite, EG, o-EG, and i-MG flakes, in which o-EG has less thermal stability and small weight loss of o-EG around 200 °C than the EG due to the decomposition of polar groups and unstable oxygen containing functional groups present on the surface of o-EG. Another partial weight loss around 700 °C can be attributed to the subtraction of more stable oxygen functionalities of o-EG. In comparison with o-EG, i-MG exhibits a maximum weight loss at around 300 °C. The TGA curves of the different PBR/SBR nanoblends are represented in Fig. 6.15b–d, which indicate that the thermal stability of the EG/CB (total-33 wt%)- and i-MG/CB (total-33 wt%)-filled PBR was found to be increased as compared to that of the only CB-loaded (40 wt%) rubber blends.

EG/CB- and i-MG/CB-filled PBR/SBR nanoblends exhibit an increase in the thermal stability because EG and i-MG sheets act as efficient heat shielders by dissipating more heat and do not permit the heat to grow up within the blends, thereby preventing oxidation at the early stage of degradation [36].

6.7 Scanning Electron Microscopy Studies

The PBR-nBs were examined by a scanning electron microscope (SEM) to qualitative the state and nature of agglomeration, to assess the nanofillers dispersion. Small pieces of the rubber nanoblends can generally placed in liquid nitrogen for 3 min and then fractured to create two fresh surfaces. These blend surfaces can be gold-coated prior to SEM analysis to study the degree of dispersion in the PBR-nBs.

Saeed et al. [1] assessed dispersion of silica in the PBR blend to select a suitable mixing time for incorporating the silica in the rubber blend. After completion of 24 h of mixing, the rubber nanoblends were examined in SEM to study surface morphology and nanofiller dispersion. Small pieces of the uncured PBR blends were placed in liquid nitrogen for 3 min and then fractured to create two fresh surfaces. These small pieces of samples were gold-coated and then examined under SEM. SEM images examined after 10 min of total mixing for the SBR and 16 min for the PBR, respectively, which were found to be sufficient to fully disperse the silica nanoparticles in the rubbers [1].

Further insights on the morphology of cryogenically fractured PBR/NR (25/75) blends with and without clay are observed in Fig. 6.16a, b. The fractured surface of pristine PBR/NR blend is clearly shown in Fig. 6.16a which is very smooth, while

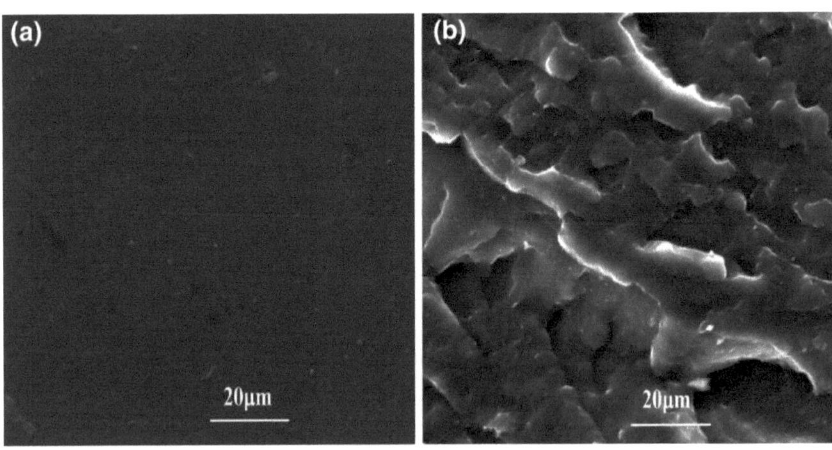

Fig. 6.16 SEM photo-micrographs of PBR/NR (25/75) blend, **a** without nanoclay, **b** 5 wt% nanoclay (by permission of [10])

the harsh surface of PBR/NR blend filled by 5 wt% nanoclay as shown in Fig. 6.16b which exemplifies an effective interaction established between rubber blend and filler. The SEM micrographs also confirm that smaller particle size of the nanofiller provides a larger surface area for a better filler dispersion and interfacial bond between the filler and rubber blend [10].

Das et al. [13] observed morphological view of PBR/SBR blend filled with CNTs. Debpriya et al. [14] studied the SEM of the tensile-fractured samples of PBR/NR/RR nanoblends filled with nanosilica to get an idea about the dispersion of silica nanoparticles. It is also reported that RR content plays a vital role in nanosilica dispersion in the ternary rubber nanoblends, which is corroborated from the evaluation of mechanical properties; two extreme compositions (20 and 60 %) of RR were taken into consideration at different sol–gel reaction temperatures [14]. Figure 6.17a–f represents micrographs of PBR/NR/RR nanoblends containing 20 and 60 % RR as well as nanosilica, respectively. Micrographs (a), (b), (d), and (e) show smooth, rubbery failure, whereas micrographs (c) and (f) indicate the rupture type of failure. Although some agglomerates of nanosilica are observed in micrographs (c) and (f), micrographs (a), (b), (d), and (e) represent almost homogeneous dispersion of nanosilica in the rubber matrix. Therefore, homogeneous and uniform dispersion of nanosilica into the rubber blend depends on increasing RR content and sol–gel reaction temperature. The absence of vacuoles in all the PBR/NR/RR blends indicates a high level of interaction between nanosilica and rubber, which is the indication of the superior mechanical properties of ternary blends [14].

As shown in Fig. 6.18, SEM micrographs show the low effect of nanoclay at lower compositions, but as the % loading of nanoclay increases, the filers get dispersed which imparts thermal resistance to PBR/PS blend of filled with Kaolinite nanoclay. If % loading of nanoclay increases further, agglomerates would be formed which can reduce the strength and affects negatively on properties of rubber nanoblend of PBR/PS filled with Kaolinite nanoclay [20].

The cryo-fractured surfaces of the PBR blends were analyzed by SEM to observe the dispersion, interfacial adhesion, and reinforcing mechanism of EG and i-MG in rubber nanoblends. SEM micrographs of the rubber blends indicated that EG flakes are randomly distributed in the different rubber blends. The lateral size of the EG was exposed for preparation of rubber nanoblends by melt blending, which indicated moderate interfacial bonding between EG and blends. But for i-MG-loaded rubber blends, the microstructure of the fractured surfaces displayed the close contact and high embedding of i-MG flakes with rubber blend due to a good interracial interaction. SEM micrographs of the EG- and i-MG-filled rubber blends show rough, tortuous fractured surface and broad tear paths; hence, EG and i-MG sheets can be dispersed in the rubber matrix in the presence of CB and can be changed the crack lines depending on their orientation in the rubber blends [36].

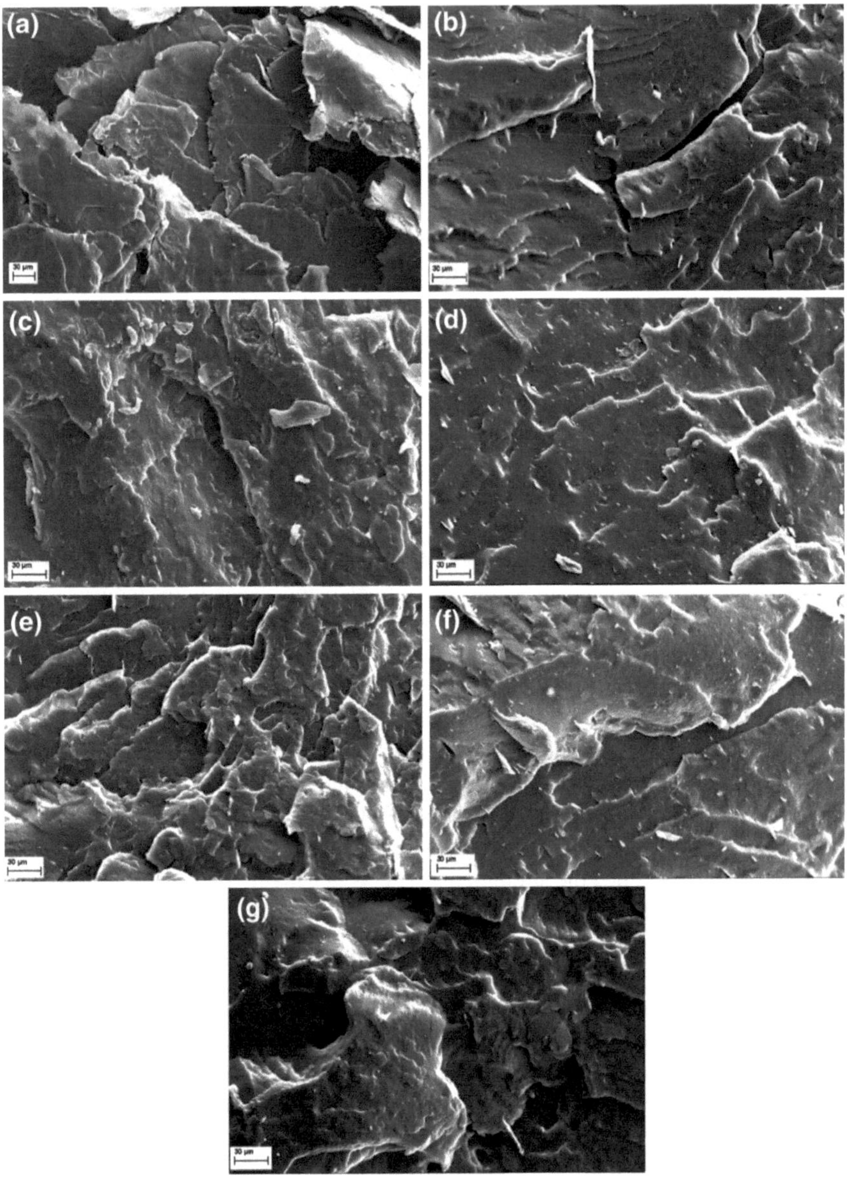

Fig. 6.17 SEM images (30 μm) of **a** i-MG **b** BEC, **c** Bi–MC, **d** SEC, **e** Si–MC, **f** SBEC and **g** SBi–MC composites (by permission of [36])

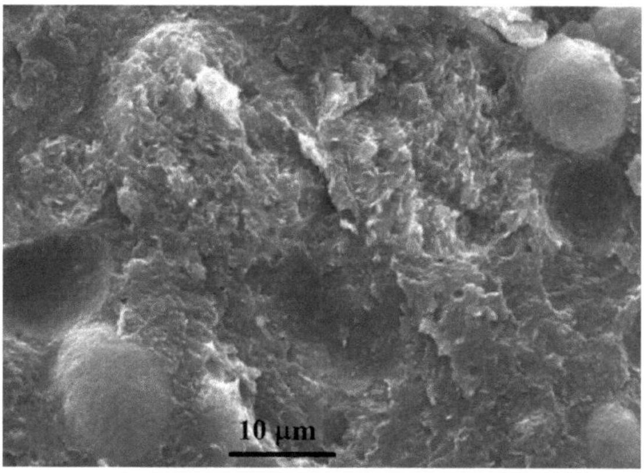

Fig. 6.18 SEM micrograph of PBR/PS blend of filled with Kaolinite nanoclay (by permission of [20])

6.8 Transmission Electron Microscopy Studies

Transmission electron microscopy (TEM) is used to study the size, shape, morphology, and crystallinity of hard blocks, structural evolution of segmented copolymers, and blends. Svoboda et al. [2] studied TEM of PBR/PP blend and reported that crystallization at 120 °C has a crosshatch lamella structure, while the blends crystallized at the 26 °C has only small fragmented lamellar structure [2]. Visakh et al. [4] reported TEM images of 5 wt% PBR in polychloroprene blend and reported mean diameter of the dispersed particles as 80 nm, with a very broad size distribution [4].

TEM micrographs of the carbon black-filled PBR/EPDM blends with various rubber blend ratios were analyzed by Go et al. [7]. He reported that CB agglomerated in PBR phase is observed due to high diene PBR and also it has a much larger affinity with CB than does a saturated elastomer such as EPDM. Morphology in Fig. 6.19 shows a good compatibility of EPDM and PBR in the presence of AAHC and also 25/75 blend of PBR/EPDM has large domain size the of PBR, which can be dispersed in the EPDM matrix, but the size of PBR domain reduces significantly when AAHC or TESPT was added to the PBR/EPDM blend [7]. In the previous sections, dynamic mechanical properties of the PBR/EPDM blend showed that the AAHC acted as a plasticizer, homogenizing agent, and compatibilizer as well as a lubricant, based on the morphological features in Fig. 6.19.

The compatibilizer is also known as interfacial agents, which have been used to enhance the compatibility of incompatible polymer blends as well as to improve interfacial adhesion between gross-phase-separated polymer pairs by reducing the interfacial energy between the phases [7]. Go et al. [7] studied TEM morphology and showed clearly the enhanced compatibility of PBR/EPDM (50/50) blend in the presence of TESPT, which indicates that the PBR phase of a large domain size is dispersed in the EPDM, but PBR domain size reduces significantly; i.e., PBR/EPDM/TESPT blend showed finer domain morphology as compared to PBR/EPDM one. The PBR/MA-g-EPDM blend also showed finer domain

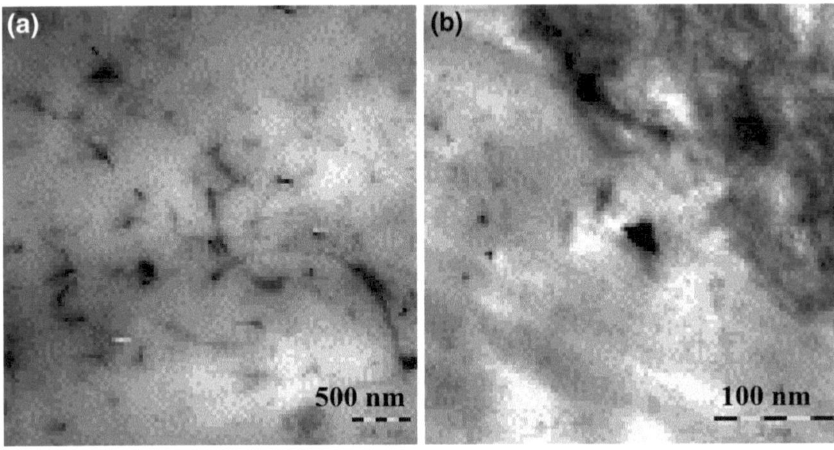

Fig. 6.19 TEM micrograph of PBR/SBR blend filled with CNTs (by permission of [13])

morphology than the PBR/EPDM blend, but less finer than the PBR/EPDM/AAHC ternary blend, so it can be said that the addition of AAHC to the PBR/EPDM blend was more effective to enhance their compatibility than the MA-g-EPDM/PBR-based blend [7]. Das et al. [13] observed TEM view of PBR/SBR blend filled with CNTs as shown in Fig. 6.20.

Figure 6.21 shows TEM images of OMMT-loaded PBR/NR-based nanoblends, which describe a combination of intercalated and exfoliated structure and also the presence of aggregated nanoclay layers [30].

The PNR/NR/OMMT-based nanoblend showed two different phases as shown in Fig. 6.22a. The dark area is the indication of PBR-dispersed phase and the light area is of NR continuous phase. The majority of OMMT layers was dispersed homogeneously in the NR matrix and at the interfaces between NR and PBR as shown in Fig. 6.22b. Only small amounts of OMMT are observed in the PBR matrix, which indicates that the OMMT layers had better compatibility with NR matrix. The OMMT particles appear to have an almost uniform thickness of 20 nm and a length between 100 and 300 nm [35].

HR-TEM images of the EG- and i-MG-filled rubber blends based on different synthetic rubbers were analyzed by Das et al. [36]. The dark lines in the images dictate the presence of EG and i-MG nanosheets with thickness at nanoscale. It is reported that the few EG sheets were agglomerated in the different rubber blends in the presence of CB due to the use of direct melt blending technique and also due to minimum basal spacing between EG sheets which restricts the rubber chains to enter the intergallery spaces. The i-MG flakes exhibited higher d-spacing as compared to the EG. Therefore, rubber chains can easily be intercalated into the gallery space of i-MG flakes. The i-MG/CB-loaded rubber blends show intercalation and partial exfoliation of i-MG sheets in the different rubber matrices [36].

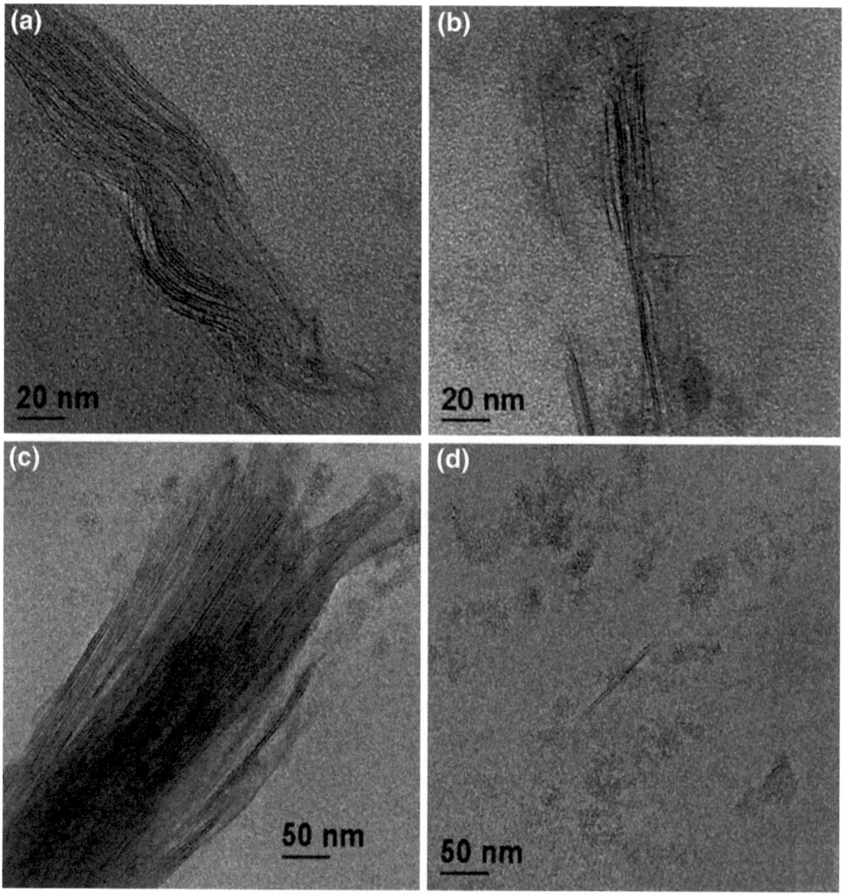

Fig. 6.20 TEM micrographs of the PBR/NR blend filled with OMMT (by permission of [30])

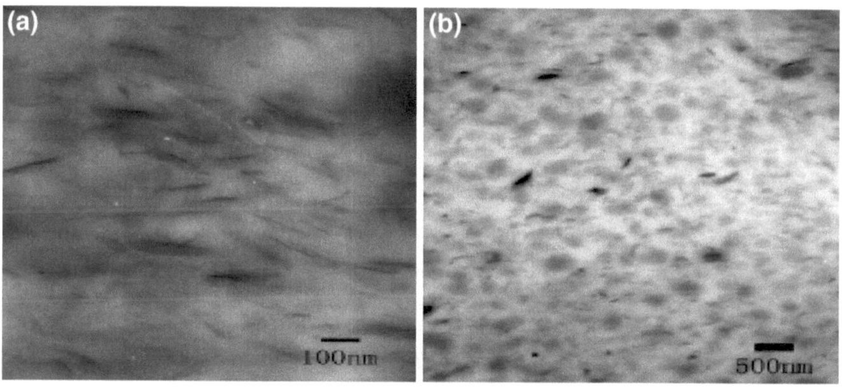

Fig. 6.21 TEM photographs of 6 wt% PBR/NR/OMMT, **a** 100 nm, **b** 500 nm (by permission of [35])

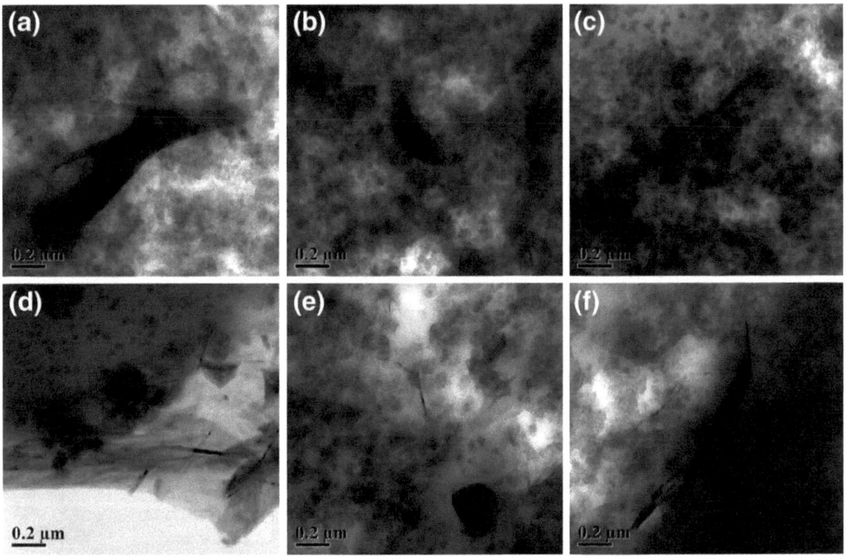

Fig. 6.22 HR-TEM images of **a** BEC, **b** Bi–MC, **c** SEC, **d** Si–MC, **e** SBEC, **f** SBi–MC (by permission of [36])

6.9 Neutron Scattering Studies

Small-angle neutron scattering (SANS) is used to characterize the size and surface properties of the filler aggregates in rubber blends. SANS setup with two triple-bounce Si crystals can be used as monochromator and analyzer for the ultra-high angular resolution. With the neutron wavelength set at 2.59 Å, the instrument generally produces ultra-low Q range data of 0.0001–0.002 $Å^{-1}$. To probe an extended range of the momentum transfer, or scattering wave vector, Q, a combination of conventional SANS and ultra-small-angle neutron scattering (USANS) measurements can be performed. Because USANS data can be taken with slit geometry and conventional SANS has generally pinhole geometry, a deconvolution can be done for the USANS data before all three sets of the data producing a wide Q range from 0.0001 to 0.2 $Å^{-1}$. As SANS measures the micro-morphology of a bulk material and also conventional SANS and USANS can be combined to probe length scales from 10 Å to microns. Therefore, the purpose of SANS experiments is generally to observe dispersion of nanofiller in rubber blends and also to characterize the size of nanofiller. In addition, interfacial properties between the particles and their surrounding polymer chains can also be probed with SANS. Zhnag et al. [34] reported the results of PBR/BIMS/CB blend, which showed that the interface behavior is more sensitive to carbon black than to silica. Prescribed amounts of CB and PBR/BIMS blend were mixed for 10 min in a Brabender at 100 rpm and 150 °C. Approximately, 0.5 g of the mixture of PBR/BIMS/CB blend having circular 1-mm-thick sheet was placed in a sample

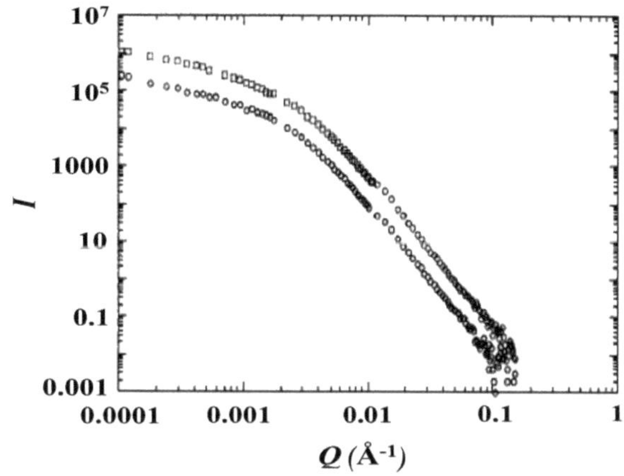

Fig. 6.23 A log–log plot of scattering intensity (I) as a function of Q for PBR blend containing (*a*) IMS (*b*) BIMS blended with CB fraction = 0.025 (*upper curve*) and CB fraction = 0.005 (*lower curve*) (by permission of [34])

holder, which was kept flat faces perpendicular to the neutron beam. This path length was chosen to ensure the best balance between the scattering intensity and attenuation. A neutron wavelength of 6 Å was used, with two collimation configurations: One has a sample-detector distance of 2.9 m, measuring the high Q range of 0.03–0.2 Å^{-1}, and the other has the sample-detector distance set at 15.2 m, measuring a medium Q range of 0.003–0.04 Å^{-1}. Standard mixture of PBR/BIMS/CB blend was used to calibrate the instrument so all intensity measured was of the absolute scale of scattering cross section per unit volume [34]. The results of two slightly different PBR blends containing CB/BIMS and CB/IMS are shown in Fig. 6.23a, b. In these figures, the plot of scattering intensity I as a function of Q is shown. The two curves in each figure represent two different CB/PBR volume fractions (upper curves for CB containing 0.025 fraction) and (lower curves for CB containing 0.005 fraction) to test concentration effects, if any. The intensity for both PBR blends containing CB/BIMS and CB/IMS was scaled with a fraction of CB indicating little concentration effects. The combination of SANS and USANS data from Q = 0.0003–0.2 Å^{-1} covers a very broad Q range of the CB aggregates [34].

As shown in Fig. 6.24, the low-Q data for PBR/silica/BIMS system do not show a rollover shape in the Guinier region, even at the USANS range approaching Q_{min} = 0.0001 Å^{-1}. Therefore, the size of the agglomerates is likely larger than several microns because silica fillers have much stronger interparticle interaction. The broad shoulder peak, which corresponds to the form factor of the primary particles appeared around Q = 0.01 Å^{-1}. On the basis of the spherical model that takes the polydispersity of the particles into account, the size or diameter of the primary particle was estimated to be 90 Å with polydispersity ($\Delta R/R$) of 30 %. In addition, the reasonable straightline shape of the curve at low Q suggests a fractal structure within the aggregates. As expected, both types of the nanofillers do not fuse with the surrounding rubbers [34].

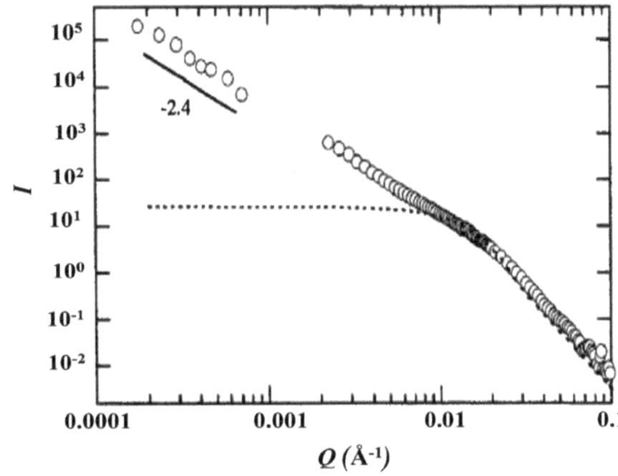

Fig. 6.24 A log–log plot of scattering intensity (*I*) as a function of *Q* for BIMS blended with hydrophilic silica with CB fraction = 0.005. *Dotted line* is the best fit of spherical model to scattering profiles (by permission of [34])

The coalescence can also be monitored from the SANS intensity; initially half of the dispersed PBR is generally deuterated, so that coalescence reduced the scattering. Despite the highly viscoelastic nature of the rubber blends and the contact time of the particles during flow, coalescence readily occurs and thus exerts a major effect on the blend morphology [4]. Most blends consist of discrete particles in a continuous phase with lower viscosity component, which are generally present at a sufficient concentration (Fig. 6.25). During mechanical mixing, domains of this low viscosity materials deform and encapsulate the higher viscosity phase which produce a globular morphology. However, blend the morphology is never at equilibrium for an immiscible. The size distribution of the dispersed phase represents a steady-state balance between the breakup of the particles and their coalescence processes that continue throughout mixing and processing [4, 39].

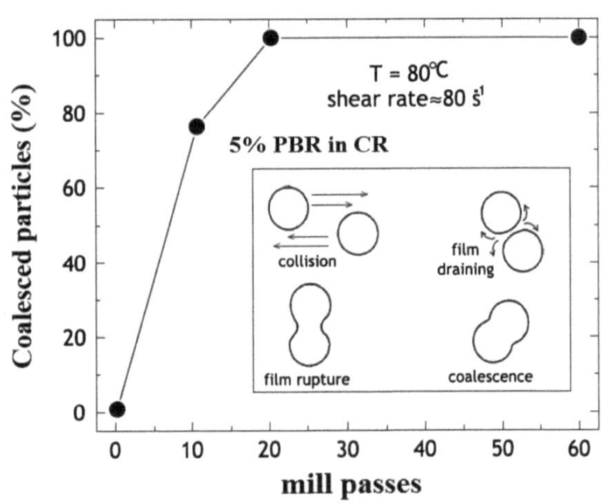

Fig. 6.25 Amount of PBR dispersed in polychloroprene that has coalesced with other particles, as a function of the number of passes through a two-roll mill (by permission of [4])

6.10 Rheological Studies

The viscosity of the PBR nanoblends (e.g., PBR/SBR [1]) can be measured at 100 °C in a single-speed rotational Mooney viscometer according to a British Standard. The scorch time is the time for the onset of cure, and the optimum cure time is the time for the completion of the cure. Both these times were determined from the cure traces generated at 140 and 170 °C by an oscillating disk rheometer with tests run for 2 h. The cure rate index measure of the rate of cure in the rubber can be calculated using a British Standard [1]. Figure 6.26 shows that Δtorque is the difference between the maximum and minimum torque values on the cure trace of the rubber and also indication of change in the cross-link density. Δtorque was plotted against the loading of TBBS/ZnO and TMTD on PBR/SBR blends. The scorch and optimum cure times were found to be increased by 60 and 54 %, respectively, while the cure rate index decreased by 35 %. However, the addition of the oil reduced torque by 43 %, which is an indication of significant reduction in the cross-link density of the rubber blends [1].

The effect of TBBS and ZnO on the Δtorque of the PBR/SBR/nanosilica-based blends showed that Δtorque was found to be increased to 22 dNm as TBBS loading was raised to 3 phr. Further increment to 9 phr had a lesser effect on the Δtorque value, which was 26 dNm. Evidently, 3 phr was sufficient to activate the rubber reactive tetrasulfane groups of TESPT and form cross-linking points between the rubber and nanofiller. To enhance the cross-link density of the PBR/SBR/nanosilica blends, 3 phr of TBBS and ZnO was added. Δtorque was found to be increased to 80 dNm with 0.2 phr of ZnO loading and showed no further improvement on the addition of 1.8 phr of ZnO in the rubber. PBR/SBR nanoblend was prepared by adding 60 phr silanized silica, 3 phr TBBS, 0.5 phr ZnO, 7.5 phr processing oil, and 1 phr anti-degradant. To improve the efficiency of TMTD, ZnO can be mixed with the PBR/SBR/nanosilica blends. Δtorque of the PBR/SBR/nanosilica blends with 5 phr TMTD raised from 82 dNm to 131 dNm on the addition of 0.1 phr ZnO,

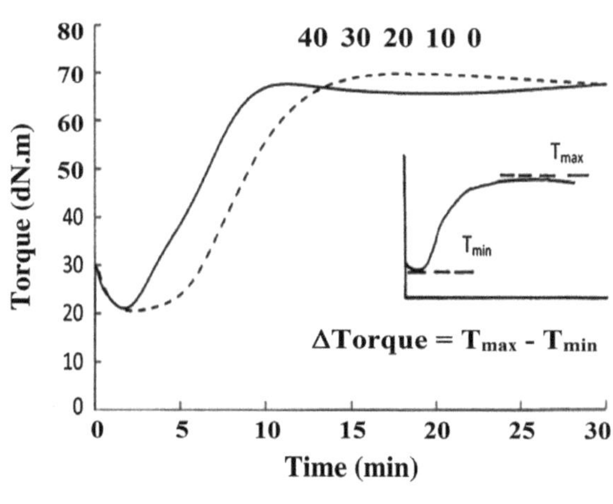

Fig. 6.26 Torque versus time traces by ODR for the PBR/SBR blends at 170 °C (*dash line*) containing ZnO and TBBS, (*solid line*) containing only TMTD (by permission of [1])

and no more ZnO was added to avoid making them too brittle. PBR/SBR/nanosilica-based blend was prepared by adding 60 phr silanized silica, 5 phr TMTD, 7.5 phr processing oil, and 1 phr anti-degradant. The addition of nanosilica, TBBS, TMTD, ZnO, anti-degradant, and processing oil can increase the viscosity of the raw PBR/SBR-based nanoblend from 40 to 98 μ, which indicates the addition of fillers can form 69 % bound rubber with increased rubber viscosity and high level of rubber–filler interaction. These rubber blends were stored at ambient temperature for 3–4 days before and one week after they were cured. These mechanical properties of the PBR/SBR rubber nanoblend had shorter scorch and optimum cure times with faster cure rate. TMTD is a fast-curing accelerator, which explained the shorter cure cycle PBR/SBR rubber nanoblend [1].

The phases of an immiscible blend can be co-continuous, or one component can be dispersed within a continuous matrix of the other. The former is favored by equal concentrations and equal viscosities; that is, η_1 and η_2 and $\Phi_1 = 0{:}5$. This is illustrated in Fig. 6.27 for a PBR/EPDM blend [4, 40].

Most blends consist of discrete particles in a continuous phase, with the latter usually the lower viscosity component at a sufficient concentration. During mechanical mixing, domains of the lower viscosity materials deform and encapsulate the higher viscosity phase, to produce a globular morphology [4]. Even in the case of Newtonian fluid mixtures, the viscosity can depend on the particle size and mixing or flow conditions.

Figure 6.28 shows the variation of the viscosity with composition for PBR/NR blend of and polypentenamer; the dependence is very complex and not obviously described by any mixing rule [4, 41]. Go et al. [7] studied the melt viscosity of the unvulcanized PBR/EPDM blends with various polymer ratios as a function of apparent shear rate and showed rheological and typical power law behaviors for all the blends. He reported that melt viscosity was found to be increased with increasing EPDM contents. In comparison with the rheological behaviors of the AAHC- or TESPT-filled PBR/EPDM ternary blend as well as with the PBR/EPDM/MA-g-EPDM blend with that of the PBR/EPDM (25/75), it is reported that the

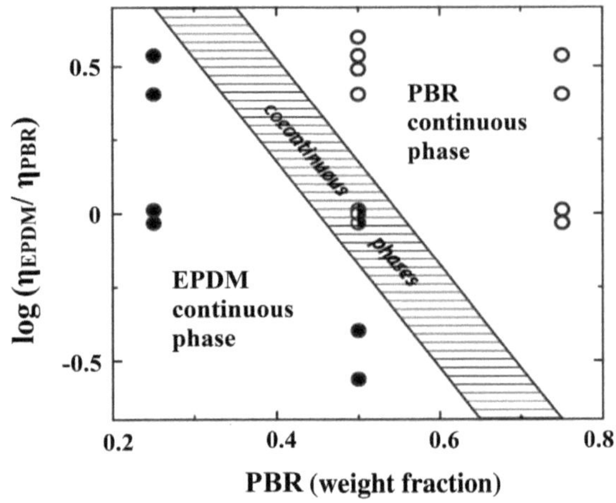

Fig. 6.27 Dependence of the phase morphology on the viscosity ratio and composition of PBR/EPDM blends. Dispersed PBR and EPDM particles are indicated, respectively, by *open* and *filled circles*; *half-filled circles* indicate a co-continuous morphology (by permission of [4])

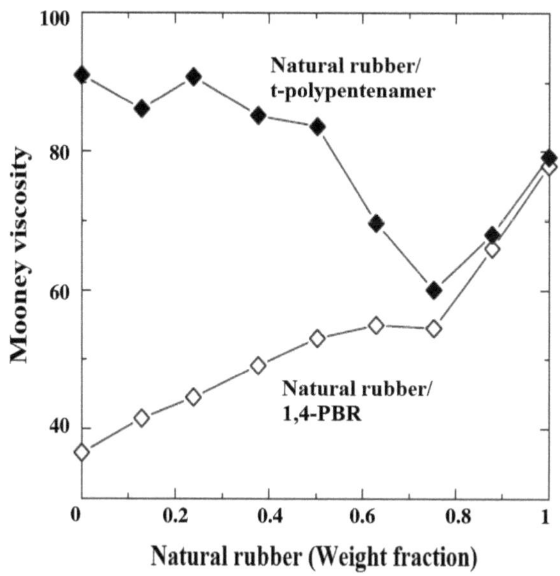

Fig. 6.28 Mooney viscosity of PBR/NR blends (by permission of [4])

AAHC acted as a lubricant that can decrease the melt viscosity of the PBR/EPDM blends. The AAHC is a blend of nonpolar aliphatic and polar aromatic resins and in polymer blending, this mixture of different polarities gives a true solvating effect in wetting out two rubbers, i.e., PBR and EPDM. The comparison of the rheological behavior of the unvulcanized ternary blend (PBR/EPDM/TESPT) with that of the PBR/EPDM binary blend indicates that the ternary blend exhibits a higher melt viscosity than that of binary blends. The result also implies that the silane-type TESPT is an effective coupling agent for PBR/EPDM/nanosilica blend due to the buildup of silica-to-rubber bond. However, PBR/MA-g-EPDM blend exhibited higher viscosity than both PBR/EPDM/AAHC ternary blend and the PBR/EPDM binary blend, which was due to high gel contents and molecular weight of the graft copolymer [7]. The rheographs of the PBR/EPDM blends with different rubber ratios are also illustrated in Fig. 6.29a, which show that the EPDM-rich blends exhibit higher torques and lower cure rates than that of PBR-rich blends.

Figure 6.29b illustrates the effect of the addition of AAHC on the curing characteristics of the PBR/EPDM (25/75) blend [7]. Go et al. [7] observed that the cure rate was not changed with the addition of AAHC but the scorch time became longer due to the reduced melt viscosity and improved processibility. It is said from viscosity behavior as shown in Fig. 6.29b that the PBR/EPDM/TESPT ternary blend exhibits higher torque and a faster cure rate than that of the PBR/EPDM binary blends without the coupling agent. The effect of coupling agents on the curing properties of PBR/EPDM (50/50) blends is more clearly observed. When TESPT was added, scorch time became shorter due to the increasing of melt viscosity. For

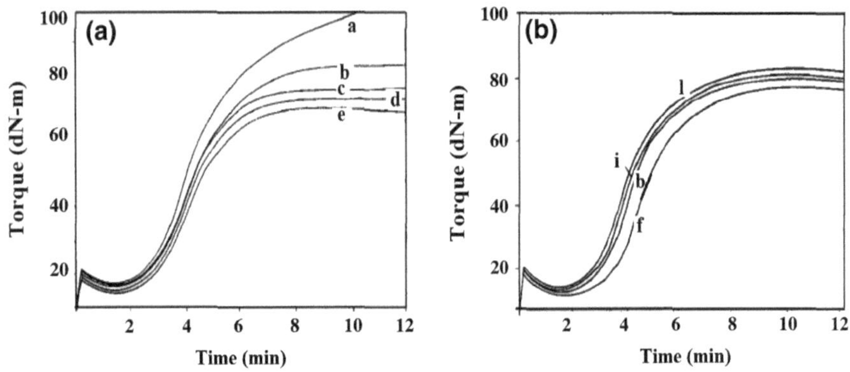

Fig. 6.29 Rheographs of the **a** PBR/EPDM blends at ratios of (*a*) 0/100, (*b*) 25/75, (*c*) 50/50, (*d*) 75/25, (*e*) 100/0, **b** 25/75 blends of (*f*) PBR/EPDM/AAHC (*i*) PBR/EPDM/TESPT; (*l*) PBR/MA-g-EPDM (by permission of [7])

the both blends PBR/EPDM/TESPT and PBR/MA-g-EPDM, the scorch time became shorter and the cure rate became faster when compared to the PBR/EPDM binary blend the PBR/EPDM/AAHC ternary blend. These might be due to co-cure characteristics in the PBR/MA-g-EPDM blends plausibly because of the presence of the functional group—MA in the graft copolymer to induce the in situ formation of compatibilizing interchain copolymers of EPDM and PBR [7].

The rheological properties of PBR/NR (25/75) nanoblend are of great interest and have been investigated [10]. The complex viscosity (η^*) and elastic modulus (G') of PBR/NR blends are shown in Fig. 6.30, which show complex viscosity sensibly higher than that of pure PBR/NR blend due to established interaction between the rubber and nanofiller. The falling trend of viscosity with respect to frequency represents the pseudoplastic nature and shear-thinning behavior of rubber nanoblends and also at low and medium range of frequencies, the complex viscosity showed an increment indicating a yield stress. The existence of yield stress has been indicated as a common characteristic of highly filled polymer melts, associated with interaction between the filler nanoparticles [10].

Vulcanization is one of the most important processes in the rubber industry, which is useful to perform a complete thermal and rheological characterization of the material during the curing reactions. This cure characterization can be conducted by a Rheometer to control the industrial production and to accelerate the quality control phase. Numerous curing tests carried out on PBR/NR blends demonstrated that these curing temperature and time (152 °C for 30 min) could be good parameters to reach the optimum vulcanization. The curing characteristics, expressed in terms of the vulcanization times: scorch time (t_2) and optimum cure time (t_{90}), and the maximum and minimum values of the torque, M_H and M_L, respectively, are deduced from the rheometric curves. These parameters, along with delta torque and a cure rate index, are expressed as Eqs. (6.3) and (6.4), respectively, and their values for various blends are listed in Table 6.2 [30].

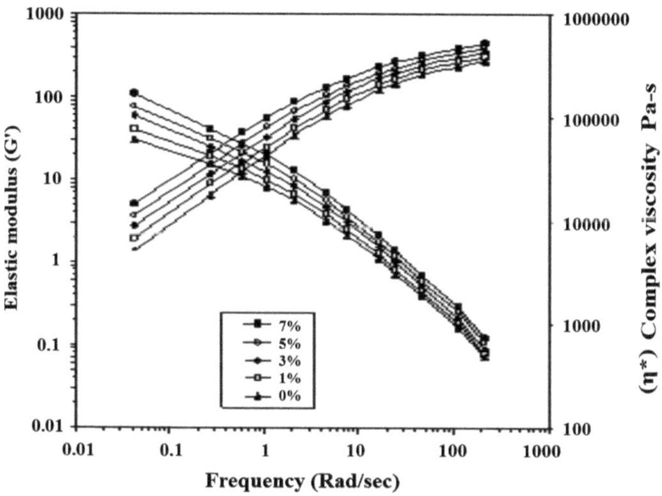

Fig. 6.30 Rheological properties of PBR/NR (25/75) blend (by permission of [10])

Table 6.2 Cure characteristics of the rubber compounds (by permission of [30])

Name of blend	M_L (dN m)	M_H (dN m)	ΔM (dN m)	t_{90} (min)	t_{90} (min)	CRI = $t_{90} - t_2$ (min^{-1})
PBR/NR	0.29	5.83	5.54	0.85	1.31	217
PBR/NR/Ca-Nanoclay	0.31	4.45	4.14	1.50	2.12	161
PBR/NR/nanoclay1	0.40	6.65	6.25	0.80	1.23	232
PBR/NR/nanoclay2	0.43	8.35	7.92	0.54	0.93	256
PBR/NR/LDH1	0.58	4.32	3.74	0.91	1.13	454
PBR/NR/LDH2	0.47	4.42	4.01	1.05	1.38	303

$$\Delta M = (M_H - M_L) \tag{6.3}$$

$$\text{CRI} = 100/(t_{90} - t_2) \tag{6.4}$$

The minimum torque (M_L) achieved during the vulcanization can be considered as a measure of the viscosity of the material. For the rubber–clay compounds, the higher (M_L) values indicate a worsening of their processibility. The maximum torque (M_L) and the delta torque (ΔM) values depend on the extent of cross-linking created and on the reinforcement induced by the filler nanoparticles. For the PBR/NR blends with MMT and LDH clays, decrement in M_H and ΔM suggests a poor interaction of the filler-matrix interphase. These results show that adding MMT and LDH clays to the pure PBR/NR blend rubber, a conventional composite at the microscale level is probably formed. On the other hand, the PBR/NR rubber blends with the OMMTs present a higher torque value with respect to the pure PBR/NR blend [30].

It is assumed that the torque value relates to the cross-link density, so PBR/NR blend becomes more cross-linked with clay. From the reported values, it can be said that the addition of unmodified MMT decreases the scorch and cure time of pure PBR/NR blend. It is reported that the pristine MMT can absorb the curatives inducing its partial inactivation, differently from OMMT which causes delay in longer optimized cure. On the contrary, t_2 and t_{90} were found to be reduced when OMMTs were added to the rubber blend. It is possible to conclude that the OMMT behaves as an accelerant agent for the rubber blend due to ammonium groups placed in the interlayer regions. The large torque difference and the faster vulcanization rate of PBR/NR/nanoclay1 compared with PBR/NR/nanoclay2 are probably due to higher modifier concentration, 125 mequiv/100 g and 90 mequiv/100 g, respectively. The PBR/NR blend filled with LDHs contents shows not only a longer scorch time (t_2) but also a high value of CRI. To explain this particular behavior, which can be explained by absorption of the curatives on the LDHs inducing a partial inactivation of the vulcanizing agent [30].

Time required to reach 90 % cross-linking (T_{90}) was almost invariant, while the scorch time T_s was much lower than that of the pure NR/BR blends (Fig. 6.31). It can be said that OMMT activated in the vulcanization reaction and the maximum torque M_H was found to be increased with the increasing OMMT content, and finally cross-link densities of NR/BR blends were reported to be increased [35]. Das et al. [36] reported cure characteristics of the different PBR vulcanizates filled with EG and i-MG in the presence of CB. Minimum torque (M_L), maximum torque (M_H), and torque difference (ΔM) are measuring parameters of viscosity, stock modulus, and degree of cross-linking of the PBR blends, respectively. EG/CB and i-MG/CB containing different PBR blend vulcanizates showed an increment in the M_H and ΔM values as compared to that of CB-filled PBR blends. EG- and

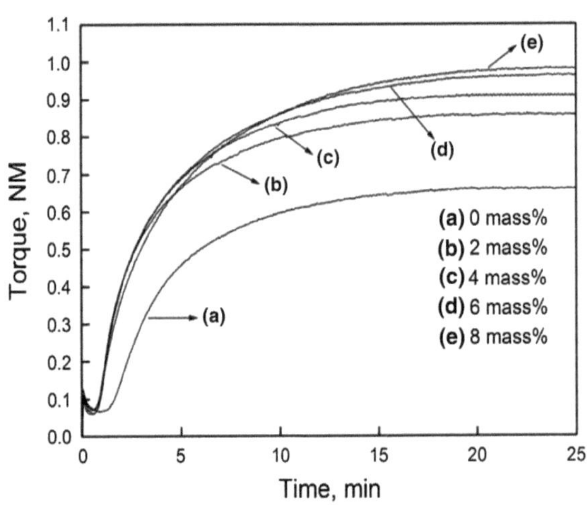

Fig. 6.31 Vulcanization curves of the nanocomposites (by permission of [35])

i-MG-filled PBR blends exhibit low scorch and cure time as compared to their respective controls. The better dispersion of i-MG increases the stiffness and modulus of the PBR blends, which can increase their M_H. The fast-curing reaction was activated by the large surface area of graphite nanoplatelets in the rubber blends and the polar groups on the surface of i-MG acted as an accelerator to speed up the ring opening of sulfur simultaneously [36].

6.11 Electrical Properties

PBR (*Cis*-1,2) has an electrical resistivity of 1×10^{15} (ρ/Ω cm) [4]. CB can be reinforced into rubber blend to provide electrical conductivity, because CB can easily transport electrons. Various theories such as conduction path, electron tunneling, and electrical field radiation are the proposed mechanisms of the electrical conduction in the polymers and rubber blend filled with CB [4–6]. The high aspect ratio of the CNTs can enable the formation of a conductive percolating network in PBR/SBR blend at below 2 wt% concentrations [13]. Figure 6.32 shows the effect of CNT loading on electrical conductivity of PBR/SBR/CNT blend, which indicates that addition of alcoholic-CNT mixture has the ability to enhance conductivity of nanoblend as compared to blend containing dry mixture of CNTs [13].

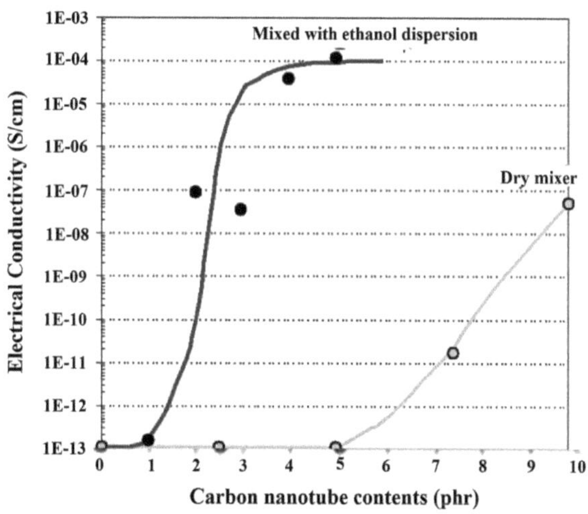

Fig. 6.32 Effect of CNT loading on electrical conductivity of PBR/SBR blend (by permission of [13])

Table 6.3 Cure characteristics of the NR/BR blends (by permission of [30])

Material composition (PBR/NR/nanoclay)	25/75/0	25/75/1	25/75/3	25/75/5	25/75/7
N_2 permeability	12.1	10.7	9.6	8.3	7.8

6.12 Permeability Studies

Properties such as gas permeability and, solvent permeability, complex viscosity as well as the storage modulus have been assessed to study the effects of the addition of nanoclays into rubber blends. For true blends and nanocomposites, the nanolayers of clay must be uniformly exfoliated in the polymer blend/matrix, to avoid aggregation or intercalation. The drastic improvement in properties can be observed when nanolayers have been exfoliated, i.e., an increment in tensile properties and decrement in solvent permeability [10, 30]. Isobutylene-based rubbers are widely used in the automotive industry because these materials possess low permeability to gases, as well as ability to control and cross-link density and mechanical properties by varying the levels of IB, PMS, and filler content [34]. Alipur et al. [10] carried out the permeation experiment of nitrogen gas on PBR/NR/nanoclay blend with gas permeability-measuring apparatus by keeping pressure at 0.57 MPa on one face of the sheet (about 1 mm in thickness and 8 cm in diameter) and zero at other face pressure. Nitrogen gas was allowed to permeate through the sheet at a rate of transmission of nitrogen at 40 °C was obtained by gas chromatography and the nitrogen permeability was calculated from it [10]. The gas permeability of PBR/NR/nanoclay blends [10] is presented in Table 6.3, which indicates that the nitrogen permeability reduces from 10.7 to 7.8 with an increasing amount of nanoclay by 1, 3, 5, and 7 wt%, respectively. It can be said that the clay nanolayers with the large aspect ratio and the planar orientation lead to the great increase of the diffusion distance by creating a tortuous path for the diffusing gas [10].

6.13 New Challenges

Still, there are many challenges in developing a new type of PBR-nBs which include health, safety, and the environmental issues related to passenger car tires that can play a vital role in the tire manufacturing and their applications. Excessive use of the curing chemicals is harmful to health, safety, and the environment and their uses are restricted by various legislations for the environment and safety in all around the world. The presence of ZnO in tread compounds has also come under growing scrutiny because of environmental concerns and also the tire industry is the largest single market for ZnO. The environmental damage can be reduced by elimination or a more restricted use of ZnO in rubber compounds [1].

6.14 Recent Advances

This chapter has already provided a thorough basic understanding of the PBR-based blends. Various polymer blends and nanocomposites such as PBR-filler compositions, PBR-elastomer blends, their formulation, compounding, and shape-memory polymers, block polymers, as well as elastomers such as butyl rubber, NBR, SBR, IR, EPDM, have been discussed. Their different processing methods such as latex blending, two-roll mixing, extrusion blending, and other processes have been reported. The synthesis, structure–property relationship, compounding, processing, and applications of each PBR blend are also discussed. In this chapter, the authors discuss the different types of compounding and their equipment, processes, and different ingredients, vulcanizing agents and their function, double networking of PBR blends. Micro- and nanostructured PBR-based rubber blends have been studied for recent developments, different manufacturing methods, and different characterization methods. The applications of these PBR/rubber blends have been reported. The current trends in the development of PBR blends with grafts/block copolymers using cross-linking agents have been discussed. Various micro- and nanofillers in PBR/rubber-based blends discussed along with compatibilization of PBR-elastomer blends by the addition of nanofillers. Preparation, fabrication of various PBR-nBs, tensile, morphological, thermal, rheological, electrical properties and study of permeability and neutron scattering of PBR-based rubber blends have also been discussed in detail.

6.15 Industrial Values

As mentioned earlier, PBR-elastomer blends are being used in most rubber products for a long time. This is evident in the competition of patents [25] on rubber blends by largest tire manufacturers. In 1984–85, at least five patents were issued to four tire manufacturers. Gen Corp (now Continental Tire) patented a blend of star-shaped solution polymerized butadiene or isoprene with regular diene polymers for lowering rolling resistance of tires. Bridgestone of Japan took patent on a blend of random PBR/SBR of specified *Cis* content and chain length to improve wet skid and rolling resistance. Another patent on a blend of isoprene and PBR copolymer with 90 % *Cis* content in PBR, which was taken to obtain better skid resistance. In view of this, it is desirable to look at the current practice briefly, to appreciate the important role of the PBR-nBs blends that can play in the rubber industry [37]. Mangaraj et al. [37] reported that PBR blend with SBR/EPDM can be used for tire tread applications. He also reported useful applications of varying PBR blends such as NBR/PBR, PBR/neoprene, PBR/epicholorohydrin, because these blends have low gas permeability, low temperature flexibility, and ease processibility. A blend of PBR/NR gives high curing temperature while that of PBR/NR/SBR gives improved heat stability. Blend of liquid PBR gives ease processing and physical properties. Addition of 10–30 % of SBR, NR, and IR to *Cis*-PBR along with alkaline salt of fatty acid improves cutting resistance and vulcanizates. Addition of liquid PBR containing 10 % carboxylic acid improves the properties of *Cis*-PBR and their various blends. Similarly, a ternary blend of PBR/NR/IR can provide good physical for giant tire treads that are useful for trucks and agricultural equipment. For black sidewalls, blending of halobutyl rubbers with PBR/NR blend provides both dynamic and static ozone resistances as well as resistance to flexural strain. Similarly, addition of chlorobutyl rubber to PBR/NR blend has increased their tearing strength and reduces outgrowth rate. Addition of CB to PBR/NR/IR blends can provide a large reduction in crack propagation rate under fatigue stress. In non-staining tire black sidewalls based on PBR/NR/EPDM blend, the addition of BIMS provides superior adhesion to carcass and improve flex resistance, ozone resistance, and rolling resistance. The addition of butyl and halo-butyl rubber to PBR/NR blends can lower resilience and improve skid resistance. When BIMS is added to carbon black-filled PBR/NR blends, the loss of tangent increases at low temperatures at −30 to 10 °C, but decreases above 30 °C and the addition of BIMS improves resistance to tread wear [37].

6.16 Applications Studies

Rubber blends can be generally divided into tire and non-tire, and approximately 70 % of rubber is generally used in application of tire and tire-related products. The non-tire products contain a variety of extruded and molded parts, including automotive hoses and belts, footwear, seals, O-rings, and roofing membrane. Pneumatic tire is a highly engineered product made of several components, including chafer, apex, black sidewall, belts, radial ply, bead, and liner,. These parts are generally prepared from PBR/NR (50/50) and (30/70) blends for passenger tire-tread applications and PBR/SBR and PBR/NR (20/80) for truck tire applications. The PBR/SBR rubber blends can provide high abrasion resistance and low rolling resistance, while NR can provide resilience and low-heat buildup. Rubber blends also find frequent applications in flat belting, i.e., SBR, IR, and NR blends with PBR. A blend of PBR/NR (20/80) is also found useful in belt cover applications, while PBR/SBR/IR blends can be useful in high-quality heel of footwears applications [37]. According to the mechanical properties of PET-recycled rubber (filled with S, Si, Zn, CB) obtained, composites based on rubber tires can be used as construction materials, such as paving slabs, while the rubber nanoblends with lower sulfur content (more elastic) can be used as sound/noise absorbing materials, for road impact damping places or for children playgrounds [4]. These rubber blends can also be recommended as pavement materials for indoor dry spaces, while randomly mixed composites can be used for outdoor applications, as construction and insulating materials such as paving slabs for outdoor landscaping, courtyards, and protection pillars [4].

6.17 Conclusions

PBR-nBs have been observed in rubber industry over a long period of time to meet the contradictory property requirements of rubber products. However, most rubber blends are immiscible and the compounding ingredients, distribute into different blend components depending on their compatibility and the rates of diffusion and reactivity in different components. As a result, most PBR blends with thermoplastics and thermosets have a heterogeneous morphology. Instrumental techniques such as UTM, DSC, TGA, DMA, SANS, SEM, and TEM are helpful in the new developments of rubber compounding.

Compatibilization is important to rubber blends, which is reflected in lower mechanical properties of incompatible blends of elastomers. Compatibilization between RR grit (ground rubber) and gum rubber is very important to successful recycling of used tires. The PBR/SBR nanoblends can be easily cured with the non-sulfur donor accelerator and zinc oxide which results superior TS, elongation at break, stored energy density at break, and modulus at different strain amplitudes. It also possesses a lower heat buildup, a higher abrasion resistance, a higher Tan δ value at low temperatures to obtain high skid resistance, and ice- and wet-grip. Optimizing the chemical bonding between the PBR/SBR and filler by addition of tetrasulfane groups can reduce excessive use of the chemical curatives significantly, which can improve health and safety at work place and reduced environmental damage.

Elastic behavior of this PBR/PP blend can be improved by annealing the blend at high temperature, when the crystal size increases, but the fragmented character remains. When the blend crystallizes at high temperature, large lamellae are formed and the blend shows very poor elastic properties. PBR/NR blends filled conductive filler carbon black can be useful for minimizing hysteresis which may depend on size, wide distribution, and structure of carbon black. Important properties such as melting viscosity, elastic modulus, and loss modulus (Tan δ) can be increased by increasing EPDM contents in PBR/EPDM rubber blends. Also AAHC and TESPT can act as a plasticizer for the PBR/EPDM rubber blend and also enhances compatibility between EPDM and PBR and vulcanization properties. PBR can also act as compatibilizer for NR/EPDM blend which decrease the domain size of the dispersed phase. PBR/PS filled with nanoclays also showed significant improvement in mechanical properties. Rubber nanoblends of PBR/NR/nanoclay showed that silicate layers can expand the distance and morphology, indicating a strong interaction between rubber and nanoclay particles in PBR/NR/nanoclay blend as compared to pristine PBR/NR blend. PBR/PS blends with epoxy resin and filled with nanosilica have improved micro-hardness properties. On incorporation of a small amount of CNTs in PBR/SBR blend, storage modulus can be increased above the glass transition temperature. These rubber nanoblends prepared by such method that can exhibit enhanced physical properties at very low concentrations and high aspect ratio of the CNTs enabled the formation of a conductive percolating network in these composites below 2 wt% concentrations of CNTs. The influence on RR content and various reaction temperatures during the sol–gel preparation of nanosilica affects physicochemical properties, mechanical behavior, thermal characteristics, and phase morphology of PBR/NR/RR/nanosilica. The thermal stability of the PBR/NR/RR ternary blend increases with the increasing RR content, while phase morphology by SEM reveals the high level of interaction between polar filler and organic matrix. A study on 3.25 wt% of PBR/PS filled with 3.25 wt% nanoclay can give improved thermal stability.

The effect of two different nanoclays (e.g., MMT and LDH with or without organically modified) has shown an improvement in mechanical properties, the abrasion resistance, and the rebound resilience. The thermal properties of PBR/NR blend were found to be improved by adding OMMTs as compared to without one. The surface modification of MMT by cationic surfactant can increase the compatibility between the PBR/NR blend and the clay, obtaining a composite with superior properties to those of neat matrix. In particular, the OMMT exhibited the highest improvement in a final performance. Probably, this clay can achieve a higher dispersion in the elastomeric matrix for its strong hydrophobic character. When both carbon and silica are incorporated into the PBR layer, the effect of carbon black is offset by nanosilica. The interfacial behavior of PBR/BIMS is more sensitive to carbon black than to nanosilica. When both carbon and nanosilica were incorporated into the PBR layer, both the surface viscosity and the interfacial broadening were similar to the PBR blends filled with only pyrogeneous silica due to segregation of the free polymer and associated silica particles to their surface and interfaces. This is analogous to the segregation of free homopolymer from a cross-linked network, which is formed by the physical cross-links of polymers adsorbed to CB particles. A blend of PBR/NR/OMMT was successfully prepared by direct mechanical blending by adding exfoliated 4 and 6 wt% OMMTs because OMMT particles dispersed homogeneously in the NR matrix. The addition of OMMT decreased the scorch time of NR/BR and accelerated the vulcanization reaction of NR, leading to simultaneous cross-linking of NR and BR. The PBR/NR/OMMT blends had excellent static mechanical properties, lower loss factor, and better wet skid resistance properties and slightly better thermal stability than that of pure NR/BR blends due to the good dispersion of OMMT in the rubber matrix.

EG/CB- and i-MG/CB-filled PBR blends based on different synthetic rubbers were fabricated by melt blending. As a result of large surface area and the presence of polar groups on the surface of the nanofillers, the physical interaction and interfacial adhesion between the nanofillers and the rubber matrices were enhanced, which resulting in an improvement in the physical, mechanical, and thermo-mechanical properties of the blends. Because of the large surface area of the nanofillers, rubber nanoblends exhibit fast-curing reaction, which result in a decrease in the scorch and optimum cure time. An improvement in storage modulus, lower rolling resistance, and better antiskid properties for the EG/CB- and i-MG/CB-filled PBR/SBR blends are compared with their respective controls. EG- and i-MG-filled PBR, SBR, and PBR/SBR blends exhibit an increase in the hardness, which resulting in an improvement in the abrasion resistance of the respective composite. Because of the higher basal spacing and exfoliated structure of i-MG sheets than EG flakes, i-MG sheets were uniformly dispersed in the different rubber matrices in the presence of CB, which resulting in a superior mechanical, dynamic mechanical, and thermal properties of the i-MG-loaded rubber blends as compared to the EG-filled rubber vulcanizates.

References

1. Saeed F, Ansarifar A, Ellis R, Meskel Y, Irfan MS (2012) Two advanced styrene-butadiene/polybutadiene rubber blends filled with a silanized silica nanofiller for potential use in passenger car tire tread compound. J Appl Polym Sci 123:1518–1529
2. Svoboda P, Saito H, Chiba T, Inoue T, Takemura Y (2000) Morphology and elastomeric properties of isotactic polypropylene/hydrogenated polybutadiene blend. Polymer 32 (11):915–920
3. Lee KH, Ryu SW (2012) Room temperature baroplastic processing of PS/PBA nano-blends. Macromol Res 20(12):1294–1299
4. Visakh PM, Thomas S, Chandra A, Mathew A (2013) Advances in elastomers-I: blends and interpenetrating networks. Adv Struct Mater 11:1–493
5. Hess WM, Chirico VE (1977) Elastomer blend properties influence of carbon black type and location, rubber. Chem Technol 50(02):301–326
6. Chough SH, Chang DH (1996) Kinetics of sulfur vulcanization of NR, BR, SBR, and their blends using a rheometer and DSC. J Appl Polym Sci 61(03):449–454
7. Go JH, Ha CS (1996) Rheology and properties of EPDM/BR blends with or without a homogenizing agent or a coupling agent. J Appl Polym Sci 62(03):509–521
8. El Sabbagh SH (2003) Compatibility study of natural rubber and ethylene propylene diene rubber blends. Polym Testing 22(01):93–100
9. Subana PS, Nair PP, George KE (2013) Studies on combined effect of nanoclay and elastomer on mechanical properties of polystyrene/polybutadiene blend. J Acad Ind Res 1(8):461–463
10. Alipour A (2012) Fabrication and characterization of nanostructured polymer composites prepared by melt compounding. Int J Biosci Biochem Bioinform 2(2):79–84
11. Mishra S, Shimpi NG (2008) Studies on mechanical, thermal, and flame retarding properties of polybutadiene rubber (PBR) nanocomposites. Polym-Plast Technol Eng 47(1):72–81
12. Pandit R, Berkessel A, Lach R, Grellmann W, Adhikari R (2012) Synthesis and characterization of nanostructured blends of epoxy resin and block copolymers. Nepal J Sci Technol 13(1):81–88
13. Das A, Stckelhuber KW, Jurk R, Saphiannikova M, Fritzsche J, Lorenz H, Klppel M, Heinrich G (2008) Modified and unmodified multiwalled carbon nanotubes in high performance solution styrene butadiene and butadiene rubber blends. Polymer 49 (24):5276–5283
14. Debapriya D, Panda PK, Bhunia S, Roy M (2014) Effect of sol-gel-derived nano-silica on the properties of natural rubber-poly butadiene rubber-reclaim rubber ternary blends/silica nanocomposites. Polym-Plast Technol Eng 53(11):1131–1141
15. Corish PJ, Powell BDW (1974) Elastomer blends, rubber, chemistry and technology. 47 (03):481–510
16. Mishra S, Baweja B, Chandra R (1999) Studies on dynamic and static cross linking of EVA-EPDM blends. J Appl Polym Sci 74(11):2756–2763
17. Mishra S, Balakrishnan S, Chandra R (1998) Thermal stability of EVA/EPDM blends. J Appl Polym Sci 70(9):1829–1835
18. Mishra S, Naik JB (1997) Mechanochemical degradation of EVA-EPDM blends. Polym Plast Technol Eng 36(2):231–240

19. Mishra S, Hazarika M, Chandra R (1999) Rheological studies of dynamic vulcanization of butyl rubber and polypropylene blends. Polym Plast Technol Eng 38(2):305–310
20. Prabhakar A, Nair PP, George KE (2013) Optimization of clay and rubber content in polystyrene based nanocomposite. Int J Innovative Res Sci Eng Technol 2(8):4082–4089
21. Chandra R, Mishra S, Parida TR (1995) Studies on dynamic behavior and flow properties of HDPE/EPDM blends by a torque rheometer. Polym Int 37(02):141–147
22. Magonov S, Heaton MG (1998) Applications of AFM for polymers. http://www.bruker.jp/axs/nano/imgs/pdf/AN084.pdf
23. Mishra S, Badgujar N, Gurav K, Patil D, Sonawane SH (2005) comparative study on mechanical and flame retarding properties of polybutadiene rubber filled with nanoparticles and fly ash. J Appl Polym Sci 96(1):6–9
24. Mishra S, Shimpi NG (2005) Mechanical and flame-retarding properties of styrene butadiene rubber filled with nano-$CaCO_3$ as a filler and linseed oil as an extender. J Appl Polym Sci 98 (6):2563–2571
25. Zanzig DJ, Yang X, Cohen MP (2010) Tyre with component comprised of a blend of polybutadiene rubber and composite of styrene/butadiene elastomer which contains exfoliated clay platelets, US 7,714,055 B2 to Goodyear
26. Mir MM (1991) Blends of crystallisable polybutadiene isomers: compatibilization by addition of amorphous diblock copolymer, PhD Theses, Massachusetts Institute of Technology
27. Gu Z, Song G, Liu W, Wang B, Li J (2009) Preparation and properties of organo montmorillonite/cis-1,4-polybutadiene rubber nanocomposites by solution intercalation. Appl Clay Sci 45(12):5053
28. Song G, Gu Z, Li P, Wang L, Gao L (2012) The properties of organo montmorillonite/cis-1,4-polybutadiene rubber nanocomposites and the effect of recovered solvent on the conversion of butadiene polymerization. Appl Clay Sci 6566:158161
29. Kim MS, Kim GH, Chowdhury SR (2007) Polybutadiene rubber/organoclay nanocomposites: effect of organoclay with various modifier concentrations on the vulcanization behavior and mechanical properties. Polym Eng Sci 47:308–313
30. Bottazzo J, Guglielmi M, Polizzi S, Brusatin G (2013) Natural Rubber/cis-1,4-polybutadiene nanocomposites: vulcanization behavior, mechanical properties, and thermal stability. Polym Eng Sci 671–678
31. Wang X, Gao Y, Mao K, Xue G, Chen T, Zhu J, Li B, Sun P, Jin Q, Ding D, Shi AC (2006) Unusual rheological behavior of liquid polybutadiene rubber/clay nanocomposite gels: the role of polymer-clay interaction, clay exfoliation, and clay orientation and disorientation. Macromolecules 39:6653–6660
32. Sadhu S, Bhowmick AK (2004) Preparation and properties of nanocomposites based on acrylonitrile-butadiene rubber, SBR and polybutadiene rubber. J Polym Sci B 42:1573–1585
33. Hansora DP, Dabke SP, Pachauri R (2012) Modeling and simulation of polybutadiene rubber process/reaction/reactor. Lambert Academic Publishing GmbH & Co. KG, Germany, pp 1–220. Year of Publication: Nov, 2012 (ISBN: 978-3-659-29246-0). https://www.lap-publishing.com/catalog/details//store/ru/book/978-3-659-29246-0/modeling-simulation-of-polymerization-reaction-process-reactor
34. Zhang Y, Ge S, Tang B, Koga T, Rafailovich MH, Sokolov JC, Peiffer DG, Li Z, Dias AJ, McElrath KO, Lin MY, Satija SK, Urquhart SG, Ade H, Nguyen D (2001) Effect of carbon black and silica fillers in elastomer blends. Macromolecules 34:7056–7065
35. Gu Z, Gao L, Song G, Liu W, Li P, Shan C (2010) Octadecyl ammonium montmorillonite/natural rubber/cis-1, 4-polybutadiene nanocomposites. Appl Clay Sci 50:143–147
36. Malas A, Pal P, Das CK (2014) Effect of expanded graphite and modified graphite flakes on the physical and thermo-mechanical properties of styrene butadiene rubber/polybutadiene rubber (SBR/BR) blends. Mater Des 55:664–673

37. Mangaraj D (2002) Elastomer blends. Rubber Chem Technol 75(03):365–427
38. Clarke J, Clarke B, Freakley PK (2001) Rubber Chem Technol 74:1
39. Roland CM, Bohm GGA (1984) J Polym Sci B Polym Phys 22:79
40. Avgeropoulos GN, Weissert FC, Biddison PH, Bohm GGA (1976) Heterogeneous blends of polymers. Rheol Morphol Rubber Chem Technol 49(1):93–104
41. Folt VL, Smith RW (1973) Rheology of elastomers blends. Rubber Chem Technol 46 (5):1193–1209

Chapter 7
Styrene–Butadiene Rubber-Based Nanoblends (SBR-nB): Preparation, Characterization and Applications

Aniruddha Chatterjee and Prashant S. Khobragade

Abstract Styrene–butadiene rubber (SBR) blends become versatile materials of polymer industry today, since their properties can satisfy a wide spectrum of customer demands. In this chapter, the fundamentals of blending, reasons for blending, ideas about the preparation, properties, application, and characterization of SBR-based nanoblends are presented. It also deals with the recent advances, new challenges, and opportunities of these SBR-based blends. A detailed review of the recent works on SBR-based nanoblended systems with their scope and objectives has been included.

Keywords SBR nanoblends · Thermal properties · Application of SBR nanoblends

Abbreviations

SBR	Styrene–butadiene rubber
Phr	Per hundred ratios
Na^+-MMT	Montmorillonite
NR	Natural rubber
NBR	Nitrile butadiene rubber
CNT	Carbon nanotubes
SiO_2	Silica nanoparticles
TMTD	Tetramethylthiuramdisulfide
TMQ	2,2,4-trimethyl-1,2-dihydroquinoline
CBS	*N*-cyclohexyl-2-benzothiazolesulfenamide
6PPD	*N*-(1,3-dimethybutyl)-*N*-phenyl-*p*-phenylenediamine)
DOP	Dioctyl phthalate

A. Chatterjee (✉)
Department of Plastics and Polymer Engineering, Maharashtra Institute of Technology, Aurangabad, Maharashtra, India
e-mail: aniruddha_chatterjee2006@yahoo.co.in; aniruddha.chatterjee@mit.asia

A. Chatterjee · P.S. Khobragade
Department of Plastic and Polymer Technology, University Institute of Chemical Technology, North Maharashtra University, Jalgaon 425001, India

G. Marković and Visakh P.M. (eds.), *Rubber Nano Blends*,
Springer Series on Polymer and Composite Materials,
DOI 10.1007/978-3-319-48720-5_7

TS	Tensile strength
EB	Elongation break
nPS	Nanopolystyrene
DSC	Differential scanning calorimetry
TGA	Thermogravimetric analysis
T_{max}	Decomposition temperature
T_g	Glass transition temperature
SAF	Super abrasion furnace
ISAF	Intermediate super abrasion furnace
SRF	Semi-reinforcing filler
NSU	Composition of NR/SBR
TEM	Transmission electron microscope
SEM	Scanning electron microscope
OC	Octadecylamine modified fluorohectorite
XRD	X-ray diffractometer
EPDM	Ethylene propylene diene terpolymer
NMR	Nuclear magnetic resonance
TiO_2	Titanium dioxide nanoparticles
VCT	Variable contact-time experiments
$T_1\rho H$	Proton spin–lattice relaxation time
FTIR	Fourier transform infrared spectroscopy
DMA	Dynamic mechanical analysis
CB	Carbon black
tanδ	Loss tangent
CNs	Nanocrystalline cellulose
RH	Hexamethylene tetramine
IR	Isoprene
t_{onset}	Online conduction value at initial time
t_{Gmax}	Online conduction value at maximum time
GP	Graphene
GO	Graphene oxide
XSBR	Carboxylated styrene-butadiene rubber
TR	Tear strength

7.1 Introduction

A styrene–butadiene rubber (SBR) is an important type of elastomeric material. It is a synthetic block copolymer (the version developed by Goodyear is called Neolite) consisting of cross-linked molecules of styrene and butadiene monomer in the ratio of 3:7 in random form and exhibits a rapid and large reversible strain in response to stress. It has good abrasion resistance and good aging stability when protected by additives than other synthetic rubbers. SBR is a suitable material for automobile tires and other applications. The styrene/butadiene ratio influences the properties of the SBR: With high styrene content, the rubber is harder and less rubbery. Generally, SBR is compounded with sulfur as vulcanizing agent, antioxidant, filler, and other different additives with respect to the final use and applications. Recently, nanoclays, inorganic nanoparticulates, and other synthetic compound such as CNTs, graphene, and their derivatives have been introduced as fillers in SBR to prepare valuable nanocomposites but at higher loading of fillers allow formation of agglomeration in elastomer composites which do not contribute to the elastic response and mechanical properties of rubber matrices. Therefore, nowadays, to maintain the desired properties with respect to the end use application, the concept of nanoblends has been introduced.

SBR nanoblends consist of a homogeneous mixing of nanopolymers, copolymers, and hybrid nanomaterials having different shapes, but at least one dimension must be in the range of 1–100 nm. SBR nanoblends are classified as miscible and immiscible depending on the constituents present in the matrix. The SBR nanoblends have attractive properties such as high thermal resistance, mechanical properties and damping at high frequency, and mechanical and viscoelastic properties than single rubber [1–5]. Among these, the rubber/polymer nanoparticles and rubber/inorganic-organic nanoparticles have drawn considerable attention and great importance due to their wide range of properties useful in the development of new low-cost products with better performance and to control properties of multicomponent rubbery systems [6, 7]. Most of these blends are immiscible due to the lack of favorable interactions between the components and SBR at the interfaces. The adhesion between the components is an essential requirement for better blending performance [8] of SBR blends with different nanopolymers and hybrid nanomaterials, and preparation and their characterization are mention below.

7.2 Preparation of Styrene–Butadiene Rubber Nanoblends

In this section, different techniques for preparation of SBR rubber nanoblends are discussed. The selection of appropriate processing technique for the preparation of SBR nanoblends depends on the constituents present in SBR. Generally, rubber nanoblends can be classified into three different areas.

(i) rubber/rubber nanoblends,
(ii) rubber/polymer nanoblends, and
(iii) rubber/hybrid nanoblends.

Following are the techniques reported for SBR nano blends.

7.2.1 Latex Blending Technique

NR/SBR/Clay Blend

Unique Latex Blending Technique

Mitra et al. [9] reported unique latex blending technique in which aqueous suspension of Na^+-montmorillonite (Na^+-MMT) clay was prepared by dispersing the clay in water (2 %w/v) with continuous stirring (700–800 rpm) for 72 h at ambient temperature. Aqueous suspension of Na^+-MMT clay and the natural rubber (NR)/ SBR latex mixture were mixed thoroughly and stirred for 2 h in a magnetic stirrer. After that, aqueous dispersions of the curing agents were added into it and the prevulcanization of the blended mixture was carried out at 80 °C for 2 h under constant stirring (300–400 rpm). The nanomaterial concentration was kept at 2, 4, 6, and 8 per hundred ratio (phr), respectively. Films of cross-linked nanoblend gel were obtained from prevulcanized latex by casting on a flat glass plate and subsequent drying at ambient temperature (25 °C) to constant weight. Finally, the films were vacuum dried at 50 °C for 12 h, prior to analyze them.

Conventional Latex Blending Technique

SBR nanoblends prepared by conventional latex blending technique where, 6 phr Na^+-MMT blended in NR and SBR latex followed by mixing and press curing. The latex–clay mixture was coagulated with dilute sulfuric acid (2 wt%) and washed with water till it becomes acid free. Then, the coagulated mass was dried in a vacuum oven at 50 °C for 24 h. After that, the dried rubber–clay mass was subjected to conventional mixing in a two-roll mill for the incorporation of various ingredients into them. Finally, the compounded mass was compression-molded into a 1-mm-thick sheet by curing till optimum cure time at 150 °C in a compression molding with pressure of 5 bar.

7.2.2 Melt Blending Technique

SBR/NBR/NR/CNT

Solution of styrene–butadiene rubber (S-SBR) (styrene content of 21 % and vinyl content of 63 %), nitrile butadiene rubber (NBR) (nitrile content of 34 %), and NR was used for the preparation of blend. NR was masticated by means of two-roll mill before the use in order to obtain a similar mooney viscosity value as the other blend partners. Multiwalled carbon nanotubes (CNTs) with an average diameter of 10 nm and a broad length distribution with several nanotubes up to 10 μm were used as conductive filler. Peroxide was used as vulcanizing agent for all the mixtures. For preparation of CNT-filled compounds and blends, an internal mixer was used at initial set temperature of 50 °C and rotor speed of 75 rpm. A conductivity sensor system was installed in the chamber to measure the electrical signal of the conductive mixtures. CNT-filled compounds were prepared by mixing 7 phr CNTs with different rubbers investigated. Ternary blend of CNT–SBR/NBR/NR from statistic mixing was prepared by mixing 5 phr CNTs with (33/33/34) SBR/NBR/NR blend in one-step mixing process. For achieving ternary blends, CNTs were mixed in each rubber, and then, it was mixed with two other fresh rubbers. The recipes were formulated in order to receive three blends (CNT–SBR)/NBR/NR, SBR/(CNT–NBR)/NR, and SBR/NBR/(CNT–NR) with the composition corresponding to that of the statistic blend. For an effective dispersion of CNTs in rubber matrix, an ethanol-assisted mixing process (wet mixing) was applied; however, no nonionic surfactant was used for stabilizing the nanotubes in the ethanolic medium. CNTs were first wetted with a certain amount of ethanol to a paste. Then, it was added into the mixing chamber with rubber for the preparation of the nanoblends and master batches. Ethanol was entirely vaporized during the mixing process. An ethanol/CNTs ratio of 3.7 was kept for a good dispersion of CNTs. At a mixing time of 46 min, 1.5 phr peroxide was added. Samples taken out at 50 min were compression-molded (at temperature = 160 °C and pressure = 100 bar) to obtain a sheet, prior to the analysis [10].

7.2.3 *Solution Blending Technique*

NR/SBR/Clay Blend

For the blend preparation, clay was first dispersed in deionized water with vigorous stirring by a special type of stirrer at a concentration of 2 wt% and an aqueous suspension of layered silicate was obtained. The 50/50 (wt%) blend of NR and SBR latexes was prepared by slowly adding the given amount of SBR latex into the NR latex under mixing for 30 min. Then, a given amount of (50/50) NR/SBR blend latex was added into the aqueous clay suspension and stirred for 30 min. Finally, the mixture was co-coagulated by a cation-type coagulating agent (dilute solution of sulfuric acid, 2 wt%), washed with water several times until the pH reached about 7 and dried at 80 °C for 24 h under reduced pressure to obtain NR/SBR/clay nanoblend. The blends and the clay-free NR/SBR blend (50/50) were mixed with the ingredients in a 6-inch two-roll mill and further vulcanized in a standard mold at 150 °C for the optimum curing time determined by a rheometer [11].

7.2.4 Two-Roll Mill Blending

SBR/SiO_2/Graphene Blend

The rubber batch materials consist of silica (SiO_2), styrene–butadiene rubber (SBR 1712), reclaim rubber, zinc oxide, stearic acid, tetramethylthiuramdisulfide (TMTD), caster oil, sulfur, 2,2,4-trimethyl-1,2-dihydroquinoline (TMQ), carbon black (CB), *N*-cyclohexyl-2-benzothiazolesulfenamide (CBS), and *N*-(1,3-dimethybutyl)-*N*-phenyl-*p*-phenylenediamine) (6PPD). The blend is prepared by mixing the materials by using the laboratory two-roll mill. The detailed procedure is as follows:

(i) First at room temperature, SBR is put through the two rolls for several times with decreasing the distance between two rolls to the extent of 0.5–1 mm.
(ii) Add reclaim rubber and masticate for 3 min.
(iii) During this process, rubber pieces were passed through the rollers in horizontal and vertical states alternatively for several times for obtaining the homogenous materials.
(iv) Add zinc oxide and continue the mixing for 2 min, followed by the addition of stearic acid.
(v) Add CB and dioctyl phthalate (DOP) alternatively and mixing continued for 7 min.
(vi) Add CBS and mix for 5 min followed by the addition of antioxidant.
(vii) Add the SiO_2/graphene-hybrid-reinforcing filler blended with coupling agent and repeat mixing.
(viii) Finally add sulfur with preblending.
(ix) Continue the mixing process for more time in order to get a good homogenization and decrease mill opening to 0.5–1 mm for increasing homogenously the efficiency of mixing [12].

7.3 Tensile Characterizations of Styrene–Butadiene Rubber Nanoblends

Mechanical properties of SBR nanoblends play a vital role in most of the load-bearing applications. In recent years, lighter, stiffer, and stronger materials are in demand which can be used in different sectors such as aeronautics and civil engineering and in various structural applications. In recent years, there has been an increased application of cellulose nanocrystals (CNs) as the load-bearing component in the development of advanced and low cost polymeric nanoblends. Cao et al. [13] prepared nanoblends, where CNs fabricated by acid hydrolysis of cotton linter with concentrated sulfuric acid and reinforced the resulting CNs in to carboxylated styrene-butadiene rubber (XSBR) matrix [a copolymer of butadiene, styrene and a small amount of acrylic acid with excellent chemistry stability, good physical and cohesional strength] with improved physical properties. As compared to other inorganic reinforcing fillers, CNs have number of additional advantages, including low density, positive environmental benefit, low energy consumption in manufacturing, wide variety of fillers available throughout the world, high sound absorbing property, high combustion with ease recycling, and reasonably easy processability due to their smooth nature, which allows high filling levels, in turn resulting major cost savings. A required amount of CNs suspension was mixed with XSBR latex and stirred well for 30 min at room temperature followed by drying under reduced pressure at 60 °C until a constant weight was obtained. The compounding of XSBR/CNs nanoblends with other ingredients was carried out on a two-roll mill. The rubber compound sheets were then compressed and vulcanized at 170 ºC in an electrically heated hydraulic compression moulding machine up to their optimal cure time. Dumbbell-shaped specimens (6 mm width in crosssection) were cut from the moulded sheet, and the tensile testing machine was used to analyze the tensile strength (TS) according to ASTM D412. Tear strength (TS) was determined according to ASTM D624 method on an un-nicked 90° angle test pieces. Shore A hardness was measured as per ASTM-D2240 method.

The mechanical properties of the XSBR/CNs nanoblends are shown in Figs. 7.1, 7.2 and 7.3, respectively. Previous literature studies reported that strong interaction by hydrogen bonding between the functionalized host matrix and hydroxyl groups on cellulose increased the polymer strength. Similar experience was also observed here. The TS (Fig. 7.1) and TR (Fig. 7.2) of the XSBR/CNs nanoblends increased with increasing CNs in the XSBR matrix. The elongation at break (EB) and shore A hardness were also observed to be improved by addition of CNs and shown in Fig. 7.3a, b. The TS and TR were increased from 16.9 to 24.1 MPa and from 43.5 to

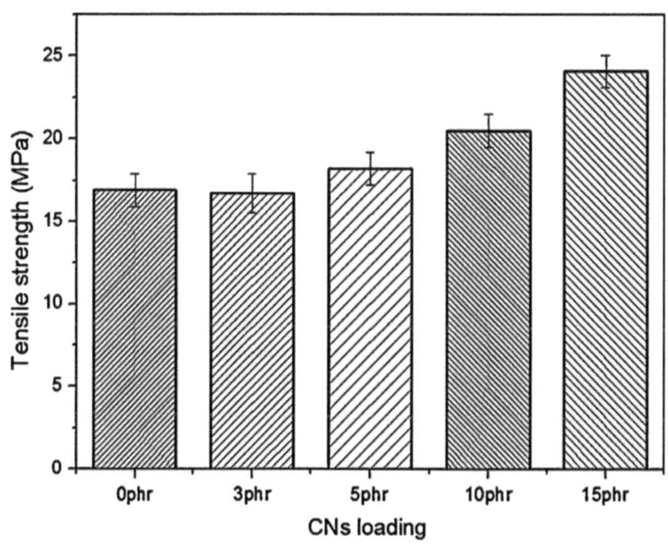

Fig. 7.1 Tensile strength (TS) of XSBR/CNs nanoblends [13]. *Source* Elsevier

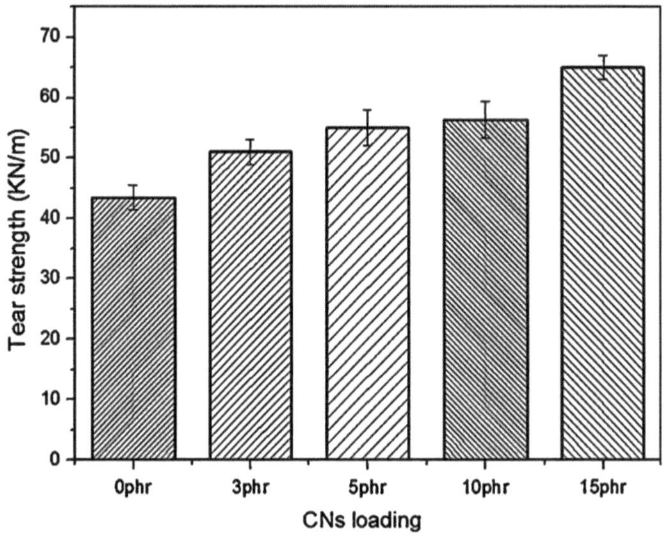

Fig. 7.2 Tear strength (TR) of XSBR/CNs nanoblends [13]. *Source* Elsevier

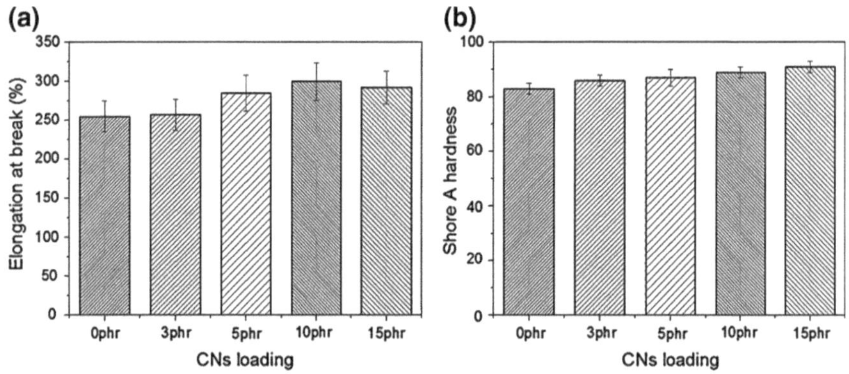

Fig. 7.3 **a** Elongation at break (%) (EB) and **b** shore A hardness of XSBR/CNs nanoblends [13]. *Source* Elsevier

65.2 MPa, respectively, with increasing CNs loading from 0 to 15 phr. This was revealed that CNs indeed showed a good reinforcement on the XSBR. The improved mechanical properties were related to the excellent homogeneous distribution of CNs in the XSBR matrix and better interfacial interaction between the CNs and the XSBR matrix.

In other words, maximum improvement in the tensile properties of nanoblends can be achieved only when all of the layered silicates are separated into single layers [11]. Mitra et al. [9] reported the tensile properties of nanoblend gels. The

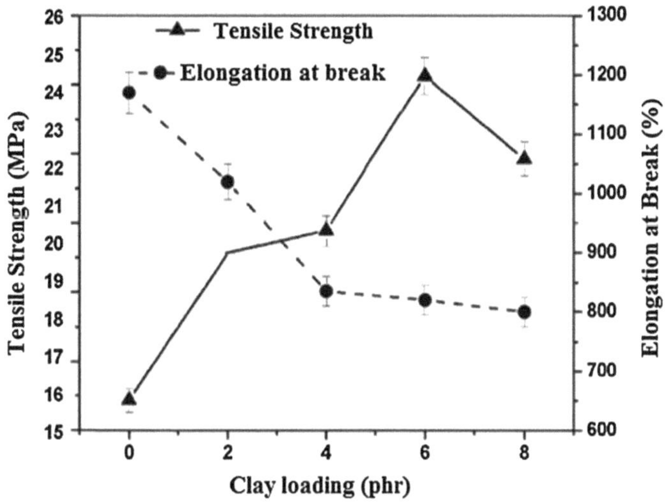

Fig. 7.4 Tensile properties of NR/SBR/Na-MMT (50/50) nanoblends [14]. *Source* John Wiley and Sons

high level of dispersion and exfoliation has a direct bearing on the mechanical properties of these nanoblend gels. As shown in Fig. 7.4, the TS registered a substantial increase in Na-MMT loadings. For example, the TS values of samples 4 and 6 are 29 and 54 % higher than the control virgin rubber. In the case of SBR nanoblend gels, the TS gives maximum value at 6 phr of Na-MMT loading for both NR and SBR nanoblend gels. The EB values generate different trends for NR and SBR nanoblend gels. In the NR nanoblend gels, EB decreases consistently with the addition of Na-MMT (Fig. 7.4), whereas in the case of SBR nanoblend gels, EB increases steadily up to 6 phr clay loading and then falls off (Fig. 7.4). This has been attributed to the higher degree of exfoliation of nanoclays in the SBR matrix. The impressive level of improvement in the TS of the nanoblends can be ascribed to (a) uniformly dispersed structure of clay at nanolevel, (b) high aspect ratio, and (c) improved polymer–filler interaction. At relatively higher clay loading (>6 phr in the present case), due to reduction in the level of exfoliation and prevalence of nanoclay agglomeration, the mechanical strength decreases significantly irrespective of the rubber matrix.

7.4 Thermal Properties of Styrene–Butadiene Rubber Nanoblends

The thermal behavior analysis of the prepared SBR nanoblends with various ratios of nanomaterial loadings was investigated by differential scanning calorimetry (DSC) and thermogravimetric analysis (TGA and DTG) in the various temperature

Table 7.1 Effect of CNTs loading ratio on the thermal properties of prepared SBR nanoblends [14]. *Source* Elsevier

SBR nanoblends	Melting temperature (T_m) (K)	Destroying temperature (T_d) (K)
Sample 1 (0.0 wt%)	365.95	539.95
Sample 2 (4.0 wt%)	370.95	541.65
Sample 3 (8.0 wt%)	373.25	542.55

ranges and heating rates under nitrogen atmosphere. Thermal properties show the thermal stability and thermal degradation characteristics of the blends. In this section, the effect of different nanomaterials on thermal properties of SBR-based nanoblends was discussed [15, 16]. Hosseini et al. have studied the effect of different wt% loadings of CNTs on thermal behavior of SBR nanoblends [14] using DSC. The obtained results (Table 7.1) showed that the thermal stability of prepared SBR nanoblends was improved by the increase in CNTs loadings. This is due to local heterogeneity in prepared blends with well-dispersed CNTs which make favorable molecular interactions between the CNTs particles and binder and create an interfacial zone of polymer segments. Table 7.1 shows the effect of CNTs loadings ratio on thermal properties of prepared SBR nanoblends.

NR/SBR blends with and without nanosilica loadings were prepared by Prasertsri et al. [15] and characterized by thermal experiments. TGA experiments were carried out using TGA/SDTA and DSC, where approximately 8 mg of NR/SBR blends, with and without nanosilica, were heated from 40 to 600 °C at a rate of 20 °C/min under the nitrogen atmosphere. Then, the gas was automatically changed to oxygen and heating of the sample was continued up to 800 °C. Both TGA and derivative TGA (DTGA) results were used to identify the composition of NR/SBR blends. About 15 mg of the samples were capsulated in aluminum pans. Each experiment was done under the nitrogen atmosphere at a heating rate of 10 °C/min within the temperature range from −80 to 0 °C. Data were plotted as heat flow versus temperature. The inflection point of the slope in the thermogram was taken as the glass transition temperature (T_g).

Figure 7.5a, b shows TGA and DTGA results of various NR/SBR blends without silica. From TGA curves (Fig. 7.5a), there are two main steps of decomposition in all samples. The first weight loss in region 350–500 °C is about 98.5 % of the total weight loss, corresponding to the decomposition of the blended rubbers. The small decomposition having weight loss of about 0.7 wt% occurs at about 600 °C after switching to oxygen atmosphere caused by oxidative degradation of carbon char from the rubber decomposition. The residue about 0.8 wt% relates to ash and inorganic substances. In Fig. 7.5b, only one decomposition peak is observed at approximately 390 and 473 °C, denoted as decomposition temperature (T_{max}) of neat NR and SBR, respectively. SBR contains partially aromatic styrene groups; it possesses higher thermal stability than NR. When NR blended with SBR, two decomposition peaks are observed and corresponds to T_{max} of the individual rubbers. In the decomposition region of the rubbers (350–500 °C), it can be noted that (i) the weight loss at a specified temperature reduces when SBR content in the blends is increased

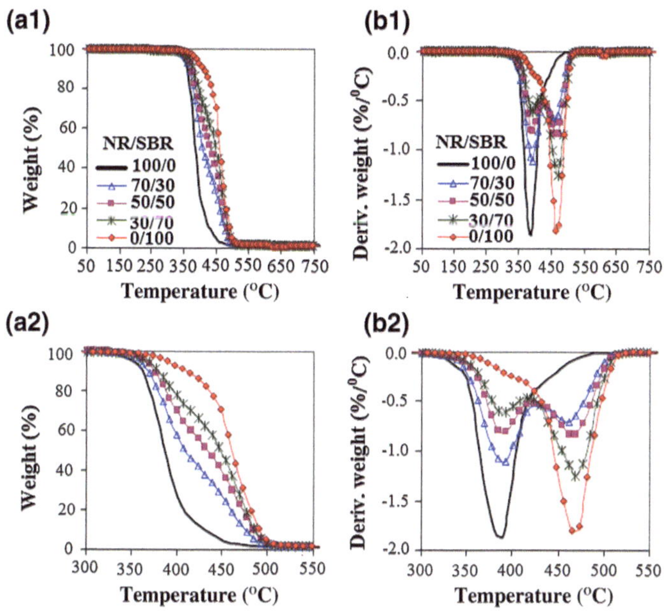

Fig. 7.5 **a** TGA and **b** DTGA curves of various NR/SBR blends without silica [subscripts 1 and 2 represent the temperature range from 50 to 750 and from 300 to 550 °C (*enlarged curve*)] [15]. *Source* Elsevier

(Fig. 7.5a) and (ii) the peak height at T_{max} strongly depends on the blend composition (Fig. 7.5b). TGA and DTGA curves of NR/SBR blends containing silica are illustrated in Fig. 7.6. After adding silica, T_{max} of the blends does not significantly change when compared to that of unfilled blends, but its peak height reduces with the increasing silica content due to the dilution effect with respect to volume fraction of rubber. Furthermore, the residue weight from the TGA curve is found to increase with respect to the combination of ash and silica. Thus, it is necessary to separate them by subtraction in order to know the actual amount of silica in the coagulated blends. Since the T_{max} of NR and SBR phases in the blends is significantly different (a difference of about 80 °C), the ratios of their peak heights are criteria to distinguish them apart. [T_{max} of 473 and 390 °C of SBR and NR nanoblends].

DSC was employed to determine glass transition temperature (T_g) of NR/SBR blends. Heat capacity ratios at transition steps of NR and SBR in the blends were also correlated with the blend composition. T_g values of unfilled NR/SBR blends at various blend ratios and silica-filled NR/SBR blend (50/50) at different silica contents are tabulated in Table 7.2. T_g of pure NR and SBR appears at −62 and −50 °C, respectively. After blending, two transitions are observed in the thermograms at around T_g of each elastomer. They are coded as T_{g1} and T_{g2} for NR and SBR phases, respectively. It can be noted that neither T_{g1} nor T_{g2} shows any deviation from the values of neat NR and SBR. This infers that there is only physical interlocking without any chemical interaction between the molecular

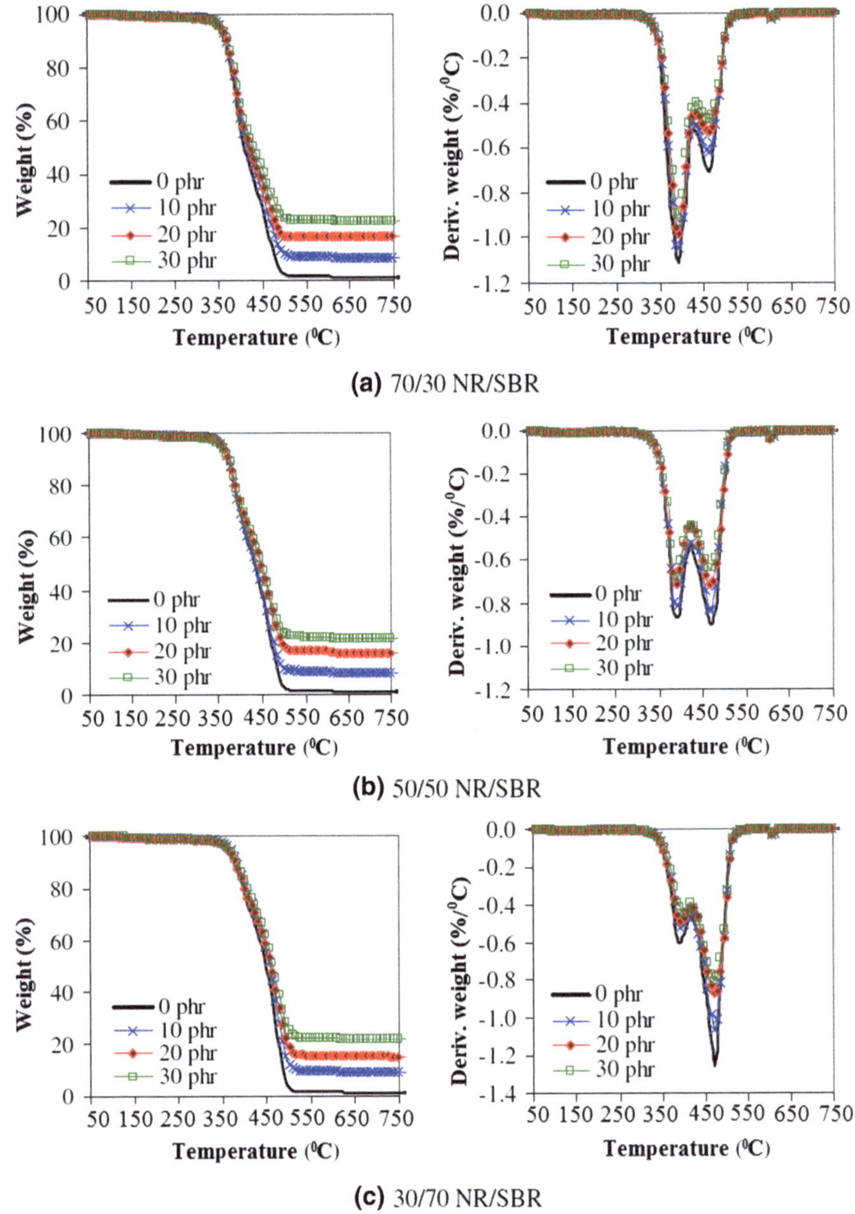

Fig. 7.6 TGA and DTGA curves of NR/SBR blends containing different silica contents [15]. *Source* Elsevier

chains of NR and SBR in the coagulated blends. Similarly, by adding silica into the NR/SBR blend (50/50), two transitions can be seen approximately at the same temperatures as for the neat NR and SBR, indicating immiscible blends.

Table 7.2 Glass transition temperature of various NR/SBR samples measured by DSC [15]. *Source* Elsevier

NR/SBR (%w/w)	Silica content (phr)	Glass transition temperature (°C) T_{g1}	T_{g2}
100/0	0	−62.3	–
70/30	0	−62.1	−50.5
50/50	0	−62.9	−52.8
30/70	0	−62.7	−51.8
0/100	0	–	−49.9
50/50	10	−62.8	−53.9
50/50	20	−63.4	−53.7
50/50	30	−63.5	−53.8

Pal et al. [16] also prepared the high styrene–butadiene rubber (HSR) and NR-based nanoblends and studied their thermal response by DSC and TGA. The compounding formulation for the HSR and NR blends with its various ingredients was mixed in a two-roll mill at a friction ratio of 1:2 following the standard mixing sequence. Compounding formulations based on the change in HSR and NR contents are shown in Table 7.3. Also, three types of CBs and one semi-reinforcing filler (SRF) are used, such as super abrasion furnace (SAF N110), intermediate super abrasion furnace (ISAF N231 and N234), and semi-reinforcing filler (SRF N774). For vulcanization, the amounts of additives such as sulfur, process oil, and CBS were based on 100 wt% of rubber, and the samples have the code name 'NSU-1,' 'NSU-2,' 'NSU-3,' 'NSU-4,' 'NSU-5,' and 'NSU-6,' respectively.

DSC measurements were carried out by taking 10 mg samples (sealed under aluminum pans) scanned in the temperature range of −80 to 80 °C. The heating rate was 10 °C/min under the nitrogen atmosphere with a flow rate of 40 ml/min. The DSC curves for all rubber vulcanizates are shown in Fig. 7.7. The criterion for compatibility or incompatibility in polymer blends is the presence of a single T_g for the material, which is intermediate between the T_g of the pure components and the existence of two $T_{g's}$ in the DSC thermograms for incompatible polymer blends. All traces show a reasonable sharp T_g transition depending on the composition; however, the graph for blend DSC traces is not shown for brevity. From Fig. 7.7, all show the T_g at around −60 °C, but in case of NR/HSR/Ultrasil-VN3 grade nanosilica-1 (NSU-1), the T_g shifted to the higher side. All the samples show only one melting peak on the DSC curves; this is attributed to the same backbone structure of the matrix and the CB as reinforcing filler. Since miscible polymer blends should exhibit a single T_g between the $T_{g's}$ of pure components. This is consistent with the effect of blend composition on miscibility when the chemical structure and the molecular mass of components are fixed. But in case of 80 wt% of NR and 20 wt% of HSR with different grades of CB, the melting peak appreciably shifts to the higher side. This may be due to the dilution effects of the matrix with the fillers.

Table 7.3 Compound formulation based on the change in HSR and NR contents [16]. *Source* Elsevier

Compounds	Sample codes					
	NSU-1	NSU-2	NSU-3	NSU-4	NSU-5	NSU-6
	Weight (wt%)					
NR	80	80	80	70	70	70
High styrene rubber (HSR)	20	20	20	30	30	30
Carbon blacks						
SAF (N110)	20	–	–	20	–	–
SRF (N774)	20	–	–	20	–	–
ISAF (N234)	–	40	–	–	40	–
ISAF(N231)	–	–	40	–	–	40
Ultrasil VN3	5	5	5	5	5	5
Si-69	1	1	1	1	1	1
Sulfur	2	2	2	2	2	2
Stearic acid	2	2	2	2	2	2
IPPD	1	1	1	1	1	1
CBS	0.7	0.7	0.7	0.7	0.7	0.7
ZnO	5	5	5	5	5	5
Wax	1	1	1	1	1	1
Elasto-710	2	2	2	2	2	2

SAF – Super Abrasion Furnace
SRF – Semi Reinforcing Filler
ISAF – Intermediate Super Abrasion Furnace

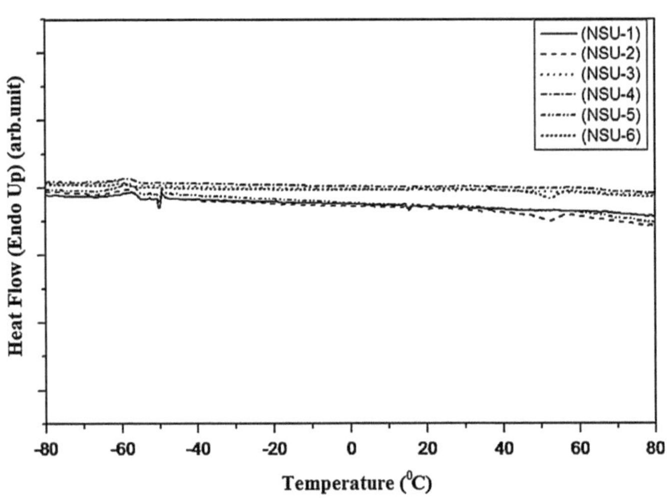

Fig. 7.7 DSC study of high styrene rubber (HSR) and natural rubber (NR) with nanosilica [16]. *Source* Elsevier

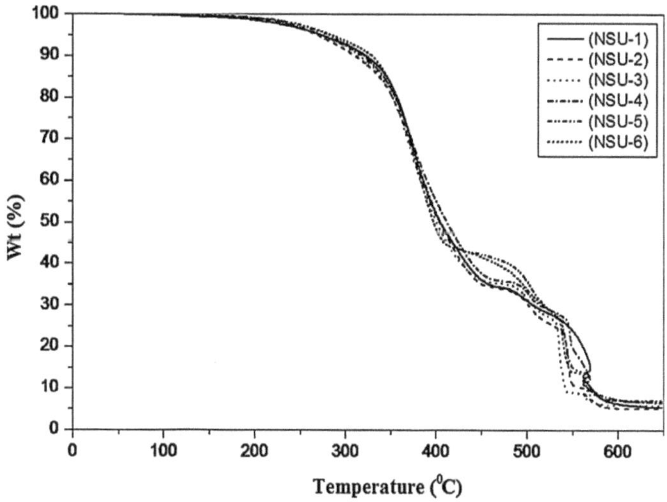

Fig. 7.8 TGA plots of high styrene rubber (HSR) and natural rubber (NR) with nanosilica [16]. *Source* Elsevier

TGA studies were carried out in the presence of air at a rate of 10 °C/min from 50 to 650 °C. Degradation temperatures of all samples were calculated by TGA plot. The temperature for the onset of degradation (T_1), the temperature at which 10 % degradation occurred (T_{10}), the temperature at which 50 % degradation occurred (T_{50}), and the temperature at which 90 % degradation occurred (T_{90}) were calculated from the TGA plots (Fig. 7.8). It was observed that the onset degradation temperature was more or less same for all the samples. Thereby, the onset degradation temperature probably decreased in the case of rubber samples due to the decrease in cross-linking density. Cross-linking increased the rigidity of the system, which in turn increased the thermal stability. The rate of degradation was almost the same up to 90 % degradation for all the samples. Also, it is found that for samples NSU-2 to NSU-3, the degradation temperature is decreased which is due to the decrease in concentration of HSR with intermediated super abrasion furnace type of CB.

7.5 Morphological Properties of Styrene–Butadiene Rubber Nanoblends

Stephen et al. [17] prepared SBR-based nanoblends with differently layered silicate nanomaterials and characterized their morphological properties by transmission electron microscope (TEM). The TEM images of layered silicates that reinforced latex nanoblends were taken with an acceleration voltage of 120 keV. The specimens were prepared using an Ultracut E cryomicrotome. Thin sections of about 100 nm were cut with a diamond knife at −120 °C.

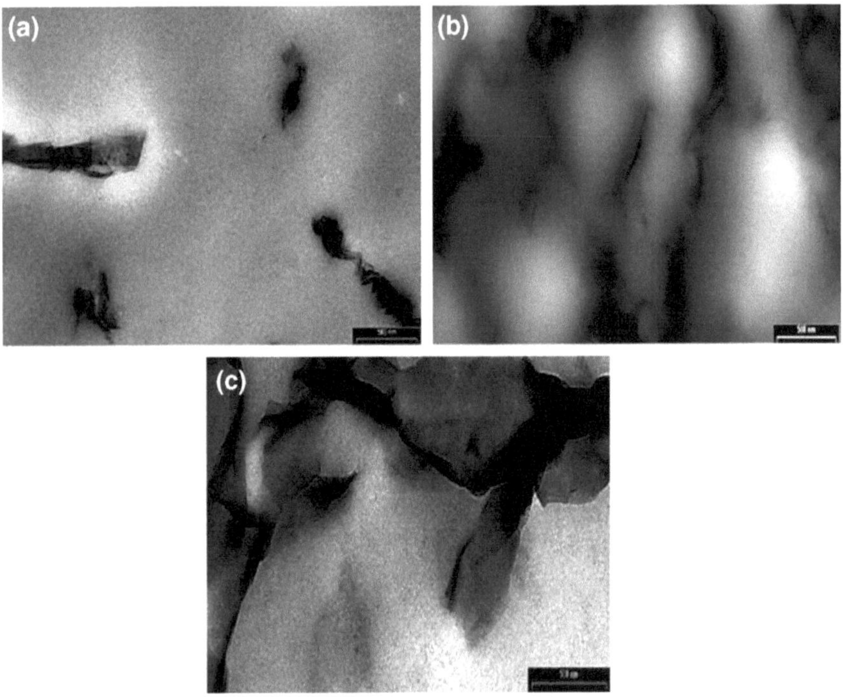

Fig. 7.9 TEM images of **a** commercial clay, **b** sodium bentonite, and **c** sodium fluorohectorite-filled latex nanoblends (7.5 phr) [17]. *Source* Elsevier

Figure 7.9 shows the morphology of latex nanoblends where the dark lines indicated the dispersion of silicates in rubber matrix. Fig. 7.9a shows that the commercial clay exists as large aggregates and is unable to undergo exfoliation. Figure 7.9b, c shows the TEM image of sodium bentonite- and fluorohectorite-filled NR latex nanoblends. On comparing the dispersion of two silicates in latex, fluorohectorite exhibits more exfoliated structure than the bentonite clay. In the latter case, it can see partially exfoliated and intercalated structure. Most of the dispersed units of bentonite-filled nanoblends are not individual layers but layer of bundles. The silicates having the ability to swell in the aqueous phase of the latex resulted in the separation of layers. The aqueous swell ability is higher for fluorohectorite, and therefore, it exhibits more exfoliated structure and increases the properties.

Pal et al. [16] characterized the morphological properties of HSR/NR nanoblends by scanning electron microscope (SEM). The compounding formulation for the HSR and NR blends with its various ingredients is already shown in Table 7.3. The tensile fracture surface of the samples was studied at 1000 times zooming and scanned after gold coating. For all rubber samples, smooth fracture surfaces with uniform filler dispersion and unidirectional tear path orientation along the direction of flow is shown in Fig. 7.10. The micrograph of the NSU-3 rubber sample is

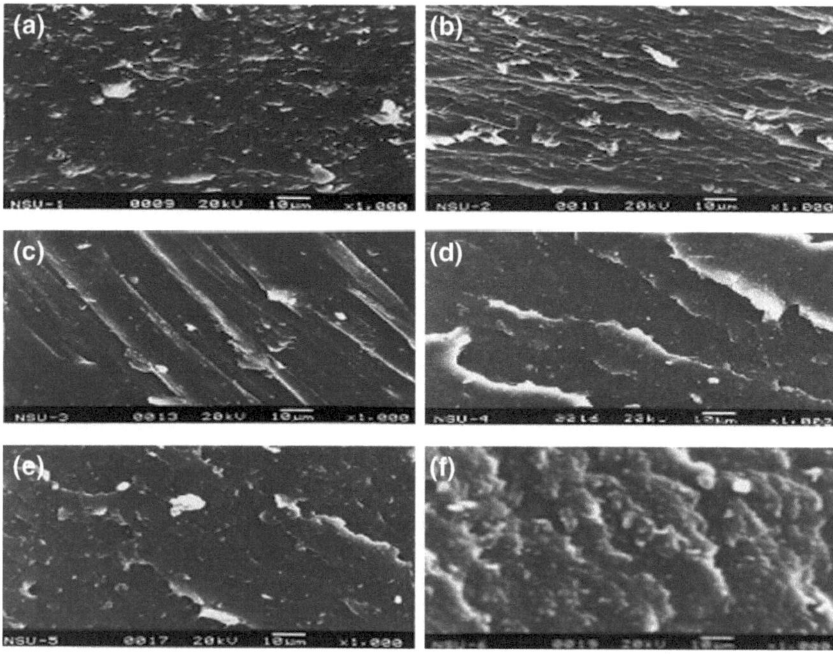

Fig. 7.10 SEM images of different types of NSU blends: **a** NSU-1, **b** NSU-2, **c** NSU-3, **d** NSU-4, **e** NSU-5, and **f** NSU-6 [16]. *Source* Elsevier

characterized by a smooth, rubbery failure (which is a smooth failure in the case of rubber samples without the formation of necking) where the additives are clearly seen; the appearance is associated with a low tensile strength. But for NSU-1, NSU-2, NSU-4, NSU-5, and NSU-6, fatigue type of failure is clearly observed. It may also conclude that rough surface has been seen for the sample containing intermediate super abrasion furnace (ISAF) type of CB. Some samples such as NSU-1, NSU-2, NSU-5, and NSU-6 have also shown rupture type of failure. This may be due to the disproportionate breaking of the CB structure and reformation of inter-aggregate bonds of CB. The compound NSU-4 showed lesser heat buildup compared to NSU-5 and NSU-6 because it contains combination of SAF N110 type of CB and SRF N774, whereas NSU-6 contains low-structured CB and NSU-5 contains high-structured CB. The combination of SRF and CB may be responsible for low heat buildup. In order to further verify the effect of the nanomaterials on the morphology of SBR nanoblend, Zhang et al. [18] studied the existence of octadecylamine-modified fluorohectorite (OC) dispersion in SBR matrix. Figure 7.11 shows TEM micrographs of SBR/OC nanoblend, where intercalated platelets are mainly observed due to its smaller interlayer distance of the intercalated structure. These are in good agreement with the XRD results, where polar interaction between polymer and OC can promote the creation of exfoliated structure in between OC and SBR. The inter-gallery environment of OC is still

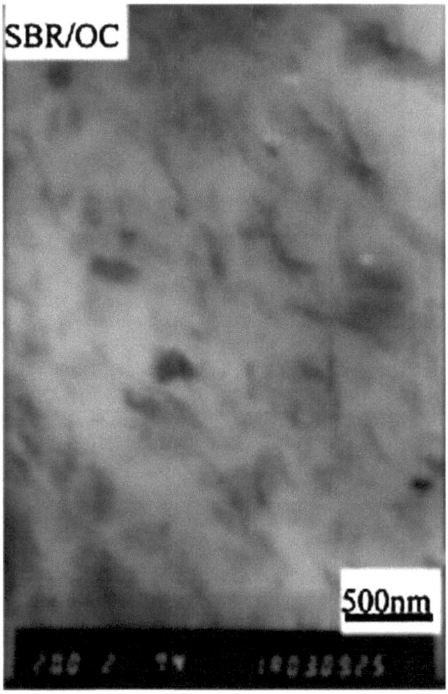

Fig. 7.11 TEM micrographs of SBR/OC [18]. *Source* John Wiley and Sons

polar, although the polarity of clay is reduced by organophilic intercalate. The polarity of polymer is closely related to chemical structure of polymer chain. As far as the molecular chemical structure is concerned, SBR contains less number of unsaturated double bonds.

7.6 X-ray Diffraction Studies of Styrene–Butadiene Rubber Nanoblends

Figure 7.12 shows tchemical shiftshe XRD diffraction patterns of SBR/OC nanoblend containing 10 phr OC. The results show that the intercalation behavior of OC evidently depends on the characteristics of rubber. The diffraction peak of OC at 2θ is 3.938 corresponding to a basal spacing (001) of 2.24 nm. Compared with OC, the intensity of the intercalated basal reflection of SBR/OC nanoblend increases and the full width at half maximum decreases, indicating a well-ordered intercalated nanoblends.

Similar phenomenon is observed in the SBR/OC system before and after curing (Fig. 7.13). Before curing, XRD peaks appear at 1.28, 1.54, 2.15, 3.29, and 6.30 nm, respectively. Compared with the initial interlay spacing 2.24 nm of OC, the peaks of 1.28 and 1.54 nm suggest that a part of OC is confined. The

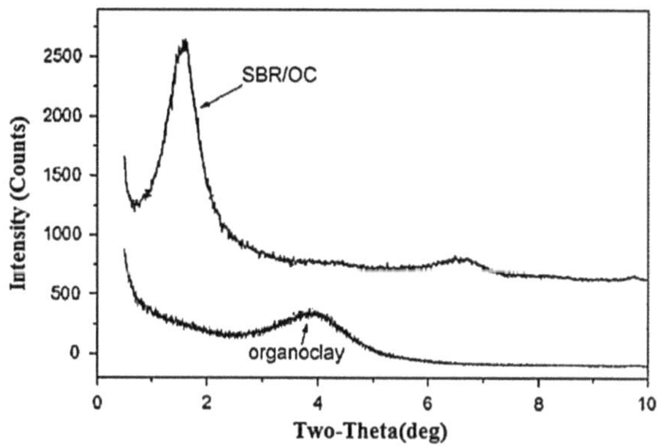

Fig. 7.12 XRD diffraction patterns of organoclay (OC), SBR nanoblend containing 10 phr OC [18]. *Source* John Wiley and Sons

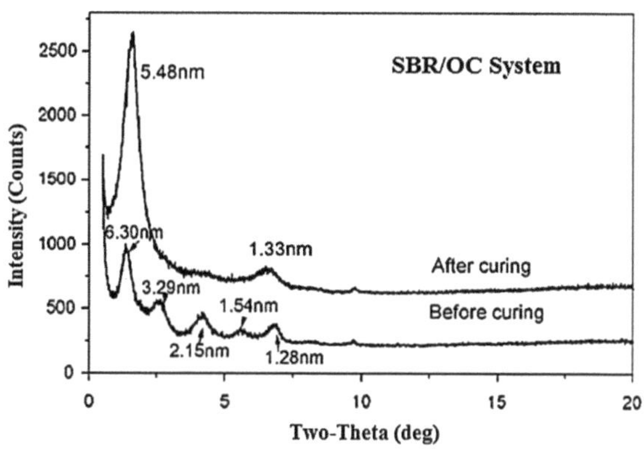

Fig. 7.13 XRD patterns of the above SBR/OC nanoblend before and after curing [18]. *Source* John Wiley and Sons

inter-gallery distances of 6.30 and 3.29 nm indicate that SBR molecules intercalate into the clay interlayer under the shear force. After curing, the SBR/OC only shows two peaks at 5.48 and 1.33 nm corresponding to inter-gallery distances, respectively. Moreover, the intensity of the peak at 5.48 nm increases, indicating the formation of well-ordered intercalated structure. Compared with the peak at 6.3 nm of SBR/OC before curing, the gallery height decreases by 0.82 nm, which is comparable to the change in gallery height of the melt intercalation of PS into fluorohectorite modified by octadecyl ammonium. Since the configuration of SBR molecules is determined by styrene monomer, and the entropy is increased during the process of expulsion of SBR, the reason for the decrease in gallery height should be ascribed to the displacement of SBR molecules from the interlayer during the process of curing. In the cases of the SBR/OC systems, the mixing process produces the intercalated and even exfoliated structures [18].

Durandish et al. [19] prepared the SBR/EPDM nanoblend based on the organoclay and characterized their X-ray diffraction patterns using 50 kV, 40 mA Cu-Kα radiation (λ = 0.1540598 nm). The samples were sheets with a smooth surface and dimensions of 10 mm × 10 mm × 1 mm. The basal spacing of silicates was

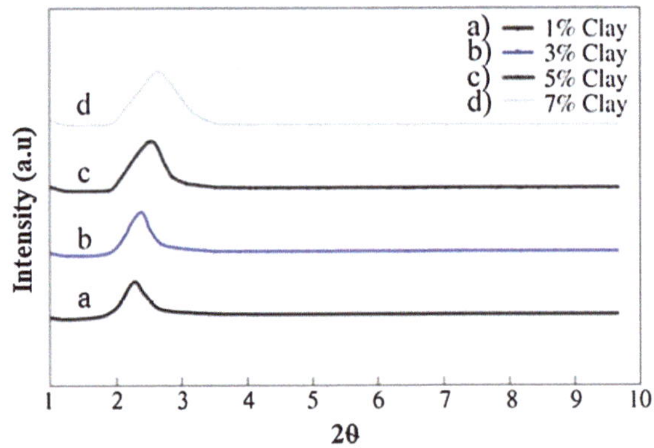

Fig. 7.14 XRD patterns of SBR 60/EPDM 40 containing 1, 3, 5, and 7 wt% Cloisite 20A [19]. *Source* John Wiley and Sons

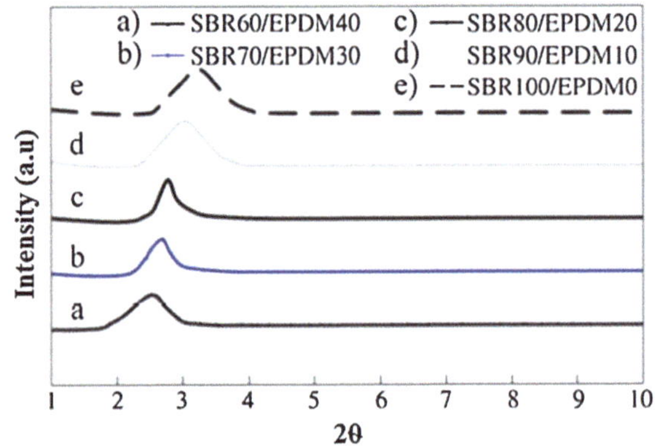

Fig. 7.15 XRD patterns of different compositions of SBR/EPDM containing 5 wt % nanoclay [19]. *Source* John Wiley and Sons

estimated and calculated on the basis of plane peak positions in the WAXD intensity profile using Bragg's law, $d = \lambda/(2\sin\theta_{\max})$. XRD patterns of the SBR/ethylene propylene diene terpolymer (EPDM) (60/40) samples containing 1, 3, 5, and 7 wt% organoclay are typically shown in Fig. 7.14. As we know, organoclay (Cloisite 20A) presents a peak at $2\theta = 3.52°$, which as deduced by the Bragg's equation corresponds to an interlayer distance of 2.42 nm. Shift of the organoclay characteristic peak to lower 2θ values (see Fig. 7.14) indicates that elastomer chains have intercalated between silicate layers. This clearly shows that inter-lamellar spacing of the clay would enlarge after melt compounding. As seen, the intensity of the peaks increases by clay loading. Broad peaks with lower intensities are usually indicative of more intensive intercalations or even partial exfoliations.

Moreover, XRD results related to the different compositions of SBR/EPDM containing 5 wt% organoclay are shown in Fig. 7.15. It is shown that as the EPDM content within the samples increases, the organoclay diffraction peak shifts toward lower angles along with decreasing peak intensity. It can be attributed to the higher mooney viscosity of EPDM, resulting in a higher shear stress undergone by the matrix which might help the deagglomeration of the clay stocks within the elastomer matrix.

7.7 Spectroscopic Studies of Styrene–Butadiene Rubber Nanoblends

7.7.1 Nuclear Magnetic Resonance (NMR)

The nuclear magnetic resonance (NMR) technique on polymers is generally confined to studying the spin–spin and spin–lattice relaxation processes as a function of temperature and composition. This spectroscopy has been applied effectively to crystalline systems where the spin–spin relaxation time for the protons in the amorphous phase is much greater than that in the crystalline phase.

Arantes et al. [20] prepared the TiO_2/SBR nanoblends and analyzed by NMR using magic angle spinning (MAS) with high-power proton decoupling technique, carried out to observe the mobile region using a short delay between 90 pulses. Figure 7.16 shows the NMR spectra of pure SBR elastomer, SBR/TiO_2 nanoblend, and SBR/TiO_2-photodegraded nanoblend. The obtained spectra are shown in Fig. 7.16 and tabulated in Table 7.4. Despite the fact that the spectrum of the

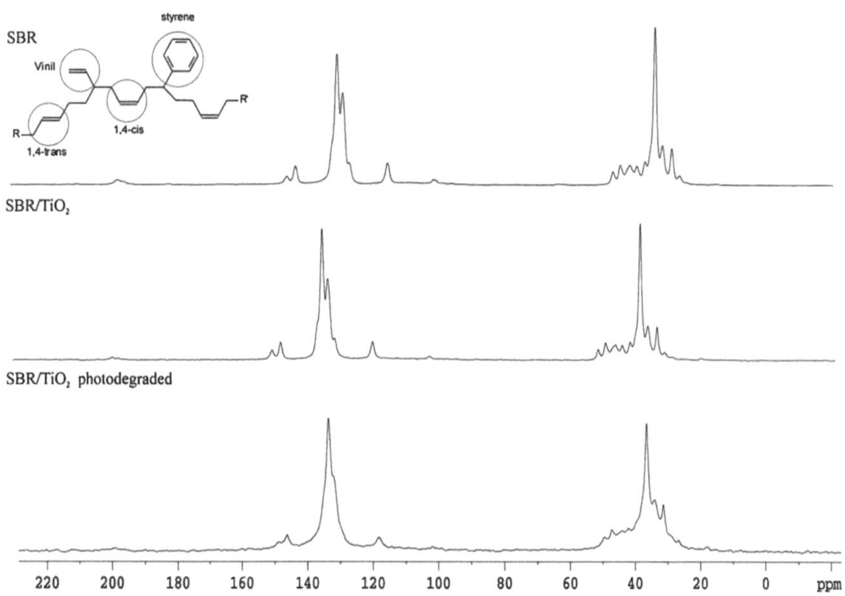

Fig. 7.16 Solid-state ^{13}C NMR MAS spectra: **a** SBR; **b** SBR/TiO_2 (0.2 %) nanoblend, and **c** SBR/TiO_2 (0.2 %)-photodegraded nanoblend [20]. *Source* Elsevier

Table 7.4 ^{13}C NMR chemical shifts (d = ppm) of SBR and SBR/TiO_2 nanoblend [20]. *Source* Elsevier

Structure	SBR (d/ppm)	SBR/TiO_2 (d/ppm)
C_1 aromatic	143.8	148.9
CH=CH vinyl	141.2	146.3
C_2 and C_4 aromatic	130.0	135.1
C_3 and C_3 (S) and CH=CH 1,4 (T)	128.6	133.8
CH=CH 1,4 (C)	126.7	132.1
$=CH_2$ vinyl	113.1	118.2
CH_2 styrene	44.3	49.5
CH_2 styrene	41.9	47.2
Cά styrene	39.0	44.1
CH_2 1,2 vinyl	36.8	42.1
CH_2 1,4 trans	34.5	39.6
CH_2 1,4 cis	32.8	37.8
CH_2 1,2 vinyl	31.4	36.5
CH_2 1,4 cis	28.9	34.0
CH_2 1,2 trans	26.2	31.3
CH_2 1,2 cis	23.8	28.1

non-degraded SBR/TiO_2 blend is identical to the pure SBR elastomer as regards the number, shape, and relative intensities of the peaks, it is possible to observe that the nanoblends spectrum is dislocated 5 ppm to higher chemical shift, possibly due to the interaction of the TiO_2 nanoparticles in between SBR chains. This effect is a good inference of the strong action of nanoparticles on the organization of the chains in nanoblends. It is also possible to note a broadening of NMR signals after the photodegradation and the absence of the characteristic assignments of butadiene, indicating that the butadiene chain was effectively photodegraded. Peaks with chemical shifts in 99.0 and 98.2 appear due to the impurities present in the SBR. The broadening observed in the NMR spectrum indicates that the molecular mobility of the SBR chains is reduced after photodegradation. This is mainly due to the fact that the resonance line width is related to the segmental polymer motion. If the molecule is free to move in any direction, then the spectrum line width is narrow. On the other hand, if the motion is restricted, then the line width increases. In terms of rubber degradation processes, it seems to be consistent with the formation of extra cross-links restricting the motion of molecules, since the chains lose their ability to move as individual polymer chains. Analyzing the pure SBR and the SBR/TiO_2 nanoblend spectra according to the variable contact-time (VCT) experiments (Fig. 7.17), two domains with very different molecular mobilities were encountered. The signal distribution profile shows clearly a large and flexible domain for the pure SBR and SBR/TiO_2 nanoblend, but a rigid domain for the SBR/TiO_2-photodegraded nanoblend. The intensity of the NMR signals during the cross-polarization experiment depends on the rate of cross-polarization

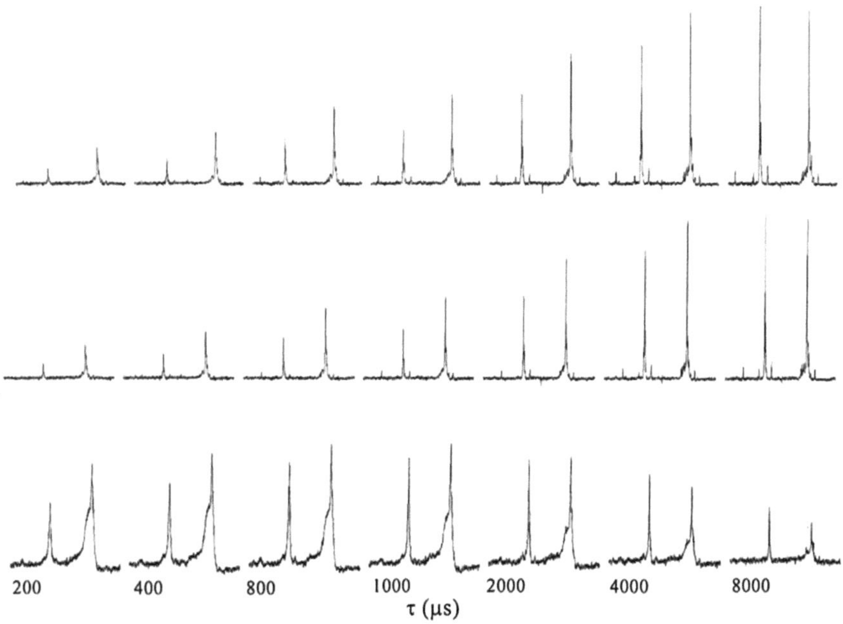

Fig. 7.17 ^{13}C NMR CP/MAS spectra with variable contact-time experiment (VCT): **a** SBR, **b** SBR/TiO_2 nanoblend, and **c** SBR/TiO_2-photodegraded nanoblend [20]. *Source* Elsevier

from the ^{1}H nuclei to near-neighbor ^{13}C nuclei, which makes it possible to determine the proton spin–lattice relaxation time in the rotating frame ($T_1\rho H$). In an atactic and amorphous flexible polymer, such as polybutadiene rubber, the molecular motion reduces the interaction of the dipole moments, and consequently, the cross-polarization is a slow process. Spectra a and b (Fig. 7.17) show peaks located at long contact times, in agreement with rubbery materials. On the other hand, the spectrum c of SBR/TiO_2-photodegraded nanoblend shows a rigid domain in the variable VCT decay profile.

7.7.2 Fourier Transform Infrared Spectroscopy (FTIR)

Over the past few decades, FTIR has proved to be a valuable technique for characterizing polymers. With this technique, the identification of many polymeric materials is easy and quick. From infrared data, it is feasible to determine many qualitative and semiquantitative polymeric features, such as stereochemistry, types of additives, degree of degradation, presence of a copolymer, chain length, orientation, and crystallinity [21].

The FTIR spectra of the NBR/SBR and (NBR/SBR)/ginger powder (GP)/crys-nanoclay (CN) nanoblends are shown in Fig. 7.18. FTIR spectra indicate that the

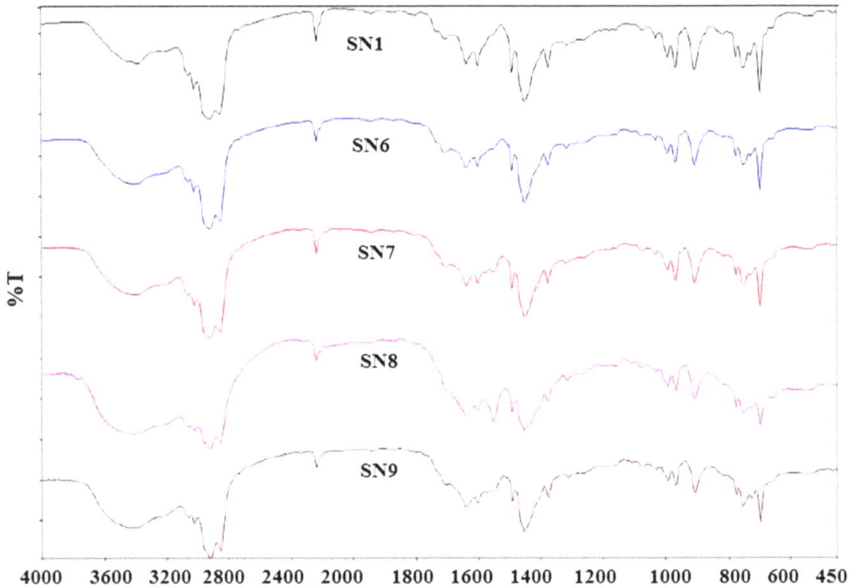

Fig. 7.18 FTIR spectra of NBR/SBR blend and its nanoblend with different wt% of crysnanoclay [22]

peak at 3730–3400 cm^{-1} is the stretching peak of –OH bond of stearic acid, but it has become broader after incorporating GP and CN due to overlapping of the –OH groups of lignin, cellulose, and Si–OH groups. The broad and strong band ranging from 3000 to 3600 cm^{-1} indicates the overlapping of Si–OH (silanol) and R–OH (hydroxyl) groups of GP-stretching vibrations. The peak visible at 2913 cm^{-1} is due to the symmetric stretching of the C–H band. The peaks observed at 2922 and 2851 cm^{-1} can be assigned to asymmetric and symmetric –CH_2 groups, respectively. The peak at 2325 cm^{-1} is due to the C–H band of SBR and at 2260–2240 cm^{-1} is due to nitrile linkages of the blend. The peak at 1600 cm^{-1} is due to the C=C band of butadiene in the system. The peak at 1636 cm^{-1} corresponds to the C=C stretching attributed to the aromatic group, and the peak at 1021 cm^{-1} is due to the presence of styrene in SBR. There is no peak at 1730 cm^{-1} (>C=O stretching peak), which indicated that the oxidation of main polymeric chain did not occurred at the time of rubber milling with the help of cracker-cum-mixing mill at high temperature. The wave numbers from 1400 to 1600 cm^{-1} are assigned to aromatic skeletal vibration [22].

Figure 7.19 shows the FTIR spectra of unmodified SBR particle and various photografting degrees of PP-g-SBR particles obtained by extracted PP-g-SBR systems by boiling xylene. The intensities of the two wide absorption bands at about 2926 and 2855 cm^{-1} increase with the increasing UV irradiation time, which have been assigned to the antisymmetry and the symmetry-stretching vibration of –CH_2-group, respectively. Similarly, the intensities of two absorption bands at 1450

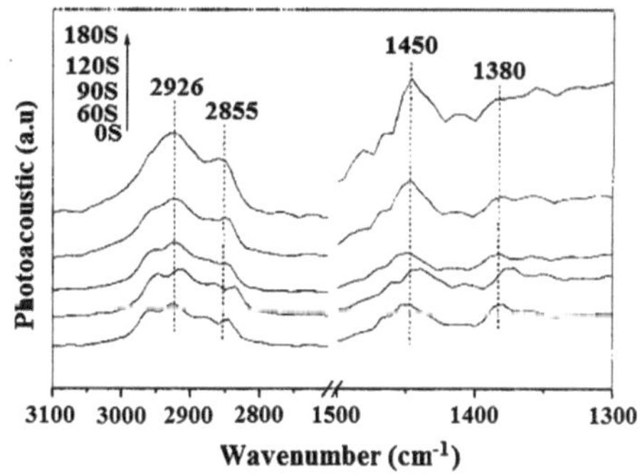

Fig. 7.19 FTIR spectra of unmodified SBR nanoparticles and photografted PP-g-SB nanoparticles for different UV irradiation times [23]. *Source* John Wiley and Sons

and 1380 cm^{-1} increase with increasing the irradiation time due to the antisymmetry and the symmetry bending vibration of $–CH_3$ group, respectively. These results indicate that the $–CH_2$- and $–CH_3$ groups in the UV-irradiated samples increase with increasing irradiation time due to PP molecular chains grafted onto the SBR nanoparticles [23].

7.8 Rheological Studies of Styrene–Butadiene Rubber Nanoblends

Rheology is the study of flow and deformation of materials under applied forces which are routinely measured using a rheometer of all materials from fluids such as dilute solutions of polymers. Rheological properties can be measured from bulk sample deformation using a mechanical rheometer or on a microscale by using a microcapillary viscometer or an optical technique such as microrheology. The viscosity and viscoelasticity can vary depending upon the external conditions applied, such as stress, strain, timescale, and temperature. The type of rheometer required for measuring these properties is often dependent on the relevant shear rates and timescales as well as sample size and viscosity [24]. The rheological behaviors of polymer nanoblends have been studied by several researchers, because it is essential for the polymer processing and the end product designing. The dynamic oscillatory and steady-shear rheology studies of polymer nanoblends revealed the enhanced viscosity values when compared to pristine and can be attributable in terms of the formation of an intercalated or exfoliated network structure by clay platelet. The flow properties of sodium bentonite- and sodium fluorohectorite-filled NR, XSBR, and 70/30 NR/XSBR lattices have analyzed by Stephen et al. [25] with reference to shear rate, filler loading, and temperature. Figure 7.20a–f shows the shear rate versus viscosity curves of sodium bentonite and sodium fluorohectorite latices and their blend at various filler loadings. In all the systems, the viscosity of layered silicate-reinforced latex samples decreases with the increasing shear rate, representing pseudoplastic nature, i.e., shear-thinning behavior. The change in networks of latex nanoblends with shear rate is shown

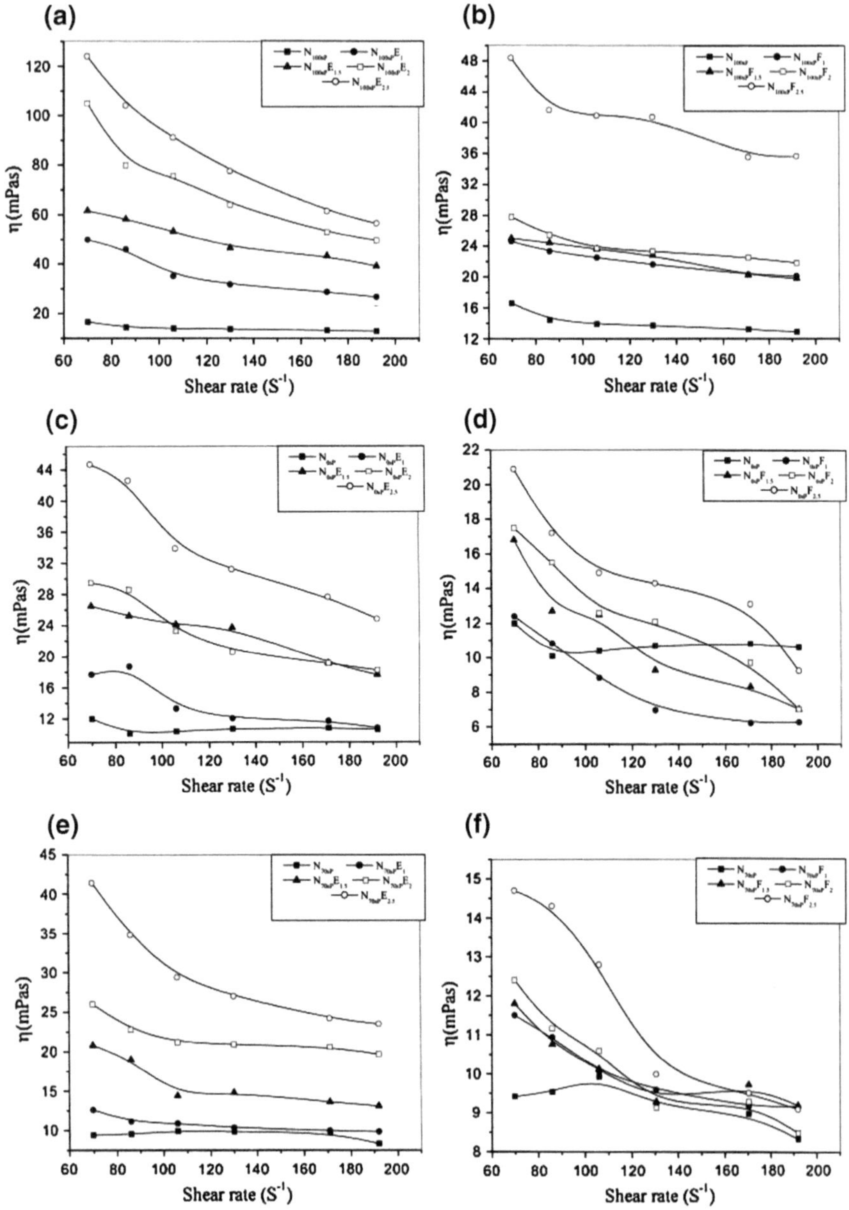

Fig. 7.20 Effect of shear rate of the viscosity of **a** NR+ sodium bentonite (E), **b** NR+ sodium fluorohectorite (F), **c** XSBR + E, **d** XSBR, **e** 70/30 NR/XSBR + E, and **f** 70/30 NR/XSBR + F [25]. *Source* John Wiley and Sons

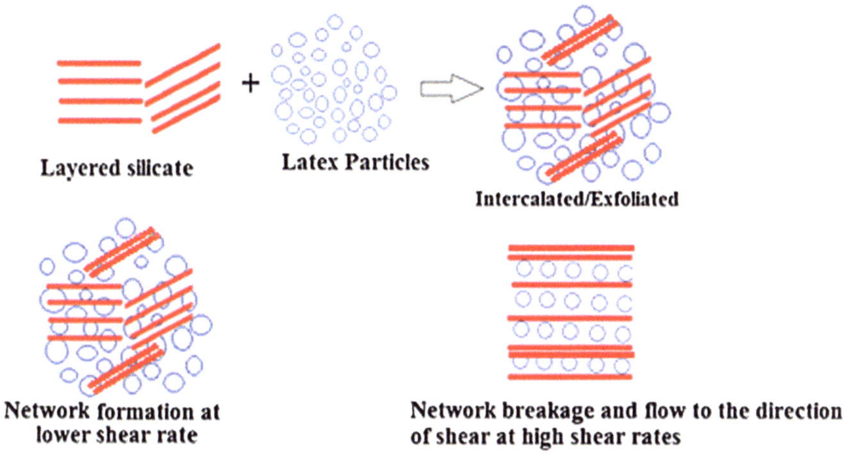

Fig. 7.21 Schematic representation of the influence of shear rate on network formation and breakage

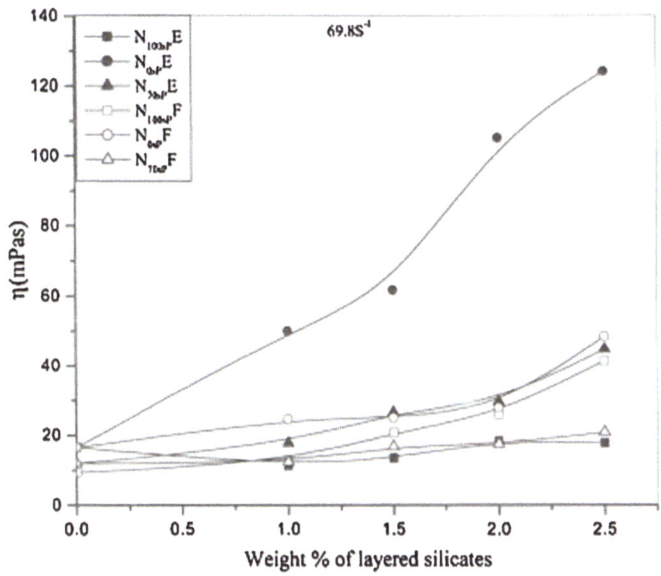

Fig. 7.22 Variation in viscosity with concentration of layered silicates in NR, XSBR, and 70/30 NR/XSBR [25]. *Source* John Wiley and Sons

schematically in Fig. 7.21. On contrary, Fig. 7.22 shows that the viscosity of latices increases when the function of weight percentage of filler increases. The increase in viscosity is owing to the reinforcement of layered silicates and their interaction with rubbers. The extent of reinforcement of layered silicates can be analyzed using

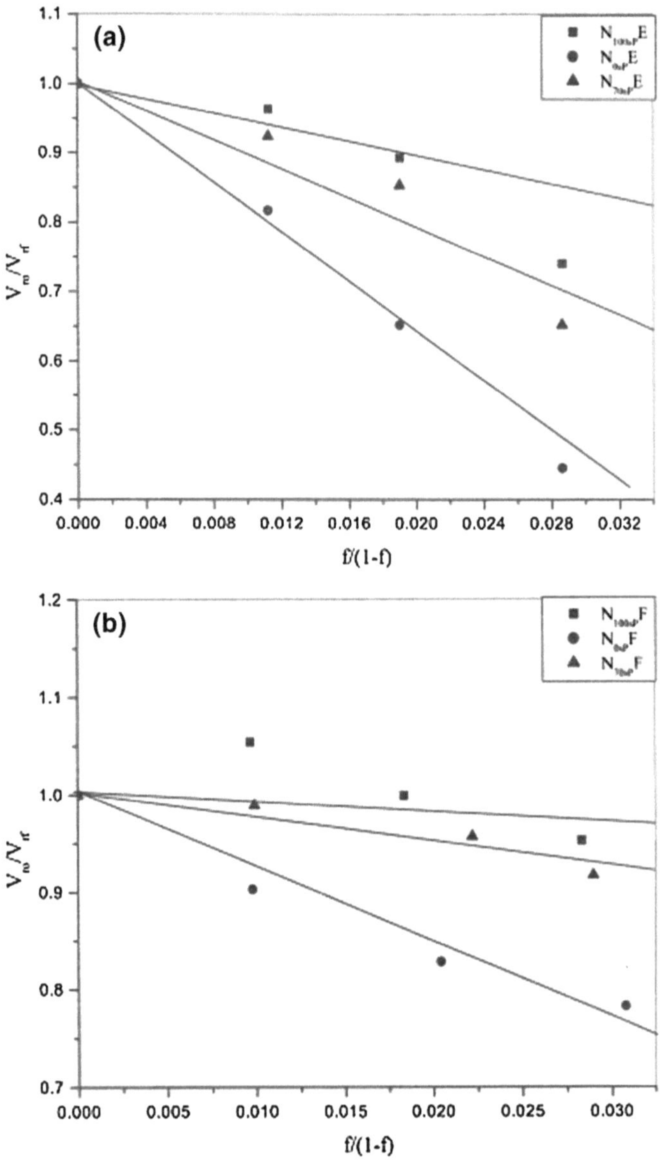

Fig. 7.23 Kraus plots of nanofiller-reinforced latex nanocomposites **a** sodium bentonite and **b** sodium fluorohectorite [25]. *Source* John Wiley and Sons

Kraus equation. Kraus plots of sodium bentonite- and fluorohectorite-filled latex nanoblends are shown in Fig. 7.23a, b. The negative slope values obtained for Kraus plots are pointing to the reinforcement of filler in the rubber matrix. Because of the breakage of networks, the viscosity decreases with increase in temperature. It is found that the zero shear viscosity of filled samples is higher than virgin

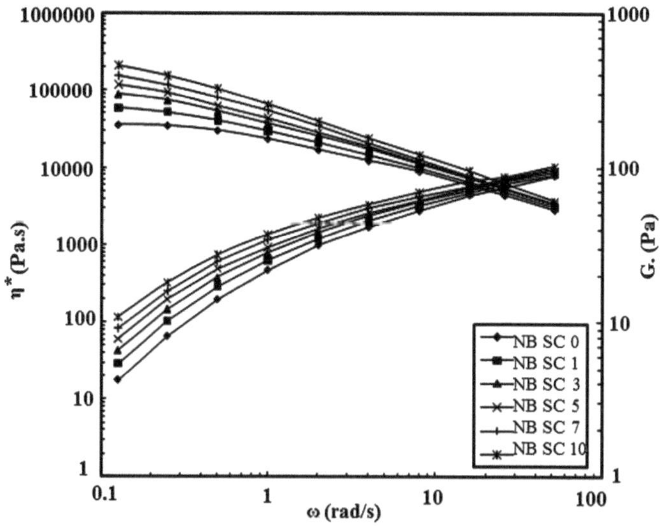

Fig. 7.24 Complex viscosity and storage modulus of NR/BR/SBR (65/20/15) nanoblends prepared with different nanoclay contents [26]. *Source* John Wiley and Sons

polymers. This can be explained in terms of the enhancement in polymer/filler interaction because of high aspect ratio of layered silicates.

Naderi et al. [26] also studied the rheological properties using oscillatory rheometer at 80 °C with parallel plate geometry of NR/BR/SBR (65/20/15) nanoblends prepared with different clay contents (Fig. 7.24). As expected, the complex viscosity of the nanoblends is higher compared to the unfilled matrix, indicating the formation of a network structures and also good matrix–clay interactions, resulting in a rise in the complex viscosity. The falling trend of viscosity with frequency is observed in all the clay loadings, representing the pseudoplastic nature and shear-thinning behavior of the compounds. The remarkable rise in complex viscosity at low frequencies could be compared to the materials exhibiting a yield stress, indicating that a yield stress may be present for nanoblends compared to the pure matrix. The elastic modulus of nanoblends also increases with clay loading which would be ascribed to the good interactions established between the matrix components and organoclay functional groups.

7.9 Viscoelastic Properties of Styrene–Butadiene Rubber Nanoblends

The viscoelastic properties such as storage modulus, loss modulus, and loss tangent of SBR depend on the structure, crystallinity, and extent of cross-linking. To determine the viscoelastic properties of SBR nanoblends, the dynamic mechanical

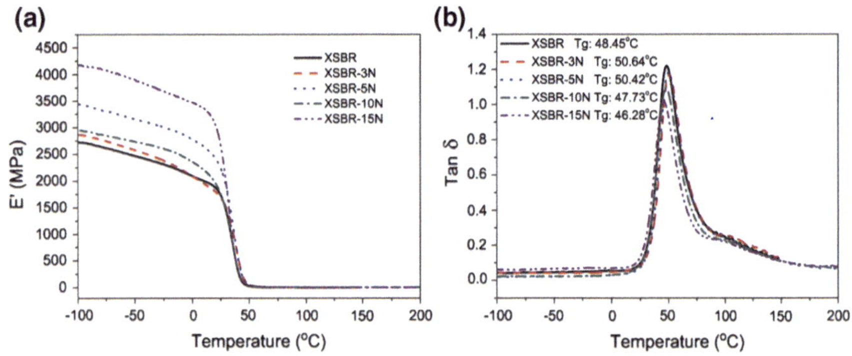

Fig. 7.25 $E'-T$ and tan δ–T curves of NR/BR/SBR blends reinforced by NCC and CB [13]. *Source* Elsevier

Table 7.5 Data of the E'–T and tan δ–T curves of NCC-loaded NR/BR/SBR composites [27]

NCC/CB/RH (T_g)	NCC/CB/RH (T_g)	NCC/CB/RH (T_g)	NCC/CB/RH (T_g)	NCC/CB/RH (T_g)
0/45/0	−55.2	6.867	0.1942	0.1151
10/35/0	−55.8	7.751	0.1906	0.1144
10/35/4	−53.9	9.023	0.1761	0.0887

analyzer (DMA) is used in tension mode. The temperature program was run from −100 to +150 °C using a heating ramp of 5 °C min−1 at a fixed frequency of 10 Hz. The samples were prepared as a cut strip with the size of 8 mm × 1 mm × 6 mm (length × thickness × width). Dependency of storage modulus (E') as a function of temperature of XSBR/CNs nanoblends was shown in Fig. 7.25. All nanoblends show a sharp decrease of E' value at the temperature range between 25 and 45 °C, along with a rubbery plateau (Fig. 7.25a). As the amount of CNs increased, a successive increment of E' values were observed (e.g. the storage modulus at −100 °C increased from ~2700 MPa to 4200 MPa with the addition of CNs from 0 to 15 phr in the XSBR matrix). The improvement in modulus below glass transition temperature (T_g) is an excellent confirmation for the strong reinforcing tendency of CNs in the XSBR matrix. The compatibility between CNs and XSBR, as well as sufficiently strong filler–filler interactions was playing a major role in XSBR matrix. Fig. 7.25b shows the variation of tanδ as a function of temperature for composites with various concentrations of CNs. The tanδ is the ratio of dissipated energies and storage energies. It is the value to characterize internal friction. The higher the tanδ, the internal system will suffer more friction and the more heat is produced. XSBR/CNs nanoblends exhibit a maximum tanδ value at a temperature around 50 °C, which can be altered by changing the CNs amount. Generally, T_g of polymer could be shifted to higher temperature after incorporation of nano-fillers and it could be possible due to the segmental motion of polymer chain segments hinder by nano-fillers. As shown in Fig. 7.25b, T_g was shifted from 48.45 °C to 50.64 °C with the addition of only 3 phr CNs in the XSBR matrix. On the other hand, T_g was decreased from 50.64 to 46.28 °C with increasing CNs amount to 15 phr. It is well studied that the polar atraction can improve T_g of polymer matrix and therefore the carboxyl groups in XSBR matrix can easily form strong hydrogen bonds with the hydroxyl groups on the CNs surface. This may act as an internal plasticization which results in the shifting of T_g to a lower temperature with increasing CNs. However, further investigation is still needed to study the essential reason for this particular phenomenon. No significant broadening of the tanδ peak was observed in Fig. 7.25b. Finally, the intensity of the relaxation process decreased with increasing amount of CNs, and it is attributed that the decrease of matrix material amount is responsible for particular damping properties [13].

Chen et al. [27] also prepared a series of CNs reinforced natural rubber/butadiene rubber/styrene-butadiene rubber (NR/BR/SBR) blends via mastication of NR/CNs, BR/SBR and other ingredients. According to the DMA analysis, CNs achieved comparable reinforcing effect with carbon black (CB), and the modulus was improved with modification of hexamethylenetetramine (RH). It was observed that there were obvious reductions in storage modulus at the temperature ranging from −80 to −45 °C, followed by a rubbery plateau and the tanδ reached the maximum at about −55 °C, illustrating the glass transition (T_g) of nanoblends. Comparison of different nanoblends revealed that the replacement of CB by CNs did not impair the reinforcing effect, which was reflected by the slightly increased value of storage modulus, and RH modification further facilitated this tendency. The reinforcement of CNs was due to the high aspect ratio and modulus. So a certain amount of CNs could perform comparable reinforcing effect with CB. Additionally, RH enhanced the CNs network by hydrogen bonding and linked this network to rubber molecules. Besides, the contribution of RH to rubber cross-link also played a major role. According to the characteristics of tanδ at the temperature above 0 °C decreased slightly upon the replacement of CB by CNs. The existence of RH also decreased the tanδ as the RH facilitated the cross-linking of rubber. This trend suggested a low rolling resistance and probably reduced heat buildup.

7.10 Electrical Properties of Styrene–Butadiene Rubber Nanoblends

Recently Le et al. [28] reported the electrical properties of CNT-filled SBR/NR and SBR/polyisoprene (IR)-based nanoblends and their distribution. For the preparation of CNT blends, an internal mixer was used. Peroxide was used as vulcanizing agent for all the mixtures. For an effective dispersion of CNTs in rubber matrix, an ethanol-assisted mixing process (wet mixing) was applied. In order to determine the suitable amount of ethanol, the online conductance was recorded by adding

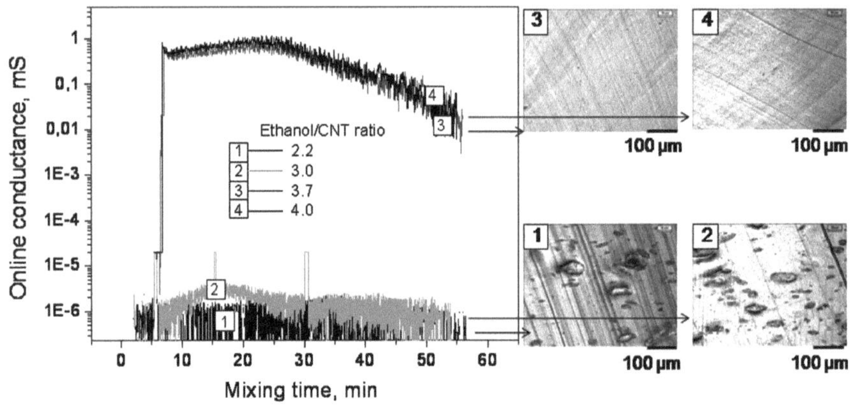

Fig. 7.26 Online conductance measured during the mixing process of CNT-filled SBR/IR blends with different ethanol/CNT ratios and images made by optical microscopy for different blends [28]. *Source* John Wiley and Sons

different ethanol/CNT ratios into a 75/25 SBR/IR blend. A conductivity sensor system was installed in the chamber of the internal mixer to measure the electrical signal of the conductive mixtures. Up to an ethanol/CNT ratio of 3.0, the conductivity signal is still very weak as shown in Fig. 7.26. At a ratio of 3.7, the online conductance curve was recorded at a high level due to the better dispersion of CNTs in blend system. On the images made by optical microscopy, some large CNT agglomerates are seen in blends with ethanol/CNT ratio of 2.2 and 3.0. They disappear when the ratio increases to 3.7 and 4.0.

In order to further determine the online conductance of CNT-filled SBR/NR (S/N) and SBR/IR (S/I), the online conductance curves of S/N-1 blend containing 5 phr CNTs and S/N-3 containing 4 and 3 phr CNTs, respectively, are shown in Fig. 7.27. With increasing CNT loading, the conductance curves shift to a higher level. A strong increase in conductance of several orders of magnitude is observed when CNT loading increases from 3 to 4 phr that is relating to the percolation threshold of CNTs in SBR/NR blends. All the online conductance curves show typical conductance–time characteristics with t_{onset} and t_{Gmax}. At t_{onset}, the online conductance starts to rise and reaches the maximum value at t_{Gmax}. The macrodispersion of filler and the online conductance correlate closely to each other. The largest change in the size of filler agglomerates, i.e., the dispersion of large filler agglomerate into smaller aggregates or even individual tubes, is determined in the period between t_{onset} and t_{Gmax}. Upon t_{Gmax}, the online conductance decreases slightly that is related to the better distribution of small aggregates throughout the matrix and shortening of tubes. The online conductance of S/I blend containing 5 phr CNTs presents t_{onset} at 7 min and t_{Gmax} at 25 min, while S/N-1 shows t_{onset} at 15 min and t_{Gmax} at 30 min. The slower development of online conductance indicates a slower dispersion of CNTs in S/N-1 compared to S/I. Upon t_{Gmax}, both blends show nearly the same decrease in conductance. The dispersion of S/I is faster

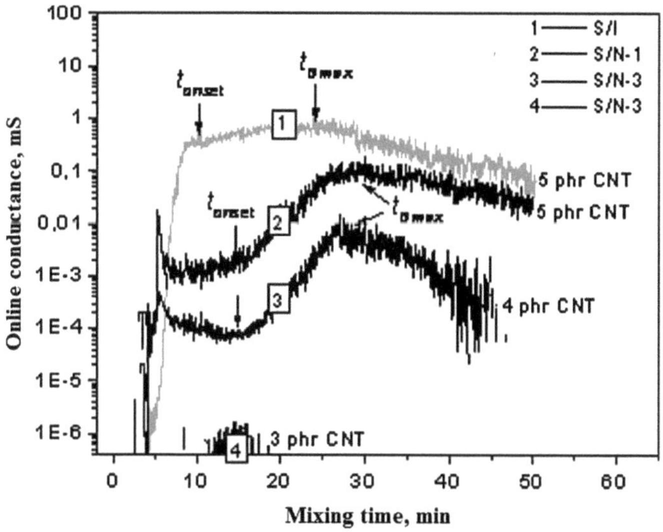

Fig. 7.27 Online conductance measured during the mixing process of CNT-filled SBR/IR blends with different ethanol/CNT ratios [28]. *Source* John Wiley and Sons

than that of S/N-1 that is corresponding well to the development of the online conductance. As shown in Fig. 7.27, at 50-min mixing time, both S/N-1 and S/I blend exhibit the similar values of the online conductance and macrodispersion.

In order to discuss the electrical properties of CNT-filled SBR nanoblends, Le et al. [29] further investigated the electrical properties of SBR, NBR, and NR filled with CNT ternary blends without and with ethanol. The method is used to determine the electrical properties of blends including online measured electrical conductance and wetting concept. Figure 7.28 a shows the online conductance curves of different rubbers containing 5 phr CNTs without and with ethanol recorded during the mixing. It is observed that in the absence of ethanol, SBR and NR show no electrical signal (curves 1 and 3), indicating that there is no percolated conduction filler network in rubbers due to bad filler dispersion. But in case of CNT-filled NBR, an online conductance curve was recorded (curve 2), which shows a typical conductance–time characteristics with $t_{onset} = 17$ min and $t_{Gmax} = 30$ min. At t_{onset}, the online conductance starts to rise and reaches the maximum value at t_{Gmax}. According to the results, macrodispersion of filler and the online conductance correlate closely with each other. The largest change in the size of filler agglomerates, i.e., the dispersion of large filler agglomerate into smaller aggregates or even individual tubes, is determined in the period between t_{onset} and t_{Gmax}. Upon t_{Gmax}, the online conductance decreases slightly that is related to the shortening of tubes and the better distribution of small aggregates throughout the matrix, which is correlated with the increasing aggregate–aggregate distance. In contrary, when CNTs/ethanol mixture is added to SBR and NR, a significant increase in the conductance of several orders of magnitude is observed (curves 4

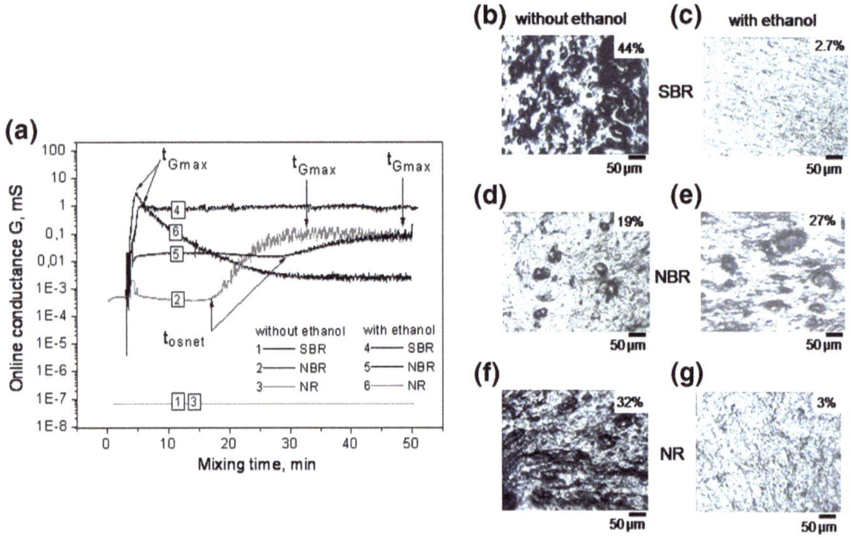

Fig. 7.28 Online conductance curves of CNT-filled rubber compounds in dependence on mixing time (**a**) and optical microscopic images of samples taken out at 30 min mixing (**b–g**) [29]. *Source* Elsevier

and 6). It is worthy to note that the presence of ethanol in ethanol/rubber mixture without any CNTs may show some conductivity signals, but they were not detectable in the measuring range of equipment. For SBR and NR compound filled with ethanol/CNTs paste, a t_{Gmax} recorded at 5 min and 6 min, respectively, is related to a fast dispersion of CNTs.

Optical microscopic images of samples taken out at 30-min mixing confirm the improvement of CNT dispersion in SBR and NR with assistance of ethanol. Without ethanol, SBR and NR show several large CNT agglomerates with a macrodispersion A/A_0 of 44 and 32 %, respectively (Fig. 7.28b, f). By the addition of ethanol, CNT dispersion is improved markedly with a macrodispersion of 2.7 and 2 %, respectively (Fig. 7.28c, g). In contrast, ethanol negatively affects the dispersion of CNTs in NBR. The online conductance curve shifts to longer mixing time with t_{Gmax} of about 50 min (Fig. 7.28a, curve 5), while the macrodispersion A/A_0 increases from 19 to 27 % according to Fig. 7.28d, e. The reason for the negative effect of ethanol in NBR may be related to the bad compatibility between ethanol and NBR. With a surface tension of 22 mN/m, ethanol shows a good compatibility with the nonpolar SBR and isoprene backbone of NR; however, it is not compatible with the polar NBR. Upon ethanol evaporation during the mixing, CNTs will be dispersed well in NBR. That is why the t_{onset} and t_{Gmax} of ethanol/CNT-filled NBR shift to the longer mixing time as shown in Fig. 7.28.

In Fig. 7.29a, the statistic CNTs–SBR/NBR/NR blend and blends from masterbatch mixing shows online conductance curves with short t_{Gmax} ranging from 5 to 7 min. When CNTs were statistically mixed into SBR/NBR/NR blend, a

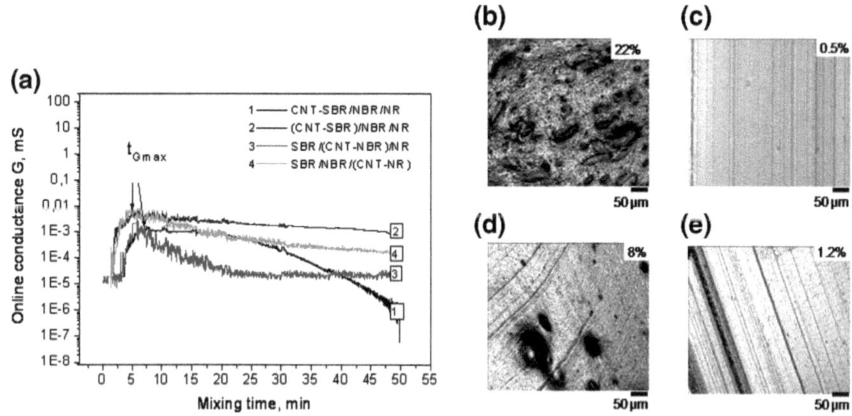

Fig. 7.29 Online conductance curves of CNT-filled blends in dependence on mixing time (**a**), optical microscopic images of the sample CNT–SBR/NBR/NR (**b**), (CNT–SBR)/NBR/NR (**c**), SBR/(CNT–NBR)/NR (**d**), and SBR/NBR/(CNT–NR) (**e**) taken out at 50 min mixing [29]. *Source* Elsevier

macrodispersion A/A_0 of 22 % is determined for the sample taken out at 50-min mixing as shown in Fig. 7.29b. It is much worse compared to that of blends from masterbatch mixing. The (CNT–SBR)/NBR/NR and SBR/NBR/(CNT–NR) show a very good filler dispersion of about 1 % (Fig. 7.29c, e), because the dispersion of CNTs in SBR and NR upon masterbatch preparation in the presence of ethanol is already good. In contrast, the bad dispersion of CNTs in NBR masterbatch results in a bad macrodispersion in SBR/(CNT–NBR)/NR blend as shown in Fig. 7.29d. The shape and level of the online conductance curves are not only dependent on the filler dispersion degree but also on the phase-specific filler distribution.

7.11 Recent Advances on Styrene–Butadiene Rubber Nanoblends

In recent years, inorganic nanomaterials have been mostly used in SBR-based nanoblends to increase the mechanical properties. But still infancy of their different applications and synthetic nanomaterials such as CNTs, graphene, and its derivatives filled SBR nanoblends have shown immense potential applications in the field of electronics, aerospace, automobile, defense industries, green energy, etc. CNTs and graphene have higher surface-to-volume ratio and make potentially more favorable for improving the properties of SBR matrices [30]. Schopp et al. [31] studied the effect of functionalized graphene (GP) on the mechanical properties, electrical conductivity, and gas barrier resistance with simultaneously improved properties. Das et al. [32] and Lorenz et al. [33] both studied the effect of CNTs and modified CNTs on physical properties such as stress–strain behavior, dynamic

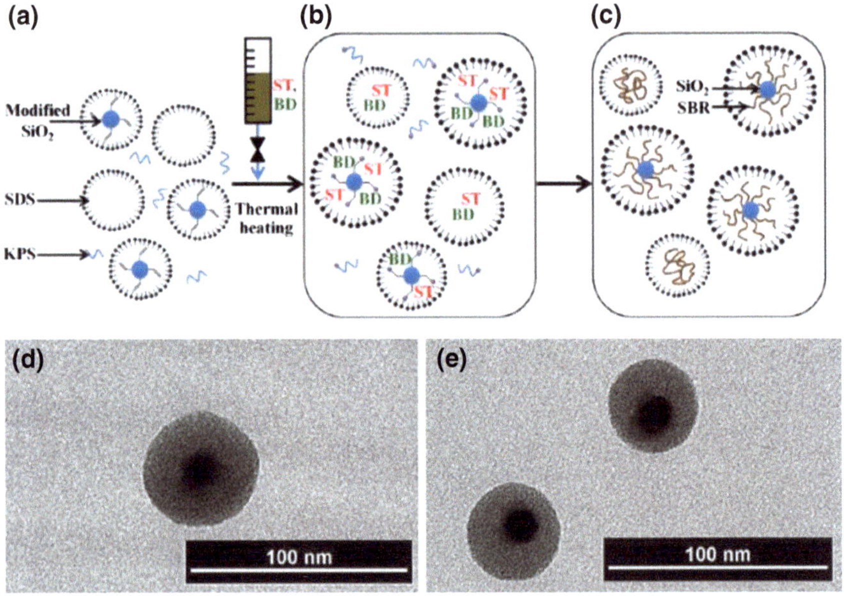

Fig. 7.30 Formation mechanism of SBR–SiO_2 core–shell nanoparticles; **a** modified silica, SDS, and KPS in water, **b** initiation stage, **c** propagation stage; TEM images of SBR–SiO_2 nanoparticles of **d** SBR–SiO_2 (10 wt% SDS) and **e** SBR/untreated SiO_2 (3 wt% SDS) at 10 wt% silica loading. Condition: SiO_2 = 10 wt%, M/H_2O = 0.2, and KPS = 2 wt% based on monomer [34]. *Source* Elsevier

mechanical, thermal diffusivity, dielectric, and fracture mechanical of SBR/butadiene (BR) blend.

Tancharernrat et al. [34] prepared the styrene–butadiene copolymer (SBR)–SiO_2 nanoparticles with core–shell morphology via differential microemulsion polymerization and studied the effect of encapsulated SiO_2 on the mechanical properties of the NR/(SBR–SiO_2) nanoblend. The formation mechanism of SBR–SiO_2 core–shell nanoparticles is illustrated in Fig. 7.30. The modified SiO_2, SDS surfactant, and KPS initiator were dispersed in deionized water to form a homogeneous solution (Fig. 7.30a). The SDS surfactant creates micelles in the system, where the hydrophobic tails form the core of organic phase and the hydrophilic heads turn toward the aqueous phase, and could stabilize the colloidal particles. For differential microemulsion polymerization, small amounts of ST and BD monomers diffused through the water to the micelle and the KPS initiator decomposed into the aqueous phase generates free radicals that produce reactive monomer radicals on the monomer molecule and silica surface for the initiation stage (Fig. 7.30b). The monomer radicals were reacted with other monomer molecules to produce the oligomeric radicals to produce growing chains in the propagation stage (Fig. 7.30c). Hence, the styrene–butadiene copolymer could graft onto the silica surface via encapsulation with a SiO_2 core and SBR shell. Figure 7.30d, e, represents the TEM images of SBR–SiO_2 core–shell nanoparticles at 10 and 3 wt% SDS concentration, respectively.

Mohamed et al. [35] studied the effect of organomontmorillonite clay (modified by hexadecyltrimethyl ammonium bromide) on the mechanical properties of acrylonitrile butadiene rubber (NBR)/SBR blend nanocomposites. The influence of gamma irradiation on the morphology and properties of the blend nanocomposites was investigated. Intercalated polar or unsaturated matrices (e.g., NBR and SBR)/ OMMT blend nanocomposites can be obtained. The clay layers are uniformly dispersed in the rubber matrix on the nanometer level. Mechanical tests confirmed that the nanocomposites had good mechanical properties when compared to the neat blends. The results also proved that the irradiated NBR/OMMT nanocomposites had higher thermal stabilities than irradiated SBR/OMMT nanocomposites.

7.12 Industrial Values and Applications

SBR-based nanoblends have high level of industrial values and application due to a high degree of cross-linking, and the particle morphology is stable even under high shear. Owing to a high cross-linking density, the particles will not dissolve, but will swell in appropriate solvents. The applications of nanomaterials are particularly beneficial in SBR compounds, which are particularly used for the production of tire treads and different tire parts. In SBR compounds, nanomaterials are used as a partial replacement for macrofillers. The major benefit of nanomaterials is in the improvement of dispersion and the reduction in filler–filler interactions. As a consequence, hysteresis is significantly reduced. For hysteresis reduction (reduction in rolling resistance), the use of nanomaterials in tire treads results in a reduction of the storage modulus in the temperature range −50 to 0 °C. As a consequence, the grip on ice and snow is remarkably improved. SBR blend with nanomaterials $T_g = -10$ °C is best used to balance rolling resistance and wet skid performance of a tire tread. In addition, the use of nanomaterials in combination with a functionalized solution SBR can result in positive synergisms. Probably as a consequence of improved dispersion, the use of nanomaterials results in the improvement of abrasion resistance. This positive effect does not necessarily show in laboratory tests but is consistently found in commercial tire tests.

7.13 Conclusions

In the SBR blends, various nanomaterials of different shapes and sizes play an important role to make the SBR blends compatible. From the review literature of recent work on SBR-based nanoblends, it is found that the increment in tensile, thermal, rheological, viscoelastic, and electrical properties of nanoblends improved considerably with an increasing % of nanomaterials loading. In the part of tensile strength, the reinforcing effect of nanofillers improves the tensile strength at lower loading, whereas the elongation at break deteriorated but still largely retained. In addition, the incorporation of nanofillers into the SBR-based blends showed improvement in thermal stability when compared to virgin blends. The rheological behavior of SBR nanoblends in the presence of nanofillers was analyzed with shear rate, filler loading, and temperature. The viscosity of the nanoblends with nanofillers increased and also exhibited pronounced shear-thinning behavior. Enhancement in viscosity upon the addition of nanofillers indicated a more uniform distribution of particles and higher extent of exfoliation on nanoscale. In the electrical properties of CNTs and graphene-filled SBR nanoblends, it was found that the distribution of CNTs and graphene in the presence of ethanol after long mixing time shows good electrical properties. From X-ray diffraction and morphological analysis by SEM, it was confirmed that at lower wt%, loading of nanomaterials goes to exfoliated structures and increases the properties of SBR nanoblends.

References

1. Shanks RA, Kong I (2013) General purpose elastomers: structure, chemistry, physics and performance. In: Advances in elastomers I. Advanced structured materials vol 11. Springer, Berlin, Heidelberg. doi:10.1007/978-3-642-20925-3_2
2. Deepalekshmi R, Visakh PM, Mathew AP, Chandra AK, Thomas S (2013) Advances in elastomers I. Advanced structured materials, vol 11. Springer, Berlin, Heidelberg. doi:10.1007/978-3-642-20925-3_1
3. Varghese S, Karger-Kocsis J (2005) Layered silicate/rubber nanocomposites via latex and solution intercalations, 77–90. In: Friedrich K, Fakirov S, Zhang Z (eds) Polymer composites from Nano- to macro-scale 1–328. doi:10.1007/b137162
4. Apostolo M, Triulzi F (2004) Properties of fluoroelastomer/semicrystalline perfluoropolymer nano-blends. J Fluor Chem 125:303–314
5. Hu GH, Cartier H (1999) Reactive extrusion: toward nano blends. Macromolecules 32:4713–4718
6. Noguchi F, Akabori K, Yamamoto Y, Kawazura T, Kawaharal S (2009) Tear energy and strain-induced crystallization of natural rubber/styrene-butadiene rubber blend. J Phys Conf Ser 184:1742–6596
7. Bendjaouahdou C, Bensaad S (2013) The effects of organoclay on the morphology and balance properties of an immiscible polypropylene/natural rubber blend. Energy Procedia 36:574–590
8. Visakh PM, Mathew AP, Thomas S (2013) Natural polymers: their blends, composites and nanocomposites: state of art, new challenges and opportunities. In: Advances in natural polymers. Advanced structured materials, vol 18. Springer, Berlin, Heidelberg. doi:10.1007/978-3-642-20940-6_1
9. Mitra S, Chattopadhyay S, Bhowmick AK (2010) Preparation and characterization of elastomer-based nanocomposite gels using an unique latex blending technique. J Appl Polym Sci 118:81–90
10. Le HH, Sriharish MN, Henning S, Klehm J, Menzel M, Frank W, Wießner S, Das A, Stöckelhuber K-W, Heinrich G, Radusch HJ (2014) Dispersion and distribution of carbon nanotubes in ternary rubber blends. Compos Sci Technol 90:180–186

11. Rahmatpour A, Abdollahi M, Shojaee M (2008) Structure and mechanical properties of 50/50 NR/SBR blend/pristine clay nanocomposites. J Macromol Sci Part B Phys 47(3):523–531
12. Ahmed JK, Al-maamori MH, Ali HM (2015) Effect of nano silica on the mechanical properties of styrene-butadiene rubber (SBR) composite. Int J Mater Sci Appl 4(2-1):15–20
13. Cao X, Xu C, Liu Y (2013) Yukun ChenPreparation and properties of carboxylated styrene-butadiene rubber/cellulose nanocrystals composites. Carbo Polym 92:69–76
14. Hosseini SM, Madaeni SS, Khodabakhshi AR (2010) Preparation and characterization of PC/SBR heterogeneous cation exchange membrane filled with carbon nano-tubes. J Membr Sci 362:550–559
15. Prasertsri S, Lagarde F, Rattanasom N, Sirisinha C, Daniel P (2013) Raman spectroscopy and thermal analysis of gum and silica-filled NR/SBR blends prepared from latex system. Polym Testing 32:852–861
16. Pal K, Rajasekar R, Kang DJ, Zhang ZX, Pal SK, Das CK, Kim JK (2010) Effect of fillers on natural rubber/high styrene rubber blends with nano silica: morphology and wear. Mater Des 31:677–686
17. Stephen R, Ranganathaiah C, Varghese S, Joseph K, Thomas S (2006) Gas transport through nano and micro composites of natural rubber (NR) and their blends with carboxylated styrene butadiene rubber (XSBR) latex membranes. Polymer 47:858–870
18. Wu Y-P, Ma Y, Wang YQ, Zhang LQ (2004) Effects of characteristics of rubber, mixing and vulcanization on the structure and properties of rubber/clay nanocomposites by melt blending. Macromol Mater Eng 289:890–894
19. Durandish M, Alipourb A (2013) Investigation into morphology, microstructure and properties of SBR/EPDM/ORGANO montmorillonite nanocomposites. Chin J Polym Sci 31(4):660–669
20. Arantes TM, Leao KV, Ines M, Tavares B, Ferreira AG, Longo E, Camargo ER (2009) NMR study of styrene-butadiene rubber (SBR) and TiO_2 nanocomposites. Polym Testing 28:490–494
21. Tancharernrat T, Rempel GL, Prasassarakich P (2014) Preparation of styrene butadiene copolymer–silica nanocomposites via differential microemulsion polymerization and NR/SBR–SiO_2 membranes for pervaporation of water–ethanol mixtures. Chem Eng J 258:290–300
22. Pashaeil S, Sadaghini AS, Siddaramaiah, Reddy GR (2014) Studies on ginger powder and crysnanoclay filled acrylonitrile-butadiene rubber/styrene butadiene rubber (NBR/SBR) green composites. Int J Chem Tech Res 6(4):2598–2610
23. Wang W, Fu M, Qu B (2004) Mechanical properties and structural characteristics of PP/PP-g-SBR nanocomposites prepared by dynamical photografting. Polym Adv Technol 15:467–471
24. Drazkowski DB, Lee A, Haddad TS, Cookson DJ (2006) Chemical substituent effects on morphological transitions in styrene-butadiene-styrene triblock copolymer grafted with polyhedral oligomeric silsesquioxanes. Macromolecules 39:1854–1863
25. Stephen R, Alex R, Cherian T, Varghese S, Joseph K, Thomas S (2006) Rheological behavior of nanocomposites of natural rubber and carboxylated styrene butadiene rubber latices and their blends. J Appl Polym Sci 101:2355–2362
26. Zarei M, Naderi G, Bakhshandeh GR, Shokoohi S (2013) Ternary elastomer nanocomposites based on NR/BR/SBR: effect of nanoclay composition. Appl Polym Sci. doi:10.1002/APP.37687
27. Chen WJ, Gu J, Xu SH (2014) Exploring nanocrystalline cellulose as a green alternative of carbon black in natural rubber/butadiene rubber/styrene-butadiene rubber blends. Expr Polym Lett 8(9):659–668
28. Le HH, Parsekar M, Ilisch S, Henning S, Das A, Stockelhuber KW, Beiner M, Anh Ho C, Adhikari R, Wiener S, Heinrich G, Radusch HJ (2013) Effect of non-rubber components of NR on the carbon nanotube (CNT) localization in SBR/NR blends. Macromol Mater Eng. doi:10.1002/mame.201300254

29. Le HH, Sriharish MN, Henning S, Klehm J, Menzel M, Frank W, Wiener S, Das A, Stöckelhuber KW, Heinrich G, Radusch HJ (2014) Dispersion and distribution of carbon nanotubes in ternary rubber blends. Compos Sci Technol 90:180–186
30. Kawazoe M, Ishida H (2008) A new concept for nano particle distribution in SBR/NBR blend solution and films via molecular confinement. Macromolecules 41:2931–2937
31. Schopp S, Thomann R, Ratzsch KF, Kerling S, Altstadt V, Mulhaupt R (2014) Functionalized graphene and carbon materials as components of styrene-butadiene rubber nanocomposites prepared by aqueous dispersion blending. Macromol Mater Eng 299:319–329
32. Das A, Stockelhuber KW, Jurk R, Saphiannikova M, Fritzsche J, Lorenz H, Kluppel M, Heinrich G (2008) Modified and unmodified multiwalled carbon nanotubes in high performance solution-styrene–butadiene and butadiene rubber blends. Polymer 49:5276–5283
33. Lorenz H, Fritzsche J, Das A, Stöckelhuber KW, Jurk R, Heinrich G, Klüppel M (2009) Advanced elastomer nano-composites based on CNT-hybrid filler systems. Compos Sci Technol 69:2135–2143
34. Tancharernrat T, Rempel GL, Prasassarakich P (2014) Preparation of styrene butadiene copolymer–silica nanocomposites via differential microemulsion polymerization and NR/SBR–SiO_2 membranes for pervaporation of water–ethanol mixtures. Chem Eng J 258:290–300
35. Mohamed RM (2013) Radiation induced modification of NBR and SBR montmorillonite nanocomposites. J Ind Eng Chem 19:80–86

Chapter 8
Polychloroprene Rubber-Based Nanoblends: Preparation, Characterization and Applications

Milena Marinović-Cincović, Gordana Marković, Vojislav Jovanović, Suzana Samaržija-Jovanović and Jaroslava Budinski-Simendić

Abstract Polychloroprene rubber-based nanoblends allow the user to access properties of the final blended and vulcanized elastomer that are not accessible from a single, commercially available rubber alone. This chapter summarizes the preparation, characterization and applications of polychloroprene rubber based nano blends. Laboratory size two roll-mixing mill (15 × 33 cm) was used to prepare mixes 55 according to ASTM D 3182-89. The cure characteristics of the mixes were determined using elastograph with 50 per 67 minute frequency of oscillation. Two different types of vulcanization techniques have been exposed to achieve the desired results, namely one-stage vulcanization method (i.e. conventional vulcanization method) and two-stage vulcanization method. A scanning electron microscope and transmission electron microscopy have been used for morphological study. Modifications of polychloroprene rubber-based nanoblends were carried out using poly(ethylene oxide) (PEO)-coated nanocalcium silicate along with nano-N-ben-zylimine aminothioformamide (BIAT) and stearic acid-coated nanozinc oxide (nano-ZOC) in the sulphur vulcanization of the NR/NBR/CR rubber blends. Polychloroprene rubber-based nanoblends are used in structural, military, coating, biomedical, pacing, and aerospace applications.

Keywords Polychloroprene rubber · Characterization · Modifications · Mechanical properties · Vulcanization · Applications

M. Marinović-Cincović (✉)
Institute of Nuclear Science Vinča, University of Belgrade, Belgrade, Serbia
e-mail: milena@vinca.rs

G. Marković
Tigar, Nikole Pašića 213, 18300 Pirot, Serbia

V. Jovanović · S. Samaržija-Jovanović
Faculty of Natural Science and Mathematics, University of Priština, Kosovska Mitrovica, Serbia

J. Budinski-Simendić
Faculty of Technology, University of Novi Sad, Novi Sad, Serbia

G. Marković and Visakh P.M. (eds.), *Rubber Nano Blends*,
Springer Series on Polymer and Composite Materials,
DOI 10.1007/978-3-319-48720-5_8

8.1 Introduction

Rubber nanoblend composites have existed for decades. The blending of rubber is to combine the desirable features of each rubber component, but the properties of vulcanized blends cannot be linearly interpolated from those of the individual rubber. Rubber blend and their ratio factors [1–3], phase morphology [4–6], and the distribution of filler between the rubbers or at the interface [7–11] have been studied. The distribution of plasticizer [12] and cross-links [13, 14] between the rubbers and the interface, interpenetration of polymer chain segments, adhesion, and cross-linking [15, 16] are special factors for blends.

The generic name of neoprene is 2-chloro 1,3 butadiene (chloroprene) and is introduced commercially by DuPont in 1931. The molecular structures are primarily *trans* 1,4 chloroprene units (88–92 %), but three other configurations, viz. *Cis* 1,4 (7–12 %), 1,2 (1.5 %), and 3,4 (1 %), also occur. The 1,2 addition of chloroprene results in polymer having some chlorine available in allylic position (one chlorine for every four carbon atom in the chain of this homopolymer). Therefore, by comparison with non-polar diene rubbers, CR has a better resistance to swelling in mineral, animal, and vegetable oils and fats. The chlorine atom also imparts to CR better flame, weather, and ozone resistance than normally encountered with diene rubbers.

There are different types of polychloroprene rubber (neoprenes): G, W, and T types. The G types are copolymerized with sulphur and contain thiuram disulphide stabilizer. W and T types are much more stable than G types. W types have more uniform distribution of molecular weight. The G types cure rapidly without accelerator, whereas the W types require accelerators for acceptable curing and vulcanizate properties. These have many processing advantages over the G types. The cross-linking of CR compounds is carried out using metal oxides (zinc oxide and magnesium oxide). A typical combination of zinc oxide (5 phr) and specially calcined magnesium oxide (4 phr) is preferred for compounding.

Blends of polychloroprene rubber are often used commercially. For example, blends of neoprene and natural rubber (NR) are often used to provide good hot tear strength and building tack. Blends of neoprene and styrene butadiene rubber (SBR) are often compound cost reduction, and blends with NBR exhibit oil and fuel resistance.

The above requirement for chloroprene product could be satisfied by blending with polyolefin elastomers such as ethylene/propylene rubber (EPR) or ethylene/propylene/diene terpolymer (EPDM) which have better heat/ozone and cut growth resistance [17, 18]. For polarity (compatibility) reasons, EPDM is normally limited to 15–30 parts in blends with polychloroprene rubber to improve non-staining static ozone resistance.

Neoprene requires the addition of appropriate reinforcing fillers to achieve the required balance of processability, hardness, and tensile or tear properties. The most widely used is carbon black, and precipitated silica, calcium silicate, china clay, and whiting are most commonly used as mineral fillers.

8.2 Preparation of Polychloroprene Rubber Nanoblends

Laboratory size two roll-mixing mill (15 × 33 cm) was used to prepare mixes according to ASTM D 3182-89 (temperature of the rolls was maintained at 45 ± 5 °C). The compounding was carried out according to ASTM D 3184-89 and ASTM D 3182-89 in the following order: activator, filler, accelerator, and curing agents. Before the addition of accelerator and sulphur, the batch was thoroughly cooled. After the completion of mixing, homogenization of the compound was carried out by passing the rolled stock endwise six times at a mill opening of 0.8 mm. The mill is opened to give a minimum stock thickness of 6 mm, and the stock was passed through the rolls four times folding it back on itself each time.

8.3 Characterizations of Polychloroprene Rubber Nanoblends

8.3.1 Determination of Cure Characteristics

The cure characteristics of the mixes were determined using elastograph with 50 per minute frequency of oscillation. A typical elastograph cure curve is show in Fig. 8.1.

From the cure curve, cure date can be obtained (Table 8.1).

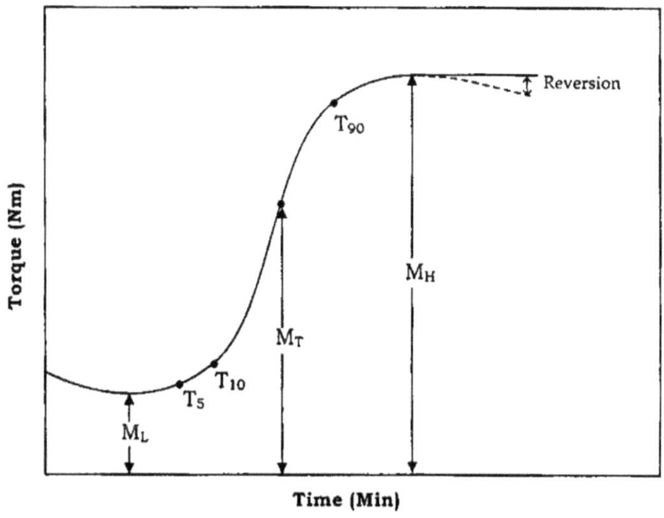

Fig. 8.1 A typical elastograph cure curve

Table 8.1 The cure date obtained from the cure curve

Minimum torque	M_L	Torque obtained by the mix after homogenizing at the test temperature before the onset of cure
Maximum torque	M_H	This is the torque recorded after the curing of the mix is completed
Torque	M_T	Torque at any time T
Optimum cure time	T_{9o}	It is the time taken for obtaining 90 % of the maximum torque
Scorch time	T_{s2}	This is the time for attaining 10 % of the maximum torque
Induction time	T_5	It is the time taken for 5 % vulcanization
Cure rate index	CRI	It is calculated as $100/(T_{90} - T_{s2})$ where T_{90} and T_{s2} are the time corresponding to optimum cure time and scorch time, respectively

For determining physical and mechanical properties, test specimens were moulded in standard moulds with a pressure of 11.764 MPa on the mould up to their optimum cure times and at specified temperatures for 24 h.

(A) Tensile Properties
Modulus, tensile strength, and elongation at the break of the vulcanizates were determined on a universal testing machine, using a crosshead speed of 500 mm/min as per ASTM D 412-87 (method Al). All the tests were carried out at room temperature (25 ± 2 °C).

(B) Tear Resistance
Tear resistance was carried out as per ASTM D 624 (1981) using uniqued, 90° angle test pieces. The samples were cut from the compression moulded sheets parallel to the mill grain direction. The test was carried out on the universal testing machine. The speed of extension was 500 mm/min, and the test temperature was 25 ± 2 °C.

(C) Hardness
The hardness (Shore A) of the moulded samples was determined using hardness tester in accordance with ASTM D 2240 (1986). The tests were performed on unstressed samples of 30 mm diameter and 6 mm thickness. A load of 12.5 N was applied, and the readings were taken after 10 s of indentation after firm contact had been established with the specimen.

(D) Compression Set
The compression set was determined according to ASTM D 395 (1982 method B). The samples (6.25 mm thick and 18 mm diameter) compressed to constant deflection (25 %) were kept for 22 h in an air oven at 70 °C. After the heating period, the samples were taken out and cooled at room temperature for half an hour, and the final thickness was measured. The compression set was calculated as follows:

$$\text{Commpression set}\,(\%) = 100 \times \frac{t_0 - t_1}{t_0 - t_s}$$

where t_0 and t_1 are the initial thickness and final thickness of the specimen, respectively, and t_s is the thickness of the spacer bar used.

(E) Abrasion Resistance

The abrasion resistance of the samples was determined using a DIN abrader (DIN 53516). Sample having a diameter of 6 ± 0.2 mm and a thickness of 6 mm was kept on a rotating sample holder, and ion load was applied. Initially, a prerun was given for the sample, and its weight was taken. The weight after the final run was also noted. The difference in weight is the abrasion loss. It is expressed as the volume of the test piece abraded by its travel through 42 m on a standard abrasive surface. The abrasion loss was calculated as follows:

$$V = \frac{\Delta M}{\rho}$$

where ΔM = mass loss, ρ = density of the sample, and V = abrasion loss. The result is expressed in cm^3/h.

(F) Rebound Resilience

Rebound resilience is measured using a vertical rebound resilience tester according to ASTM D 2632-88. A plunger weighing 28 ± 0.5 g is dropped from a height of 40 cm to the sample of thickness 12.5 mm, and the rebound height is measured.

(G) Density

Density of the samples was determined according to ASTM D 297 (1981). In this method, the weight of the specimen in air was first noted, and then, specimen was immersed in water and its loss of weight in water was determined. The density of the sample was calculated as follows:

$$\text{Density} = \frac{\text{Weight of specimen in air} \times \text{density of water}}{\text{Loss of weight specimen in water}}$$

Density of water is taken as 1 g/cm^3.

(H) Ageing Studies

Ageing studies were carried out according to ASTM D 573-88. Dumbbell samples were punched out from the vulcanized sheets and kept in air oven at predetermined temperatures for specified periods. Physical properties such as tensile strength, elongation at break, and modulus were determined after ageing. Knowing the values of these tests before ageing, the retention of these properties was calculated for assessing the effect of ageing.

(I) Chemical Test Methods

Determination of Chemical Cross-links of the Vulcanizates

The cross-link density of the vulcanizates was determined from the equilibrium swelling data as follows:

Approximately 0.2 mg of sample was punched out from the central portion of the vulcanizate, and the accurately weighed sample was allowed to swell in suitable solvent for 24 h. The outer portion of the swollen sample was then dried using a filter paper and then weighed. The solvent was removed in vacuum by placing in an oven at 50 °C for 22 h. The swollen weight was determined. The volume fraction of rubber, in the swollen network, was then calculated from the following equation.

$$V_r = \frac{(D - FT)\rho r^{-1}}{(D - FT)\rho r^{-1} + Ao\rho s^{-1}}$$

where T is the weight of the test specimen, D the deswollen weight, F the weight fraction of the insoluble components of the vulcanizates, Ao the weight of the absorbed solvent corrected for the swelling increment, and ρ, and ρ_s the density of rubber and solvent, respectively.

Knowing the value of $V_{r,}$ the total cross-link density, 1/2Mc was calculated using Flory–Rehner equations [19],

$$-[\ln(1 - V_r) + V_r + \chi V_{r^2}] = \frac{\rho r V_s (V_r)^{1/3}}{\mathrm{Mc}}$$

where V_s = molar volume of solvent, χ = the parameter characteristic of the interaction between rubber and solvent, Mc = the number average molecular weight of the rubber chains between cross-links. The values of the parameter χ taken for calculation were the following:

$$\text{NR-toluene} = 0.42$$
$$\text{SBR-toluene} = 0.32$$
$$\text{CR-benzene} = 0.26$$

The volume fraction of rubber (V_r) here is calculated assuming that the filler does not swell. It is then converted to V_r (the value v would have had in the absence of filler) according to Cunneen and Russel as follows:

$$\frac{V_{ro}}{V_r} = a^{-z} + b$$

Here, a and b are constants characteristic of the system and z is the weight fraction of the filler in the vulcanizate. The values of a and b for HAF black-filled systems are a = 0.56 and b = 0.44. The corresponding values for silica-filled system are a = 1.41 and b = −0.419. The values of V_{ro} were then substituted in the Flory–Rehner equation in place of V_r to obtain the cross-link density.

(J) Phase Morphology

Scanning electron microscopy (SEM) studies were carried out on the fractured surfaces of specimens derived from tensile testing. Fractured surfaces were then carefully cut from the failed test pieces without touching the surfaces. These specimens were stored in a desiccator and sputter coated with gold within 24 h of testing. The SEM observations were made using scanning electron microscope.

(K) Dielectric measurements

Rubber samples were cured in a mould with a spacer approximately 0.5 mm thick and 3.8 cm in diameter to the corresponding T_{c90}. Dielectric measurements were carried out by a dielectric spectrometer at room temperature in a three terminal parallel plate with a working electrode diameter 35.6 mm, allowing adequate overlap of the guard ring by the centrally placed specimen. The dielectric constant and loss factor were measured over the frequency range 10–100 kHz.

8.4 New Challenges of Polychloroprene Rubber Nanoblends

New challenges of polychloroprene rubber nanoblends have been obtained during the covulcanization of CR and NR. The problem of curative migration and filler transfer to more polar CR part is always confronted. The problem is solved when the polarity of NR phase improves by reacting with DIPDIS (Bis(-diisopropyl)

Table 8.2 Formulation of CR/NR-based nanoblends and the cure characteristics of one-stage vulcanizates at 150 °C

Mix	1	2	3	4	5	6	7	8	9	10
CR	100	–	75	50	25	100	–	75	50	25
NR	–	100	25	50	75	–	100	25	50	75
ZnO	5	5	5	5	5	5	5	5	5	5
MgO	1	–	3	2	1	1	–	3	2	1
Stearic acid[a]	0.5	2	0.88	1.25	1.63	0.5	2	0.88	1.25	1.63
N-774	–	–	–	–	–	30	30	30	30	30
Process oil	–	–	–	–	–	5	5	5	5	5
DIPDIS[b]	3.83	3.83	3.83	3.83	3.83	3.83	3.83	3.83	3.83	3.83
S	0.5	0.5	0.5	0.5	0.5	0.5	0.5	0.5	0.5	0.5
Max	6.8	3.7	4.2	3.6	3.4	7.3	4.5	4.7	4.3	4
t_{s2}(min)	4	4.5	8	5	4	3	4	5	5	4
t_{c90}(min)	43	9.5	35	27	10.5	45	11	37	16	12

[a]Stearic acid was added 2 phr with respect to NR and 0.5 phr with respect to CR

[b]Weight corresponds to 9 m mol Bis(-diisopropyl)thiophosphoryl disulphide (DIPDIS)

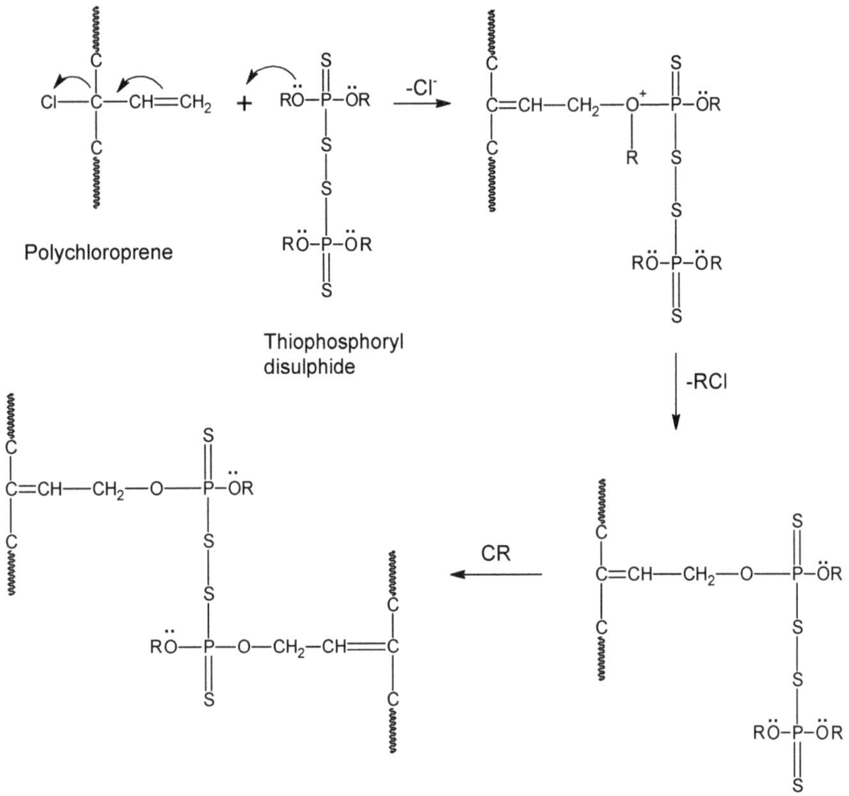

Scheme 8.1 Cross-linking of CR present with DIPDIS

thiophosphoryl disulfide) following a special technique, which is expected to improve the inter-rubber cross-linking between the two heterophases.

Two different types of vulcanization techniques have been exposed to achieve the desired results, namely one-stage vulcanization method (i.e. conventional vulcanization method) and two-stage vulcanization method.

8.4.1 One-Stage Vulcanization

The formulations and cure characteristics for CR/NR/DIPDIS rubber-based nanoblends are given in Table 8.2. The reaction mechanism [20] is presented in Scheme 8.1.

Mechanical properties of CR/NR/DIPDIS rubber blend vulcanizates are presented in Fig. 8.2a–e. The CR/NR blends (Mix 8) exhibit comparatively better

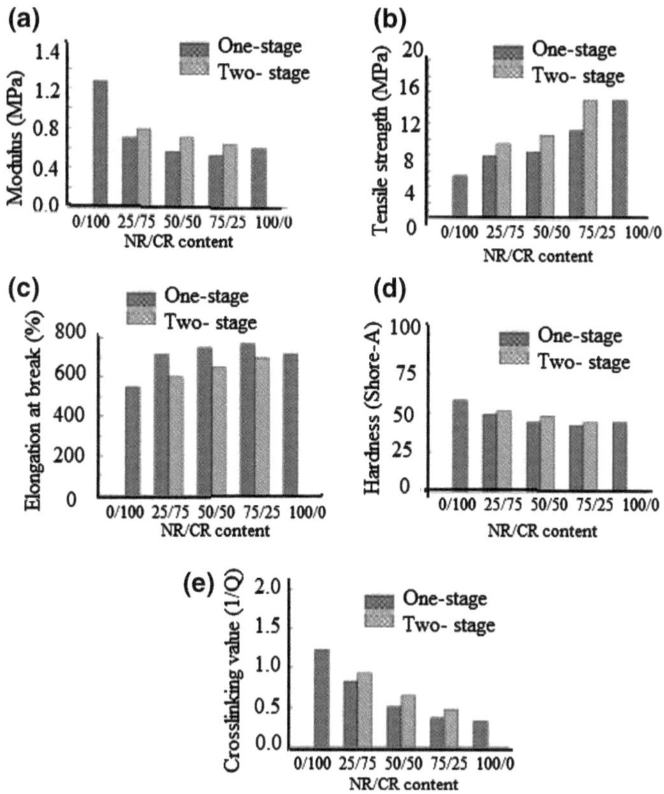

Fig. 8.2 The comparison of **a** 200 % modulus, **b** tensile strength, **c** elongation at break, **d** hardness, and **e** cross-linking values blend composition for CR and NR blends derived from one- and two-stage process

Table 8.3 Changes in physical properties after air ageing at (100 ± 2)°C during 72 h

Mix	1	2	3	4	5	6	7	8	9	10
ΔTS (%)	−31	−65	−52	−74	−85	−10	−46	−31	−36	−52
ΔM 200 %	44	−35	−6	−13	−18	77	74	139	75	65
ΔEb (%)	−29	−51	−16	−27	−46	−34	−52	−22	−30	−48
Changes in hardness	5	−8	−8	−5	−5	6	8	9	7	6

results. Cross-linking values (1/Q) in the oil immersion experiments shown in Fig. 8.2e also corroborate these observations.

Table 8.3 shows that the vulcanizates obtained from CR can retain the physical properties in the ageing process. Obtained aged values of the vulcanizates indicate a drastic fall in tensile strength and elongation at break.

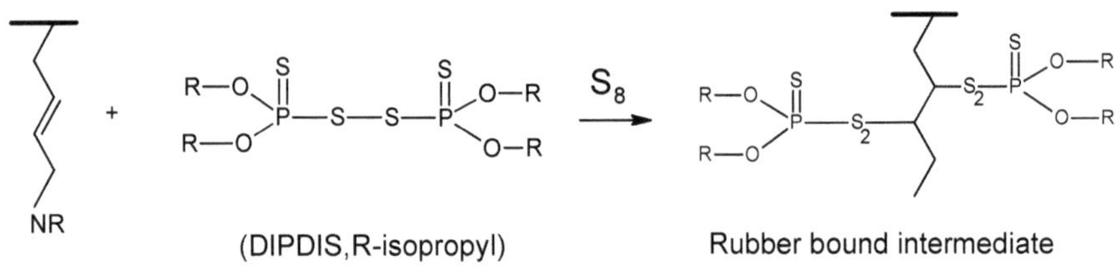

Scheme 8.2 Reaction of DIPDIS with NR in the presence of sulphur and zinc oxide

Table 8.4 Formulations and the cure characteristics of two-stage CR/NR/DIPDIS vulanizates at 160 °C

	Mix	3′	4′	5′	7′	8	9	10
First stage	NR	25	50	75	100	25	50	75
	ZnO	5	5	5	5	5	5	5
	Stearic acid[a]	0.87	1.25	1.62	2	0.87	1.25	1.62
	DIPDIS	3.83[b]	3.83[a]	3.83[a]	3.83[a]	3.83[a]	3.83[a]	3.83[a]
	S	0.5	0.5	0.5	0.5	0.5	0.5	0.5
Second stage	Preheating time	1.5	2	2.5	3	1.5	2	2.5
	N-774	–	–	–	30	30	30	30
	Process oil	–	–	–	5	5	5	5
	MgO	3	2	1	0	3	2	1
	CR	75	50	25	0	75	50	25
Max		4.8	4	3.8	5.2	5.5	4.9	4.6
T_{s2}		2	1	0.75	1.5	1	1.5	1.5
t_{c90}		32	25	7.5	5.5	32	13	7.5

[a]0.5 phr with respect to CR and 2 phr with respect to NR
[b] Weight corresponds to 9 m mol DIPDIS

8.4.2 *Two-Stage Vulcanization*

A very active cross-linking precursor and subsequently promote the interfacial cross-linking through coupling with thiophosphoryl fragments formed by the reaction of DIPDIS and NR molecules (Scheme 8.2). Inter-rubber linking is polysulphidic in nature, and DIPDIS could form di- or polysulphide linkages in the presence of elemental sulphur [21–24] (Table 8.4).

Mechanical properties of CR and NR two-stage vulcanization process are presented in Fig. 8.3a–e. Significant improvement in tensile strength, modulus, and cross-linking values are observed as compared to those obtained from one-stage vulcanization of the blends [25–28].

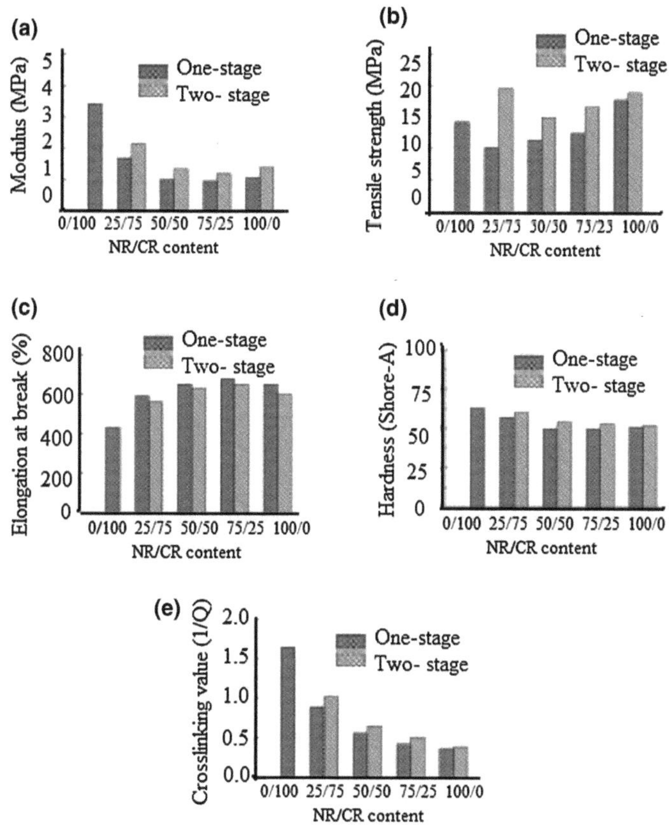

Fig. 8.3 The comparison of **a** 200 % modulus, **b** tensile strength, **c** elongation at break, **d** hardness, and **e** cross-linking values blend composition for CR and NR blends (*black-filled*) derived from one- and two-stage process

8.4.3 Dielectric Study

Dielectric constant (ε') versus frequency (f) for the pure and CR/NR blend is shown in Fig. 8.4a–c. Higher values of ε' for the one-stage blend vulcanizates are due to the increase in dipole orientation polarization and the presence of interfacial polarization [29–31].

The curves relating to dielectric loss (ε'') and log f are shown in Fig. 8.5a–c. The absorption in the higher frequency region can be due to the segmental orientation of rubber chains [32–34] (Schemes 8.3 and 8.4).

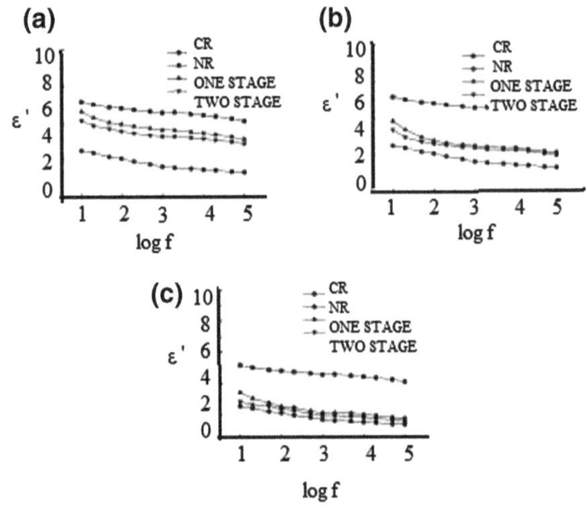

Fig. 8.4 **a** Plot of dielectric constant versus applied frequency of gum vulcanizates comparison CR, NR, and their blends (*filled circle*) NR 100 %; (*filled square*) CR 100 %; (*upward filled triangle*) CR:NR 75:25 in one stage; (*downward filled triangle*) CR:NR 75:25 in two stage; **b** plot of dielectric constant (ε′) versus applied frequency (*f*) of gum vulcanizates comparison CR, NR, and heir blends (*filled circle*) NR 100 %; (*filled square*) CR 100 %; (*upward filled triangle*) CR: NR 75:25 in one stage; (*downward filled triangle*) CR:NR 50:50 in two stage **c** plot of dielectric constant (ε′) versus applied frequency (*f*) of gum vulcanizates comparison CR, NR and heir blends (*filled circle*) NR 100 %; (*filled square*) CR 100 %; (*upward filled triangle*) CR:NR 75:25 in two stage; (*downward filled triangle*) CR:NR 25:75 in two stage

8.4.4 SEM Studies

The interfacial linking between CR and NR in the presence of di-isopropyl thiophosphoryl disulfide is due to form a compact and coherent rubber blend matrix (Fig. 8.6).

8.5 Recent Advances on Polychloroprene Rubber Nanoblends

Recent advances on polychloroprene rubber nanoblends are obtained using nanoclay as a potent compatibilizer as a reinforcing agent in the blend of CR and EPDM rubber. The nanoclay reduces the interfacial energy between the phases by accumulating at the interfaces, permits a finer intercalation–exfoliation process during mixing and vulcanization, and provides a strong reinforcement to the rubber blends [35, 36].

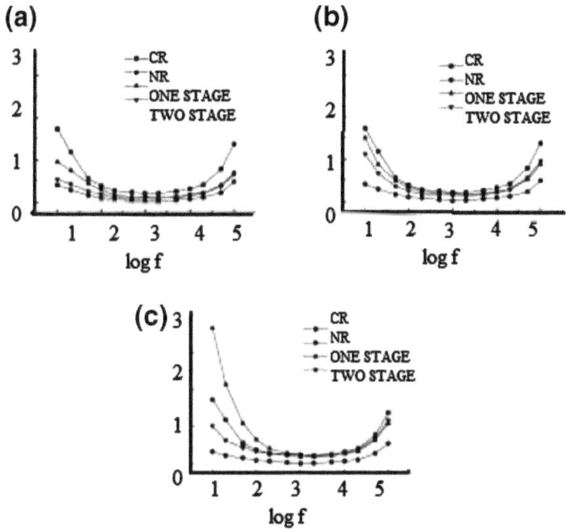

Fig. 8.5 **a** Plot of dielectric loss versus (ε'') applied frequency (f) of gum vulcanizates comparison CR, NR, and their blends (*filled circle*) NR 100 %; (*filled square*) CR 100 %; (*upward filled triangle*) CR:NR 75:25 in one stage; (*downward filled triangle*) CR:NR 75:25 in two stage; **b** plot of dielectric loss (ε'') versus applied frequency (f) of gum vulcanizates comparison CR, NR, and heir blends (*filled circle*) NR 100 %; (*filled square*) CR 100 %; (*upward filled triangle*) CR:NR 75:25 in one stage; (*downward filled triangle*) CR:NR 75:25 in two stage; **c** plot of dielectric loss (ε'') versus applied frequency (f) of gum vulcanizates comparison CR, NR, and heir blends (*filled circle*) NR 100 %; (*filled square*) CR 100 %; (*upward filled triangle*) CR:NR 75:25 in one stage; (*downward filled triangle*) CR:NR 75:25 in two stage

8.5.1 Characterization

A higher amount of stearic acid (10 phr) was added to the organically modified clay to obtain a higher d-space for the organoclay in the swollen state. Because the stearic acid-modified organoclay exhibits a higher d-space (2.98–3.96 nm), large portion of the clay will be formed and a good degree of intercalated–exfoliated structures was obtained in the rubber matrix. Figure 8.7 shows TEM images of CR/EPDM rubber blends (75/25, 50/50, and 25/75) containing 10 phr of clay.

Dynamic mechanical analysis (DMA) Fig. 8.8 is used to understand the dynamic response of the rubber blend after the addition of clay. The storage modulus obtained from oscillatory tension deformation as a function of temperature is given in Fig. 8.9.

The CR/EPDM blend system showed different Tg values. It is clear from Table 8.5 that the Tg values are closer when the blend matrix is filled with 10 phr clay, which indicates that the clays have a strong capability to increase compatibility between two heterorubbers. A shift of 9 °C was observed for the 75 CR/25 EPDM rubber nanoblend.

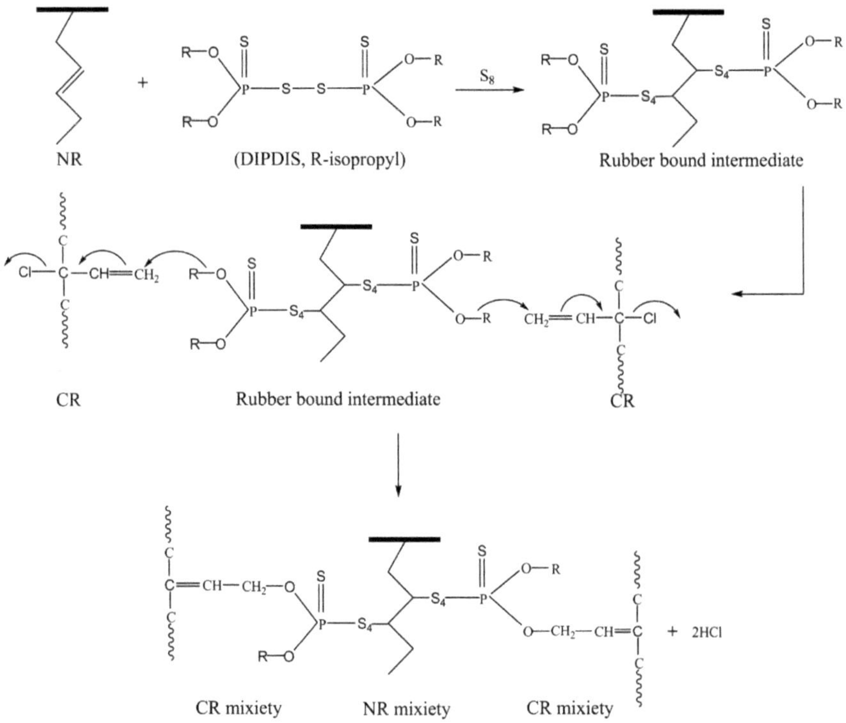

Scheme 8.3 Reaction of the rubber bound intermediates during cocuring chloroprene rubber (CR)

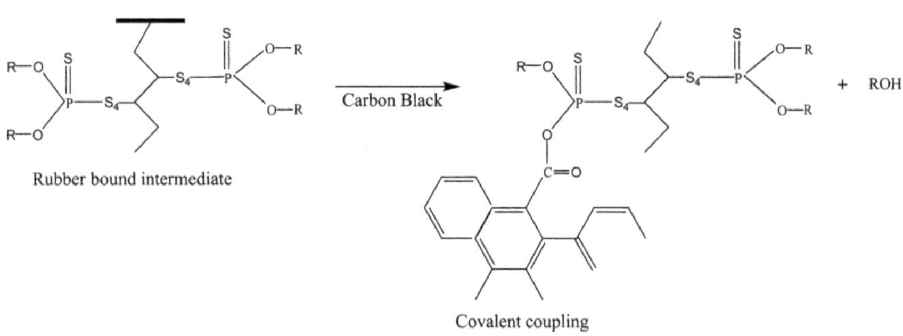

Scheme 8.4 Interaction between carbon black and pendant DIPDIS fragment

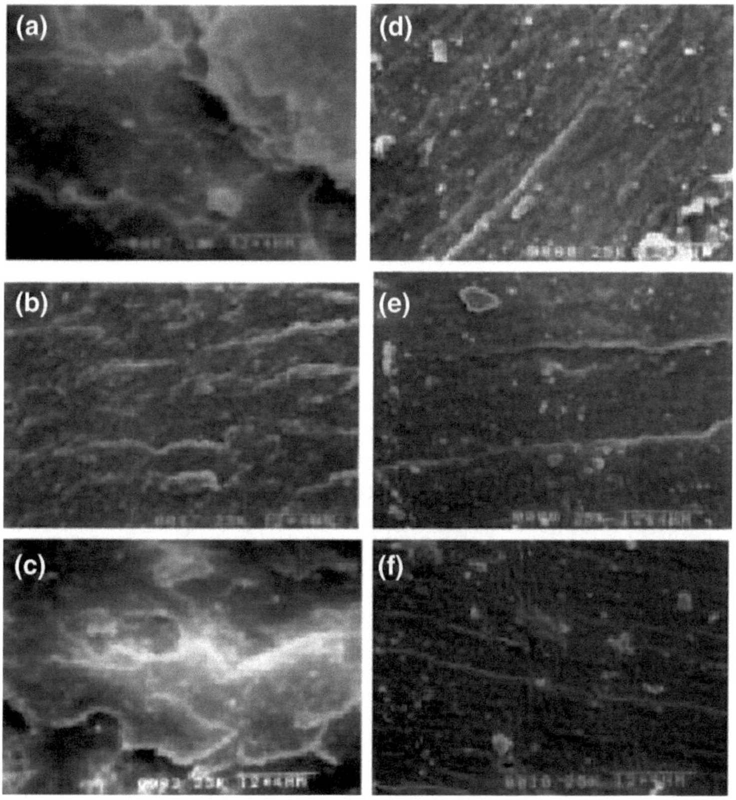

Fig. 8.6 SEM micrographs of tensile fractured surfaces of the black-filled vulcanizates cured at 160 °C: **a** 75:25 CR–NR blend (one stage) at 250× magnification; **b** 50:50 NR–CR blend (one stage) at 250× magnification; **c** 25:75 CR–NR blend (one stage) at 250× magnification; **d** 75:25 CR–NR blend (two stage) at 250× magnification; **e** 50:50 NR–CR blend (two stage) at 250× magnification; **f** 25:75 CR–NR blend (two stage) at 250× magnification

Exactly, the opposite type of behaviour was found [37–41] for an incompatible rubber blend, as revealed by the situation when the clay interferes with the compatibility between chloroprene and XNBR during the self-cross-linking reaction.

The effects of organoclay loading and chloroprene rubber (CR) concentration on the cure characteristics, microstructure, and mechanical and rheological properties of isobutylene–isoprene rubber (IIR)/CR blend were investigated, also. Different compositions of CR (10, 20, and 40 wt%) with Cloisite15A as organomodified nanoclay (1, 3, 5, and 7 wt%) were used for blends by a two roll mill. Samples were vulcanized at 175 °C using a hot press. The cure and scorch times and also the maximum torque of the composites increased with the incorporation of organoclay. Mechanical properties such as tensile strength, elongation at break, modulus (100, 200, and 300 %), and resilience improved with increasing nanoclay loading. The structure of the nanocomposites was characterized with X-ray diffraction (XRD),

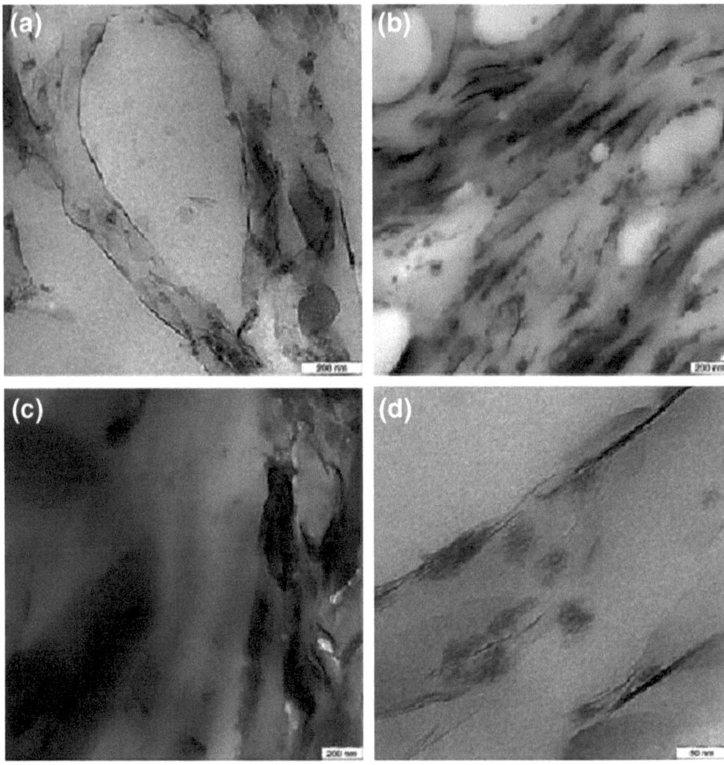

Fig. 8.7 TEM images of the blends of 25 EPDM/75 CR (**a**), 50EPDM/50CR (**b**), 75EPDM/25CR (**c**), and 25EPDM/75CR (**d**), filled with 10 phr nanoclay

SEM, and transmission electron microscopy (TEM). XRD results of nanocomposites indicated that the intercalation of polymer chains into the clay gallery was deduced from increasing the interlayer distance of silicate layers. TEM and SEM also directly confirmed XRD results.

Polychloroprene rubber blends are widely applied in the areas such as moulded goods, cables, transmission belts, conveyor belts, and profiles (Fig. 8.10a, b). According to good mechanical strength, high ozone and weather resistance, good ageing resistance, low flammability, good resistance towards chemicals, moderate oil and fuel resistance, adhesion to many substrates.

The rubber formulations for bush mounting (as shown in Fig. 8.10c) application by considering the mechanical and dynamic properties were developed. Dynamic stressing of the interface between rubber and a reinforcing material—metal, may destroy the adhesion, causing the part to fail [42]. In this project, natural rubber–polychloroprene rubber blend system were chosen to fulfil the above-mentioned requirement.

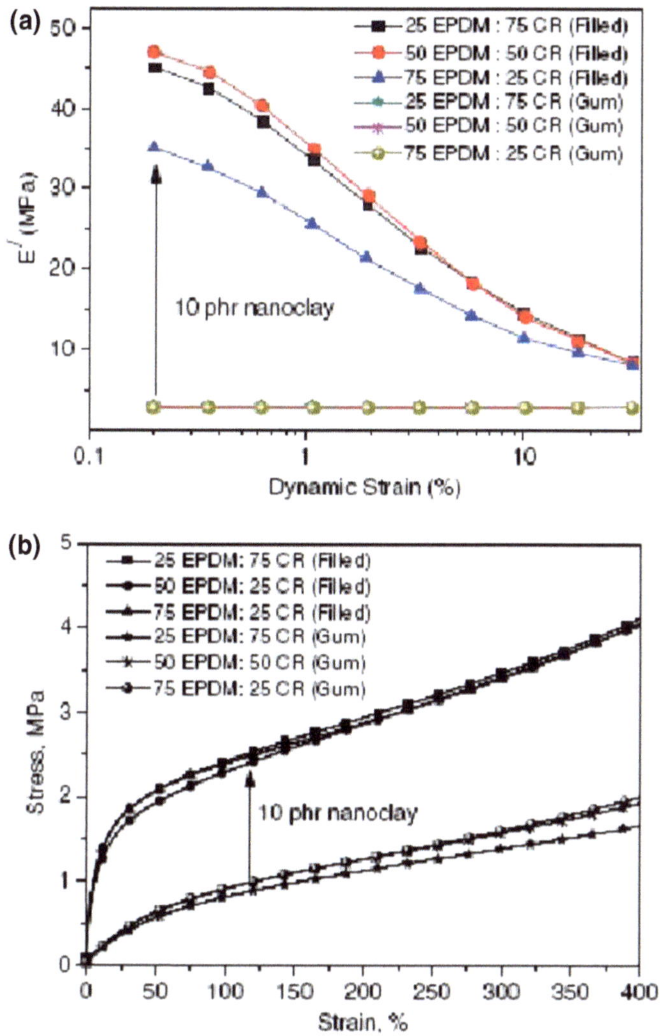

Fig. 8.8 Effect of dynamic strain amplitude on stage modulus. **a** Stress–strain behaviour of CR/EPDM blend in the absence and presence of nanoclay. **b** For this experiment, tension mode was selected for the vibration of the dynamic strain from 0.01 to 40 % at 10 Hz frequency

The vibratory system can be idealized as (a) mass, (b) spring, (c) damper, and (d) excitation as shown in Fig. 8.11. Major areas of application which employ its outstanding physical properties are used in vehicles, tyres, offshore and aerospace industries, civil engineering, and railways. The major advantage of NR which

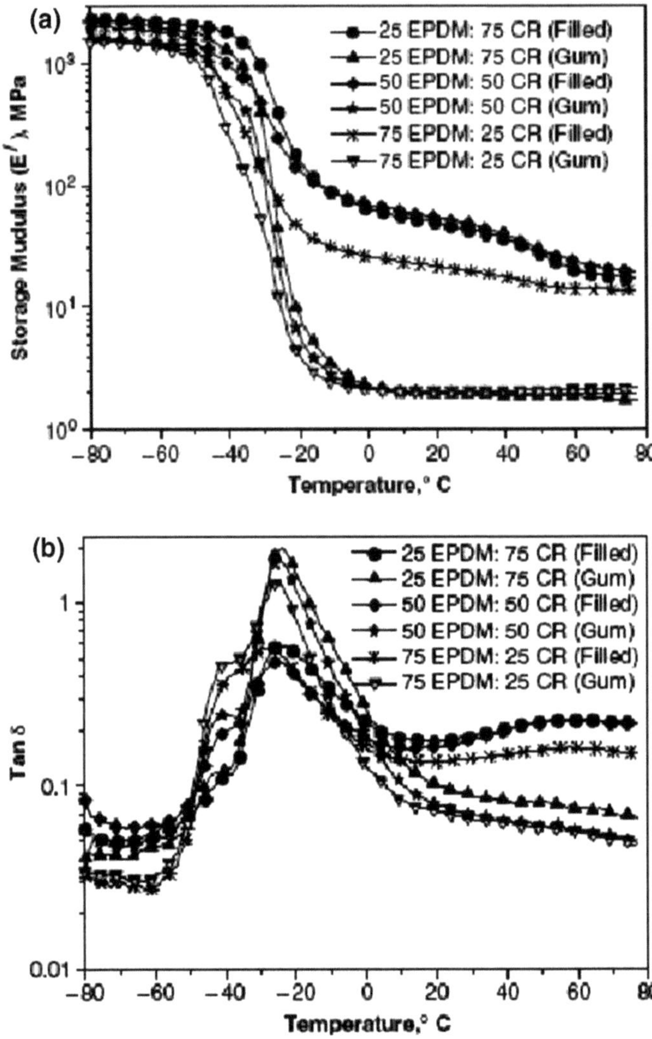

Fig. 8.9 Storage modulus (E') versus temperature (**a**) and tanδ versus temperature (**b**) plots of CR/EPDM blends in the absence and presence of nanoclay

makes it dominant in many dynamic applications is its dynamic performance and the ability of rubber to carry a high load under compression, yet function at high strains and low stiffness compared to metals [43–46].

Table 8.5 Glass transition temperature of EPDM/CR rubber nanoblends

Rubber	Glass transition temperature (°C)					
	Without clay			With clay		
EPDM: CR[a]	EPDM peak	CR peak	Separation two peaks	EPDM peak	CR peak	Separation between two peaks
100/0	−37	a	–	−37	a	–
75/25	−44	−25	19	−42	−26	16
50/50	−42	−25	17	−41	−27	14
25/75	−40	−24	16	−34	−28	7
0/100	a	−24	–	a	−24	–

[a]The curing recipe used in this study was 5 phr zinc oxide, 10 phr stearic acid, 7 phr ZDMC, and 1 phr sulfur absent for that particular case

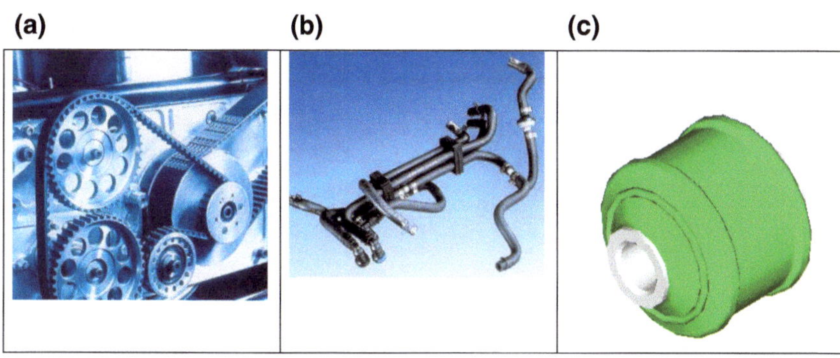

Fig. 8.10 Polychloroprene rubber-based nanoblends. **a** Belts; **b** hoses; **c** bushing

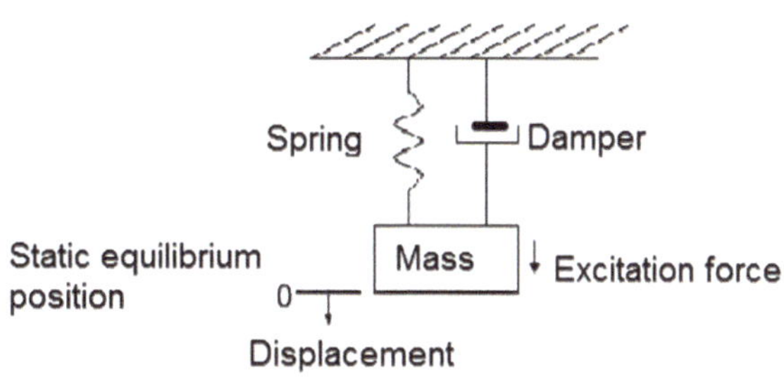

Fig. 8.11 Elements of a vibratory system

8.6 Biomedical Application

Polychloroprene rubber-based nanoblends will play a key role in medical technology in the future. Use of polymers as biomaterials is increasing day by day since many polymers having diverse properties are more similar to the body (should be biocompatible). Their appeal and acceptability are mainly due to their versatility, and the fact that they are tailor-made or modified will suit specific body functions.

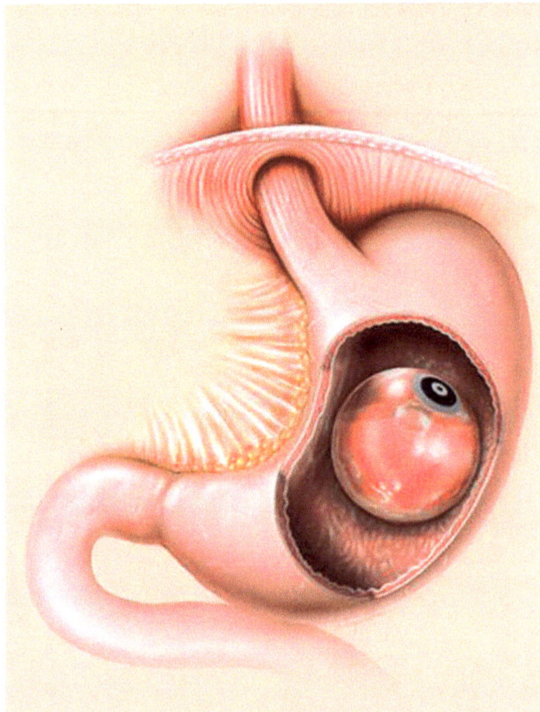

Fig. 8.12 Gastric balloons

CR nanoblend latex has been used for biomedical application such as gastric balloons (Fig. 8.12).

8.7 Pacing Application

Polychloroprene rubber-based nanoblend has several valuable properties and works well in packing applications (Fig. 8.13) requiring resistance to gasoline, sunlight, ozone, oxidation, and oil. The blend of polychloroprene and EPDM makes this material moderately resistant to oil and ozone.

Fig. 8.13 Polychloroprene rubber-based nanoblend pacing material

Besides resistance to petroleum products, polychloroprene rubber-based nanoblend sheets also withstand a higher service temperature range of −34 to 202 °C. They can be compounded for an excellent resistance to most of the acids and alkalies as well as to open flame. These sheets are widely used for various applications in refineries, food industry, and chemical and process units.

8.8 Structural Application of Polychloroprene Rubber Nanoblends

Polychloroprene rubber-based nanoblend has a wide range of structural applications (Fig. 8.14).

Chloroprene Rubber based nano blend Rubber Bush A bushing or rubber bushing is a type of vibration isolator. It provides an interface between two parts, damping the energy transmitted through the bushing. A common application is in vehicle suspension systems, where a bushing made of rubber (or, more often, synthetic rubber or polyurethane	**Chloroprene Rubber based nano blend Rubber Anti vabration** Anti vabration One of the largest range of rubber mounts includes cylindrical rubber mounts, captive anti vibration mounts, marine mounts and universal rubber mounting. The mounts are ideal Vibration damper for pumps automotive applications, generator s ets, marine use and compressors.	**Chloroprene Rubber based nano blend Rubber Door Mat** A mat is a generic term for a piece of fabric or flat material, generally placed on a floor or other flat surface, which serves a range of purposes including:providing a regular or flat surface, such as a mousepad. protecting that which is beneath the mat, such as a place mat or the matting used in archival framing and preservation of documents and paintings.	**Vibration Damping** High-mechanical damping is diametrically opposite to the requirements for high resilience. A typical application is machinery mountings in a hot and/or oily environment. Appropriate compounds are usually highly filled with soft black, china clay and aromatic oil. Hence, high Mooney Neoprene WHV or WHV-100 are indicated. ETU acceleration is required for practical cure times and minimum creep in service.

Fig. 8.14 Structural applications of polychloroprene rubber-based nanoblend

Fig. 8.15 Natural/chloroprene rubber-based nanoblend bearings

Elastomeric bearings (Fig. 8.15) can be made from a variety of materials and are very effective against most medium-to-high frequency vibration sources.

The material choice depends on the environment, but natural/chloroprene rubber-based nanoblend is the most common choice due to its high dynamic response. Bearings can be designed to provide low natural frequencies, as low as 4–5 Hz if necessary. The nature of the vibration source and the level of isolation will determine what natural frequency is appropriate. This and a number of other factors drive the bearing design.

For structural applications, polychloroprene rubbers are blended with synthetic resins to promote mechanical strength and heat stability phenolic polychloroprene (Fig. 8.16). Rate of strength development is rapid and unlike other rubber adhesives, and polychloroprene can sustain low loads at high temperatures soon after bonding. The choice of resin is determined by the type of structural loading involved. The resin addition has influence on peel and shear strength of canvas, wood, formica (melamine composite), and mild steel joints. Peel performance is maximum at 100 phr, whereas the highest shear strengths are obtained at 40 phr. Resin formulations with emulsion adhesives are similar in type to solvent blends, and their performance is comparable. The environmental risks of fire and toxicity which are a feature of solvent adhesives have spurred on the development of the

Fig. 8.16 Polychloroprene-based nanoblends adhesive

safer emulsion types. Tack retention is generally inferior to NR. Solvent or heat reactivation is less successful than other rubber adhesives. They can support loads of 200–700 kPa for extended periods soon after bonding.

8.9 Military Application

One of the most common uses of polychloroprene rubber-based nanoblends is for military (many chemicals and oil resistance, corrosion and resisting abrasion in the most extreme conditions, protects against chemical fume, splash, and salt spray) (Fig. 8.17). Elastomeric air-curing neoprene coating used in the protection of equipment on board ships. Coating neoprene rubber-based nanoblend has a long performance record of fighting. Quick and easy to apply, in-house maintenance personnel can economically protect areas most likely to be affected by corrosion and weather as well as equipment, new or old.

8.10 Coating Industry of Polychloroprene Rubber Nanoblends

A wide range of application has polychloroprene rubber-based nanoblends for coating industry (Fig. 8.18).

			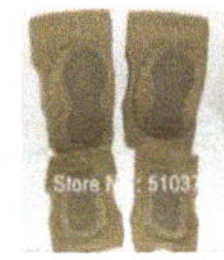
Chloroprene Rubber based nano blend cloth military	Chloroprene Rubber based nano blend shirt	Soft Black Chloroprene Rubber based nano blend Thermal Fleece Police Cop SWAT Military Army	Chloroprene Rubber based nano blend military tactical elbow pad knee pads

Fig. 8.17 Application of polychloroprene rubber-based nanoblends for military

Chloroprene Rubber based nano blend coating	spring coil with Chloroprene Rubber based nano blend coating Chloroprene Rubber based nano blend coating is generally applied to pipe support components that require flexibility such as spring coils whereas the hot dipped galvanizing coating would be applied to rigid pipe support components.	Chloroprene Rubber based nano blend Tennis arm guard,badminton wrist support	Chloroprene Rubber based nano blend swimm pac	Chloroprene Rubber based nano blend tennis arm, guard, badminton wrist support

Fig. 8.18 Application of polychloroprene rubber-based nanoblends for coating industry

		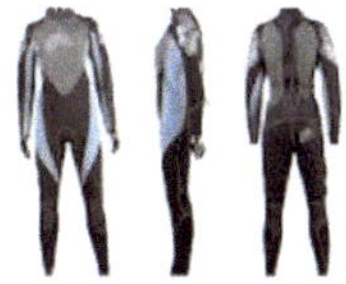	
Chloroprene Rubber based nano blend flame retardant	Aerospace heavy duty cases	Flame retardant impregnating solution	Chloroprene Rubber based nano blend plane

Fig. 8.19 Aerospace application of polychloroprene rubber-based nanoblends

Fig. 8.20 Chloroprene rubber-based nanoblend sponge and custom seat

8.11 Aerospace Applications of Polychloroprene Rubber Nanoblends

Water, flame, weather, heat, and ozone resistance are need for aerospace application of chloroprene rubber-based nanoblends (Figs. 8.19 and 8.20).

8.12 Modifications of Polychloroprene Rubber-Based Nanoblends

Modifications of polychloroprene rubber-based nanoblends were carried out using poly(ethylene oxide) (PEO)-coated nanocalcium silicate along with nano-*N*-benzylimine aminothioformamide (BIAT) and stearic acid-coated nanozinc oxide (nano-ZOC) in the sulphur vulcanization of the NR/NBR/CR rubber blends. The improvement in mechanical properties of NR/NBR blends were obtained with 5 phr of CR as a compatibilizer agent. Chloroprene rubber (CR) has a backbone structure similar to that of NR but is more polar because of a chlorine substituent. The solubility parameter value of CR is intermediate to that of NR and NBR. Also, the presence of a dipole within the repeat unit allows the possibility of the interaction of the acrylonitrile repeat unit with NBR.

The synthesis of PEO-coated calcium silicate, nano-BIAT, and nano-ZOS are given in Schemes 8.5, 8.6, and 8.7.

SEM images (Fig. 8.21) of (a) nanocalcium silicate, (b) NR/NBR/CR (20/80/5), (c) NR/NBR (20/80), and (d) NR/NBR/CR/nanocalcium silicate (20/80/5/10) are shown in Fig. 8.16. The presence of nanocalcium silicate and 3 phr of clay slightly reduce the tensile strength. The better dispersion of nanoparticles in the CR-compatibilized NBR–NBR blend caused an improvement in the technological properties.

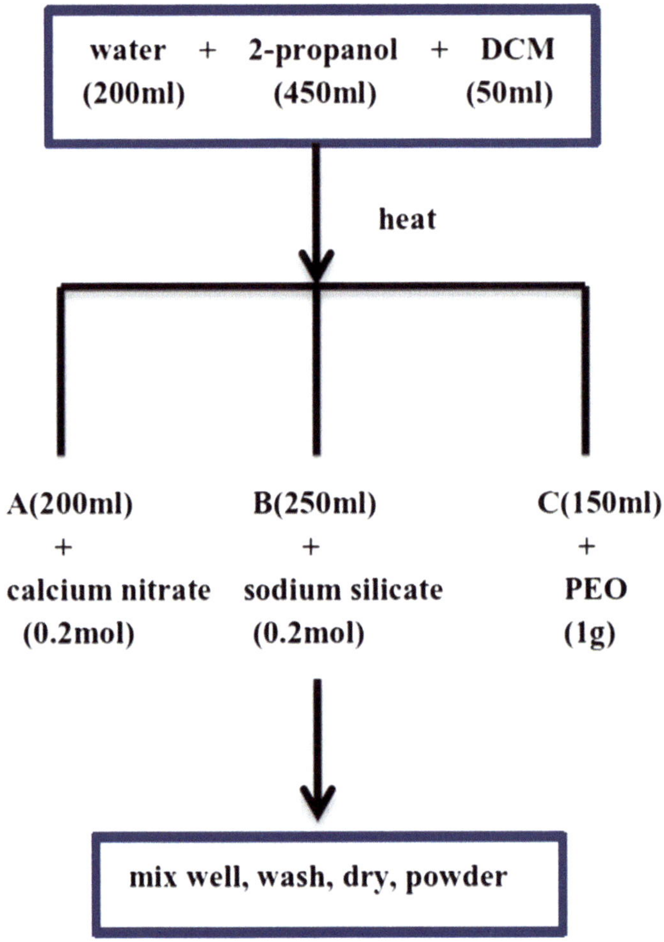

Scheme 8.5 Synthesis of nano-PEO-coated calcium silicate

Scheme 8.6 Synthesis of ZOS

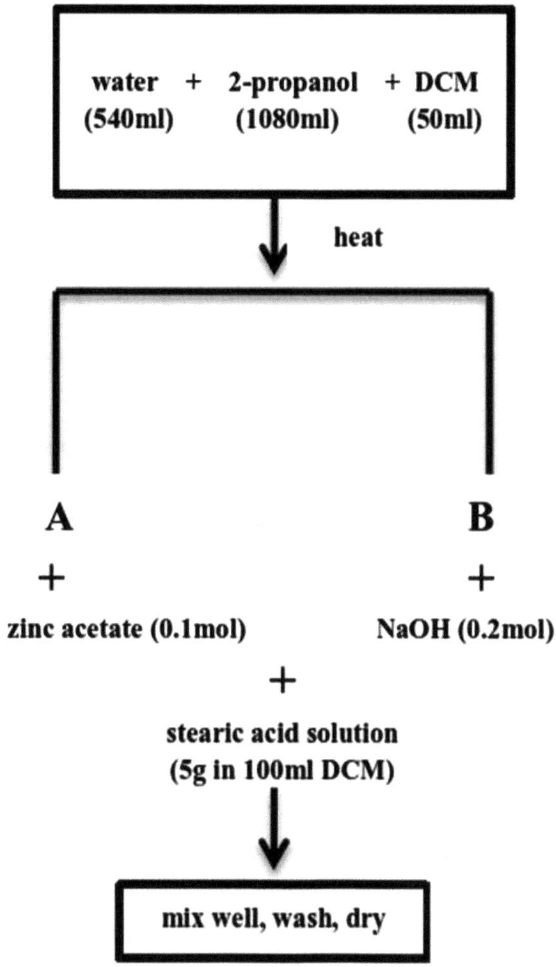

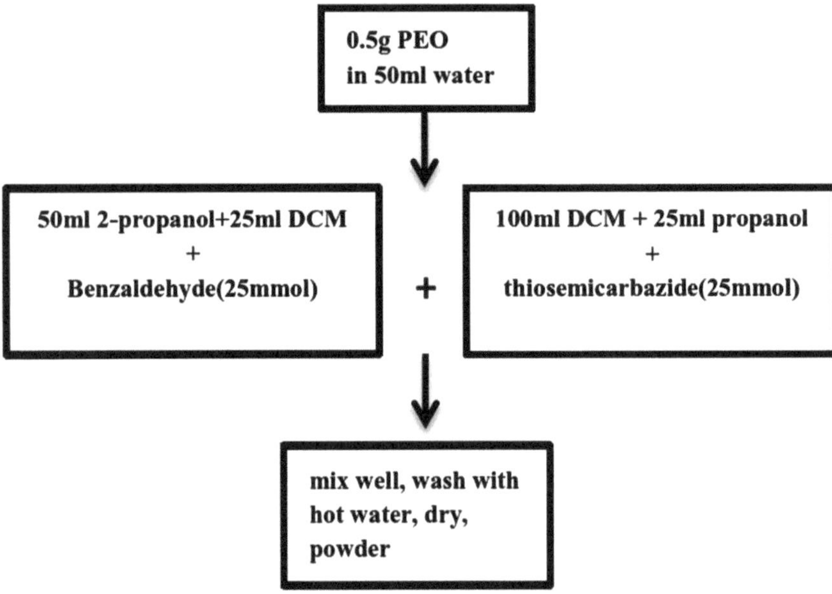

Scheme 8.7 Synthesis of nano-BIAT

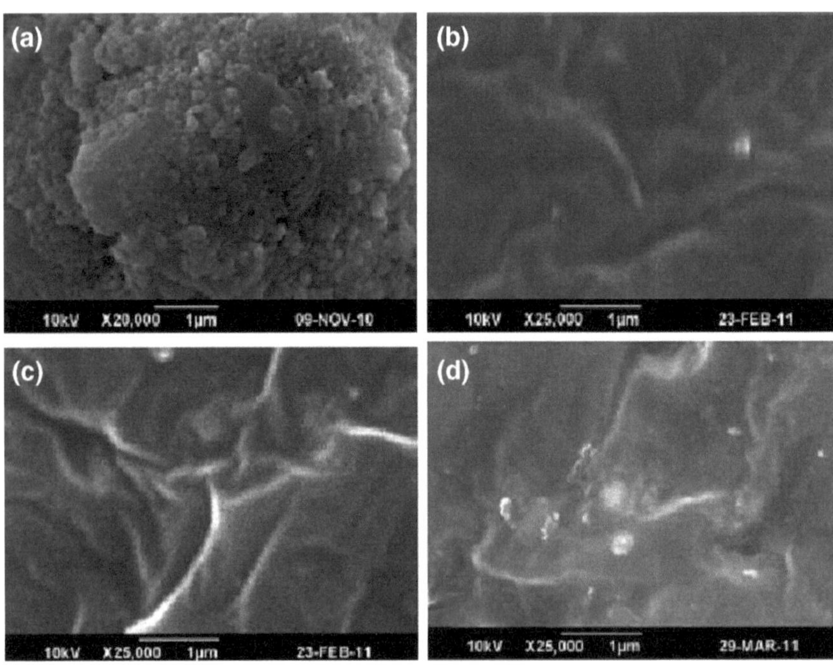

Fig. 8.21 SEM images of **a** nanocalcium silicate, **b** NR/NBR/CR (20/80/5), **c** NR/NBR (20/80), and **d** NR/NBR/CR/nanocalcium silicate (20/80/5/10)

8.13 Conclusion

Polychloroprene rubber-based nanoblends were the first commercially successful synthetic elastomer blend. Polychloroprene rubber-based nanoblends are normally used in structural, military, coating, biomedical, pacing, and aerospace applications. Latex grades of polychloroprene rubber-based nanoblends are used in waterborne adhesives and dipped goods.

Polychloroprene rubber-based nanoblends are noted primarily for its resistance to weathering, excellent combustion behaviour, ageing and heat resistance, average resistance to oils, good strength, tear and wear resistance, resistance to water, a large number of chemicals over a long period of time, good adhesion to materials, low gas permeability, resistance to soil bacteria and fungi, and good electrical properties for a large number of applications. However, it tends to be higher priced than other elastomer-based nanoblends, has only fair dielectric properties, has a mediocre resistance to low temperatures, and loses resilience above 120 °C.

In general, world demand for polychloroprene rubber has not grown like that of other synthetic elastomers because of its relatively high price and increasing substitution from competing elastomers such as EPDM (ethylene–propylene diene monomer rubber), for example. This trend has been most notable in the automotive sector, which represents about 20 % of global consumption.

World polychloroprene rubber consumption is forecast to grow at an average annual rate of almost 2 % through 2016, with both China and other Asian countries showing the highest growth at 3.5 %. Consumption growth in the more developed regions of the world is expected to be flat to slightly positive (in the USAand Japan) and slightly negative in Western Europe.

Acknowledgments The financial support for this study was granted by the Ministry of Science and Technological Development of the Republic of Serbia (Project nos. 45022, 45020 and 172056).

References

1. Zurina M, Ismail H, Chantara TR (2006) Characterization of epoxidized natural rubber/ethylene vinyl acetate (ENR-50/EVA) blend: effect of blend ratio. J Appl Polym Sci 99(4):1504
2. George S, Varughese KT, Thomas S (1999) Dielectric properties of isotactic polypropylene/nitrile rubber blends: effects of blend ratio, filler addition, and dynamic vulcanization. J Appl Polym Sci 73(2):255
3. George S, Neelakantan NR, Varughese KT, Thomas S (1997) Dynamic mechanical properties of isotactic polypropylene/nitrile rubber blends: effects of blend ratio, reactive compatibilization, and dynamic vulcanization. J Polym Sci Part B: Polym Phys 35(14):2309
4. Agarwal K et al (2011) Studies on phase morphology and thermo-physical properties of nitrile rubber blends. J Therm Analys Cal 104: 1125

5. Everaert V, Aerts L, Groeninckx G (1999) Phase morphology development in immiscible PP/(PS/PPE) blends influence of the melt-viscosity ratio and blend composition. Polymer 40 (24):6627
6. Wu S (1990) Chain structure, phase morphology, and toughness relationships in polymers and blends. Polym Eng Sci 30(13):753
7. Bokobza L (2004) The reinforcement of elastomeric networks by fillers. Macromol Mater Eng 289(7):607
8. Le HH, Ilisch S, Kasaliwal GR, Radusch H-J (2008) Filler phase distribution in rubber blends characterized by thermogravimetric analysis of the rubber-filler gel. Rubber Chem Technol 81 (5):767
9. Sudarsan M, De SK, Anil KB (1992) Quantitative estimation of filler distribution in immiscible rubber blends by mechanical damping studies. Rubber Chem Technol 65(2):293
10. Sirisinha C, Prayoonchatphan N (2001) Study of carbon black distribution in BR/NBR blends based on damping properties: influences of carbon black particle size, filler, and rubber polarity. J Appl Polym Sci 81(13):3198
11. Fenouillota F, Cassagnaua P, Majestéa JC (2009) Uneven distribution of nanoparticles in immiscible fluids: morphology development in polymer blends. Polymer 50(6):1333
12. Markovic G, Jovanovic V, Samarzija-Jovanovic S, Marinovic-Cincovic M, Budinski-Simendic J (2011) Curing and mechanical properties of chlorosulphonated polyethylene rubber blends. Chem Ind Chem Eng Q 17(3):315
13. Wootthikanokkhan J, Tunjongnawin P (2003) Investigation of the effect of mixing Schemes 8 on cross-link distribution and tensile properties of natural–acrylic rubber blends. Polym Testing 22(3):305
14. Boochathum P, Prajudtake W (2001) Vulcanization of *cis*- and *trans*-polyisoprene and their blends: cure characteristics and crosslink distribution. Eur Polymer J 37(3):417
15. Fereshteh M, Saeed Taghvaei G, Mercedeh MM et al (2013) Investigation of correlation between rheological properties of rubber compounds based on natural rubber/styrene-butadiene rubber with their thermal behaviors. Int J Ind Chem 4:16
16. Hima V, Bhagawan SS, Thomas S (1999) Effects of blend ratio, crosslinking systems and fillers on the morphology, curing behavior, mechanical properties, and failure mode of acrylonitrile butadiene rubber and poly(ethylene-*co*-vinyl acetate) blends. J Appl Polym Sci 71(14):2335
17. Muraleedharan Nair T, Kumaran MG, Unnikrishnan G, Kunchandy S (2008) Ageing studies of ethylene propylene diene monomer rubber/styrene butadiene rubber blends: Effects of heat, ozone, gamma radiation, and water. J Appl Polym Sci 107(5):2923
18. Rajasekar R, Pal K, Heinrich G, Das A, Das CK (2009) Development of nitrile butadiene rubber–nanoclay composites with epoxidized natural rubber as compatibilizer. Tech Report Mater Design 30(9):3839
19. Croll SG (2010) Application of the Flory-Rehner equation and the Griffith fracture criterion to paint stripping. J Coat Technol Res 7(1):49
20. Das A, Naskar N, Basu DK (2004) Thiophosphoryl disulphides as crosslinking agents for chloroprene rubber. J Appl Polym 91:1913
21. Makio M (2003) Study of vulcanization and degradation chemistry in natural rubber by solid dtate 13 C NMR and physical properties measurements. Chem Technol 76(5):1259
22. Biswas T, Naskar N, Basu DK (1995) Thiophosphoryl disulphides—a novel coupling agents for styrene budatiene rubber carboxilated nitrile rubber blend. J Appl Polym Sci 58:981
23. Naskar N, Biswas T, Basu DK (1994) Polymer blend—a novel method for the preparation of a natural rubber carboxilated nitrile rubber blend. J Appl Polym Sci 52:1007
24. Callan JE, Hess WM, Scott CE (1971) Elastomer blends. Compatibility and relative response to fillers. Rubber Chem Technol 44:814
25. Gonzalez L, Rodriguez A, de Benito JL, Marcos A (1996) A new carbon black-rubber coupling agent to improve wet grip and rolling. Rubber Chem Technol 69:266

26. Praveen S, Chattopathyay PK, Albert P, Dalvi VG, Chacraborty BC, Chattopathyay S (2009) Synergistic effect of carbon black and nanoclay fillers in styrene butadiene rubber matrix: development of dual structure, composite part A. Appl Sci Manuf 40(3):309
27. Donnet JB, Voet A (1976) Carbon black, physics, chemistry and elastomer reinforcement. M. Dekker, New York
28. Akanhari SS, Karnal KK (2008) Processing of styrene butadiene rubber-carbon black nanocomposites with gradation of crosslink density: static and dynamic characterization. Mater Sci Eng A 491(1–2):454
29. Canaud C, Antonio Sens M, Visconte LLY (2001) Properties of EPDM/SBR blends cured with peroxide and sulfur coagent. Kautsch Gummi Kunsts. 54:56
30. Curtis HL, McPherson AT (1925) Dielectric constant, power factor, and resistivity of rubber and gutta-percha. Technology papers of the Bureau of standards, Electrical properties of rubber
31. Gunasekaran S, Natarajan RK, Kala A, Jagannathan R (2008) Dielectric study of some rubber materials at microwave frequencies. Indian J Pure Appl Phys 46:733
32. Saad ALG, El-Sabbagh S (2001) Compatibility study of some polymer blend systems by electrical and mechanical techniques. J Appl Polym Sci 79:60
33. Abd-El-Messieh SL, Abd-El-Nour KN (2003) Effect of curing time and sulfur content on the dielectric relaxation of styrene butadiene rubber. Appl Polym Sci 88:1613
34. Sethu P (2006) Studies on the formulation and mechanical and dynamic properties of natural rubber/chloroprene rubber blend for rubber bushing application [TS1890. P987 2006 f rb]. PhD thesis, Universiti Sains Malaysia
35. Deuri AS, Bhowmick AK (1987) Aging of EPDM rubber. J Appl Polym Sci 34:2205
36. Ramesan MT, Mathew G, Kuriakose B, Alex R (2001) Polymer composite based on clorinated styrene butadiene rubber. Eur Polym J 37:719
37. Das A, Stöckelhuber KW, Heinrich G (2009) Influence of layered silicate on the self-crosslinking of polychloroprene and carboxylated nitrile rubber. Macromol Chem Phys 210:189
38. Das A, Costa FR, Wagenknecht U, Heinrich G (2008) Nanocomposites based on chloroprene rubber: effect of chemical nature and organic modification of nanoclay on the vulcanizate properties. Eur Polym J 44:3456
39. Fritzsche J, Klüppel M (2009) Molecular dynamic of silica and organoclay filled XNBR-nanocomposites. Kautsch Gummi Kunstst 62:16
40. Ganter M, Gronski W, Reichart P, Mulhaupt R (2001) Rubber nanocomposites: morphology and mechanical properties of BR and SBR vulcanizates reinforced by organophilic layered silicates. Rubber Chem Technol 74:221
41. Mohammad A, Simon GP (2006) Rubber–clay nanocomposites. In: Mai Y, Yu Z (eds) Polymer nanocomposites. Cambridge (chapter 12)
42. Gent AN, Henry RL, Roxbury ML (1974) Interfacial stresses for bonded rubber blocks in compression and shear. J Appl Mech 41(4):855
43. Mohammad TJ, Harijono DJ, Kamarul AA (2014) Vibration analysis of a cantilevered beam with spring loading at the tip as a generic elastic structure. Appl Mech Mater 629:407
44. Alberto P (1998) The analysis of dynamic capabilities in a competence-oriented organization. Technovation 18(3):179
45. Muhr AH (2005) Modeling the stress-strain behavior of rubber. Rubber Chem Technol 78 (3):391
46. Lewitzke C, Lee P (2001) Application of elastomeric components for noise and vibration isolation in the automotive industry. SAE technical paper 2001-01-1447. doi:10.4271/2001-01-1447

Chapter 9
Ethylene–Propylene–Diene Rubber-Based Nanoblends: Preparation, Characterization and Applications

Suzana Samaržija-Jovanović, Vojislav Jovanović, Gordana Marković, Milena Marinović-Cincović, Jaroslava Budinski-Simendić and Bojan Janković

Abstract In the process of macromolecule cross-linking, the choice of type and quantity of the both, components and experimental condition, are important to obtain the new cross-linked materials with better mechanical and chemical characteristics. The mechanism of cross-linking depends on of the rubber type and structure. That is an intermolecular cross-linking resulting in the elastomer network formation. The effect of different type of the fillers (carbon black-(CB) and silica) on the cure kinetics, mechanical properties, morphology and thermal stability of ethylene–propylene–diene rubber (EPDM), acrylonitrile–butadiene rubber (NBR) and EPDM/NBR rubber blends were investigated. The determination of cure characteristics was estimated by Monsanto Oscillating Disc Rheometer R-100. Mechanical properties such as tensile strength, elongation at break, modulus at 200 and 300 % elongation, hardness, have been measured at room temperature on an electric tensile testing machine (Zwick 1425) according to ASTM D 412. Morphology of the cross-linked system was carried out by scanning electron microscope (SEM). The non-isothermal degradation processes of EPDM, NBR, and EPDM/NBR rubber blends reinforced with carbon black and silica fillers (prepared in different ratios (as to parts per hundred of rubber (phr)) were investigated with

S. Samaržija-Jovanović (✉) · V. Jovanović
Faculty of Natural Science and Mathematics, University of Priština, Kosovska Mitrovica, Serbia
e-mail: suzana.samarzija@pr.ac.rs

G. Marković
Tigar, Nikole Pašića 213, 18300 Pirot, Serbia

M. Marinović-Cincović
Institute of Nuclear Science Vinča, University of Belgrade, Belgrade, Serbia

J. Budinski-Simendić
Faculty of Technology, University of Novi Sad, Novi Sad, Serbia

B. Janković
Faculty of Physical Chemistry, University of Belgrade, Belgrade, Serbia

G. Marković and Visakh P.M. (eds.), *Rubber Nano Blends*, Springer Series on Polymer and Composite Materials, DOI 10.1007/978-3-319-48720-5_9 thermogravimetric analysis (TGA) and derivative thermogravimetry (DTG), using the different calculation procedures. Model-free and model-fitting methods were combined to obtain the accurate kinetic triplets (A, Ea, and f(α)) for the degradation processes. For all investigated systems, an artificial compensation effect (art-CE) was found. It was found that the EPDM and the NBR rubber degradations represent the complex processes, where there are conversion regions with a constant value of the apparent activation energy (Ea).

Keywords Ethylene–Propylene–Diene rubber · Acrylonitrile–Butadiene rubber · Carbon black · Silica · Mechanical characteristics · Degradation process · Kinetic study · Apparent activation energy (Ea)

9.1 Introduction

In recent years, the production of nanocomposites based on polymer blends has markedly increased due to their well-balanced physical and mechanical properties, facile processability, and relatively low production cost [1]. Ethylene–propylene–diene monomer (EPDM) is a saturated, nonpolar rubber (i.e., very low—C=C content). The structure of the EPDM rubber is given in Fig. 9.1. Different grades of EPDM are available depending on the ratios of ethylene, propylene, and diene. The selection of the particular grade of EPDM depends upon the properties of the final product. Different ratios of monomers grant different characteristics to EPDM [2]. An extensive range of EPDM polymers are produced by varying the molecular weight, molecular weight distribution, ethylene/propylene ratio and level, and the type of diene ter-monomer. Elastomers are available containing from 50 % to more than 75 % ethylene by weight. Polymers with lower ethylene content are amorphous and easy to process. Higher ethylene content gives crystalline polymers with better physical properties, but more difficulty in processing [3]. EPDM with high ethylene content is the best candidate for thermal insulator in space applications [4]. The effect of the ethylene and propylene content on the glass transition temperature (T_g) and on the crystallinity of EPDM was investigated by Berdan and Verstrata et al. [5, 6]. They reported that T_g of EPDM varied with the content of the ethylene and propylene in a nonlinear way and generally increased with the increase of propylene content. Crystallinity of the EPDM increases with the increase of ethylene content and decreases with that of propylene content [5–7]. High crystallinity is associated with good mechanical properties and that is why high ethylene content provides good green strength to EPDM [8, 9]. On the other hand, high propylene content results low hardness EPDM, with more flexibility at low temperature and more elasticity [10]. Allen et al. report that ethylene content also enhances the cross-linking efficiency of the EPDM [11].

The first commercial ethylene–propylene–diene rubbers were made by the random copolymerization of ethylene and propylene in solution using Ziegler–Natta catalysts. Since these compounds were fully saturated, they were highly resistant to

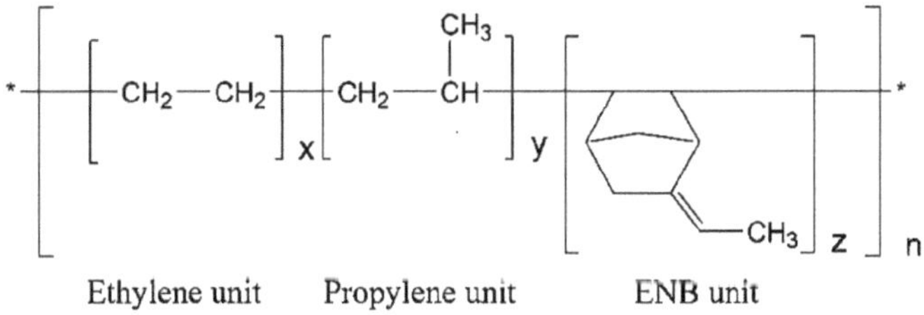

Fig. 9.1 The structure of EPDM rubber macromolecules [12]

oxidation, ozone, heat, weathering, and polar liquids. Dienes were found which were compatible with the EPM polymerization process, which had one double bond that would preferentially polymerize to leave a pendant double bond available for vulcanization [3].

In EPDM, a third monomer is incorporated to add unsaturation to the polymer. These are normally non-conjugated dienes. Non-conjugated dienes are incorporated for creating unsaturation in EPDM are 5-ethylidene-2-norbornene (ENB), 1,4-hexadiene (HD), and dicyclopentadiene (DCPD). These conjugated dienes take advantage of one double bond during copolymerization with ethylene and propylene, while it reserves the other in the side chain of EPDM for sulfur vulcanization. Generally, increase in diene content increases the rate of vulcanization. The amount of diene added ranges from 0.5 to 12 wt% [11]. Because of the difference in structures, different types of dienes provide different properties to EPDM. The most widely used one is EPDM with ENB as diene monomer (Table 9.1).

EPDM exhibits several properties, including balanced heat stability, aging resistance, elasticity (especially at very low temperatures), and water resistance; therefore, EPDM is widely applied to many rubber products (e.g., in styrene–butadiene and butadiene rubbers and as a substitute for natural rubber). One of the most important characteristic of EPDM is its ability to accept large amounts of fillers, extender oils, and other additives without affecting processability with any significant tolerance to the final properties [13]. EPDM is a rapidly growing synthetic rubber used in many industrial applications, and it is more stable than other

Table 9.1 The basic properties of diene rubber components in the EPDM rubber [12]

Diene component	Formula	Molecular mass	Physical state
(E)-5-ethylidenebicyclo[2.2.1]hept-2-ene (EN)	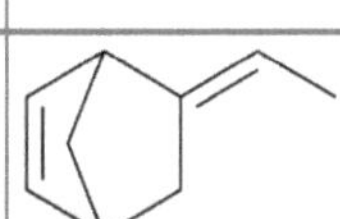	120	Fluid
3a,4,7,7a-tetrahydro-1H-4,7-methanoindene (DCPD)		132	Solid
(E)-hexa-1,4-diene (HD)		82	Solid

conventional elastomers. Unfortunately, the application of EPDM is restricted due to its poor solvent resistance and adhesion properties. Nowadays, considerable research interest is focused on new polymeric materials obtained by blending two or more polymers. The major feature of such process is that the intermediate properties are in some cases better than those exhibited by either of the single components. The majority of commercial polymer blends possess multiphase morphology and, in most cases, the satisfactory physical and mechanical properties of these materials. In addition, some modifications in terms of processing characteristics, durability, and cost can be achieved via polymer blending. Blends have been developed to meet several industrial requirements such as the need for easier processing and broadening of the properties range, either by varying the type, relative amounts, or morphology of each component. These materials can be prepared so as, for example, to combine their high mechanical strength to a better dimensional stability and thermal resistance. Elastomers based on more network precursors have become technologically important material for their diverse applications. Their physical and chemical properties recommend them as engineering materials for electric insulators, chemical industry, automotive and aircraft production, and many other areas. The blending of acrylonitrile–butadiene rubber (NBR) and EPDM rubbers was performed to achieve the best properties from each component [14]. However, blending EPDM with NBR (Fig. 9.2) can improve the aforementioned disadvantages of EPDM because polar NBR exhibits excellent solvent resistance and adhesion properties. A number of reports demonstrating the utility of EPDM/NBR blends have been previously published [15–18]. Additionally, the morphology of the polymer/polymer blend is dependent on the blend ratio, viscosity, surface property of each component, and mixing process [19, 20]. Investigators have studied the solubility properties of water in elastomers based on EPDM/NBR for use as packing materials, biomedical devices, and materials for marine applications [21]. Furthermore, EPDM, halobutyl, and NBR rubbers have been compounded for a variety of automotive applications [22]. The effect of vulcanizing systems on EPDM and NBR properties has also been studied [23, 24].

Elastomers that do not contain a filler compound are practically insignificant in any application; therefore, in practice, an elastomer is commonly combined with fillers to form a mixture, whose filler content is usually approximately 30–50 %. In the last two decades, a new class of material has found application as filler in polymer composites, namely nanofiller. Nanofillers have at least one dimension in the nanorange, i.e., 10^{-9} m. Carbon black (10–30 nm) and precipitated silica (30–

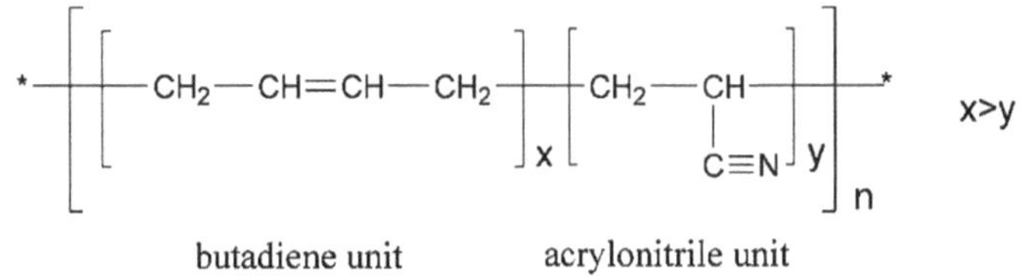

Fig. 9.2 The structure of NBR rubber macromolecules

100 nm) still remain the conventional fillers for this purpose. Obviously, the key factors for reinforcement by fillers are good dispersion and enhanced interaction with the rubber matrix. Nanofillers are classified into three types based on their dimensions in polymer matrices. The three dimensions of the particle can be in the order of nanometers (isodimensional nanoparticles) and hence appear as spherical particles. Carbon black, silica, aluminum oxide, titanium dioxide, zinc oxide, and silicon carbide are examples for nanoparticle fillers. Only two dimensions are in the nanometer scale, and the third is larger, forming an elongated structure, as in nanotubes or whiskers or nanofibers. The fillers generate stronger elastomers and are of great importance from a practical point of view. Organic and inorganic fillers, such as carbon black and silica, are often added to the elastomers.

The most popular filler for rubber modification is by far the carbon black, covering all its variants (thermal, furnace, channel, and acetylene blacks). Carbon black means elemental carbon (C) in the form of spherical particles of colloidal size. Its particles are coalesced into aggregates and agglomerates as carbon black is obtained by the combustion or thermal decomposition of hydrocarbons. The carbon black grades differ from one another regarding their particle size, aggregate form, and aggregate shape. About the terminology: Agglomerates are comprised of a large number of aggregates which are physically held together, as opposed to the continuous graphitic structure which links the particles within the aggregates (Fig. 9.3).

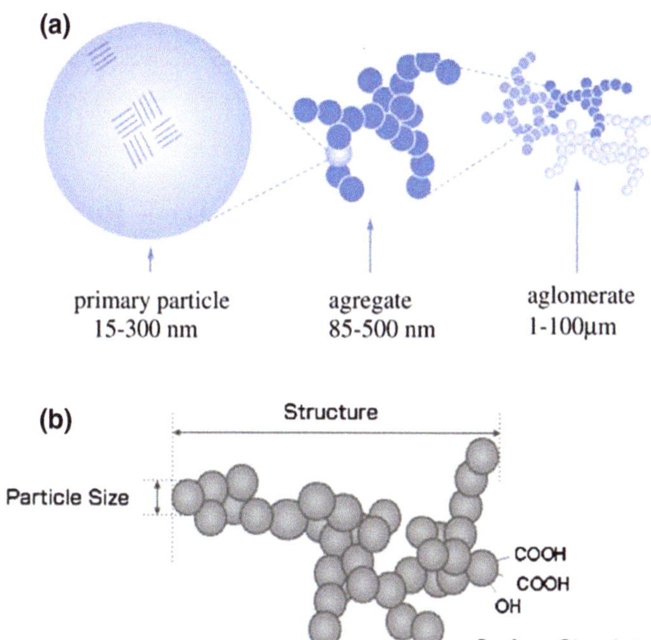

Fig. 9.3 Dimension of carbon black primary particle size and dimension of aggregate and agglomerate which have been made in cross-linking reaction of elastomer and filler (**a**) and basic properties of carbon black (**b**) [25]

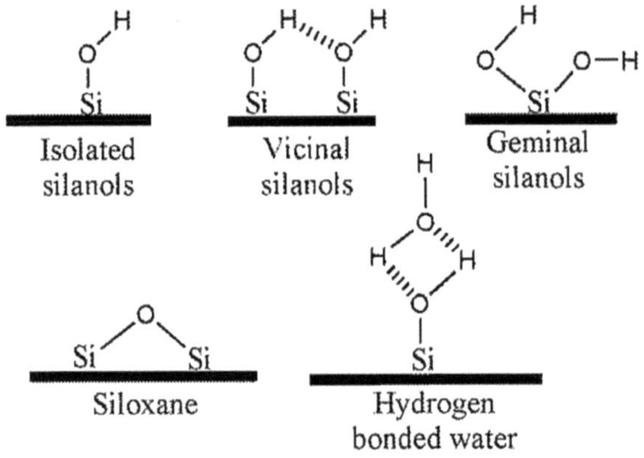

Scheme 9.1 Types of silanol group in the surface particle of SiO_2 [26]

Silica, which is amorphous, consists of silicon and oxygen arranged in a tetrahedral structure of a three-dimensional lattice. In silica, there is no long-range crystal order, and only short-range ordered domains may be found sporadically. The presence of the silanol groups (-Si-OH) on the silica surface (Scheme 9.1) induces strong particle–particle interaction. By this way, the tendency for filler agglomeration rises.

The surface silanol concentration influences the degree of surface hydration. High levels of hydration can adversely affect the physical properties of the final compound [27].

9.2 Preparation of Ethylene–Propylene–Diene Rubber Nanoblends

The materials used in this work are: network precursors: (a) ethylene–propylene–ethylidene norbornene rubber (EPDM), Vistalon 2504, ("Exxon," Belgium) with 4 % ethylidene norbornene, Mooney viscosity (ML (1 + 4)) at 125 °C, 37, ethylene content 58 % (Fig. 9.1b). Acrylonitrile–butadiene rubber (NBR), Kraynac 34–50, ("Polysar," Canada) with acrylonitrile content 33.2 %, Moony viscosity at 100 °C = 50, specific gravity 1.17 g cm^{-3}, ash content 0.5 % (Fig. 9.2). Fillers: (1) CB-High abrasion furnace carbon black, N-330 ("Degussa," Germany): primary particle size 46 nm; specific gravity 1.78–1.82 g cm^{-3}. CTAB surface area 83 ± 6 m^2/g. (2) Precipitated silica PERKASIL KS 408 ("Degussa," Germany) particle size of 22 nm, specific gravity 1.78–1.82 g cm^{-3}. BET surface area 180 m^2/g; (3) Precipitated silica SIL-1 ("Degussa," Germany) particle size of 12 nm, specific gravity 1.70–1.78 g cm^{-3}. BET surface area 200 m^2/g; Accelerators: *N*-cyclohexyl-2-benzothiazole sulfenamide (CBS): pale gray; non-hygroscopic powder; melting point 95–100 °C, specific gravity 1.27–1.31 g cm^{-3}. Tetramethylthiuram disulfide (TMTD), non-hygroscopic white powder (melting point 140 °C and specific gravity 1.33–1.40 g cm^{-3}); Antioxidants: *N*-isopropyl-*N*-phenyl-*p*-phenylenediamine (Vulkanox 4010 NA), and specific gravity 1.14–1.18 g cm^{-3}. Curing agent: Sulfur, pale yellow powder of purity 99.9 % with melting point 112 °C and specific gravity 2.04–2.06 g cm^{-3}; Activators: Zinc oxide: fine powder; purity 99 %, and specific gravity of 5.6 g cm^{-3}; Stearic acid with melting point 67–69 °C, with specific gravity of 0.838 g cm^{-3}.

Formulation of the composites with silica and carbon black is given in Tables 9.2 and 9.3, respectively. The compounds were prepared using a laboratory-mixing roll mill of dimensions 400 × 150 mm at a speed ratio of the rollers $n1/n2 = 28/22$, at a roller temperature of 40–50 °C. NBR was masticated for 3 min and blended with EPDM. After the homogenization of the rubber blend (for about 7 min), the other ingredients were added (Table 9.2). The processing time after each component addition was about 2 min. The compound rubber was allowed to stand overnight before vulcanization. The rheometric characteristics were assessed by a Monsanto oscillating disk rheometer R-100, according to the ASTM D2084-95 standard testing method. The optimum curing time (t_{c90}) was determined at 150 °C. The compound blends were molded using an electrically heated hydraulic press (Indexpell, Kerala, India) under a pressure of 60 MPa at a optimum temperature curing time. These cured sheets were conditioned before testing (24-h maturation at 25 °C). The cure characteristics, M_l (minimum torque), M_h (maximum torque), t_{c90} (optimum cure time), and t_{s2} (scorch time), were recorded, and cure rate index (CRI) was calculated according to Eq. 9.1.

$$\mathrm{CRI} = \frac{100}{t_{c90} - t_{s2}} \tag{9.1}$$

Usually, fillers are mechanically introduced into a polymer via milling [28–30]. This milling is a slow, energy-intensive process that may cause chemical degradation of the polymer chains. Moreover, undesirable particulate agglomerates often remain even after aggressive milling. Small, separated particles are ideal for reinforcing a polymer matrix. Therefore, it is desirable to explore alternative methods of incorporating inorganic fillers into elastomer mixtures.

Table 9.2 Formulation of the silica-reinforced nanocomposites based on EPDM and NBR rubbers and EPDM/NBR rubber blends [1]

Components (phr)							
EPDM/NBR	Z_nO	Stearic acid	SIL-1	4010 NA	CBS	TMTD	Sulfur
100/0	5	1	50	1	1.5	0.5	0.8
80/20	5	1	50	1	1.5	0.5	0.8
60/40	5	1	50	1	1.5	0.5	0.8
50/50	5	1	50	1	1.5	0.5	0.8
40/60	5	1	50	1	1.5	0.5	0.8
20/80	5	1	50	1	1.5	0.5	0.8
0/100	5	1	50	1	1.5	0.5	0.8

Table 9.3 Formulation of nanocomposites based on NBR/EPDM rubber blends reinforced with CB and silica [31]

Component (phr)								
NBR/EPDM	ZnO	Stearic acid	Fillers		4010 NA	CBS	TMTD	Sulfur
			CB	Silica				
80/20	5	1	50	0	1	1.5	0.5	0.8
80/20	5	1	60	0	1	1.5	0.5	0.8
80/20	5	1	70	0	1	1.5	0.5	0.8
80/20	5	1	80	0	1	1.5	0.5	0.8
80/20	5	1	90	0	1	1.5	0.5	0.8
80/20	5	1	100	0	1	1.5	0.5	0.8
80/20	5	1	35	35	1	1.5	0.5	0.8
80/20	5	1	0	70	1	1.5	0.5	0.8

9.3 Rheometric Characteristics of Ethylene–Propylene–Diene Rubber Nanoblends

Figure 9.4 presents the torque–cure time plots generated from the silica-reinforced nanocomposites based on EPDM and NBR rubbers and their EPDM/NBR rubber blends (Table 9.2). The initial decrease in torque is due to the softening of the matrix. The torque then increased due to the cross-linking between the macromolecular chains. It is shown in Fig. 9.4 that as the percentage of NBR increased in the blend systems, the torque also increased.

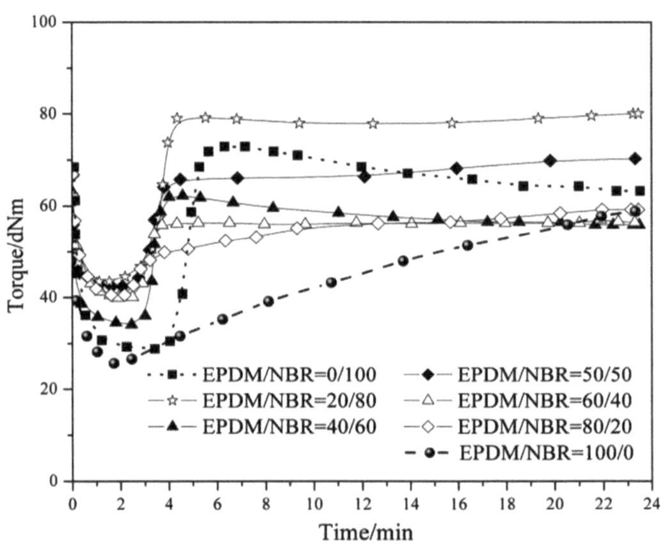

Fig. 9.4 The relationship between torque and time during the curing of silica-reinforced nanocomposites based on EPDM and NBR rubbers and EPDM/NBR rubber blends with 50 phr of silica

Table 9.4 shows the rheometric characteristics of EPDM/NBR rubber blends with 50 phr of silica. Regular variations in maximum torque (M_h) and optimum cure time (t_{c90}) have been observed for the reinforced blends. EPDM/NBR = 100/0 system shows maximum optimum cure time. Optimum cure time decreases with increasing NBR content in the blends. The t_{c90} and rate of vulcanization (CRI) values are lowest for the blend with the higher percentage of NBR. Scorch time (t_{s2}) is the time taken for the minimum torque (M_l) value. It increases by two units. It is a measure of the premature vulcanization of the material. The maximum values of the scorch time and maximum scorch safety has EPDM/NBR = 0/100 blend systems.

The maximum torque increases along with the NBR rubber content, indicating the enhancement in cross-linking density. The calculated values for cure rate increase along with the percentage of NBR content in the blend. This indicates that NBR rubber is a cure-activating component in EPDM/NBR rubber blends. This is due to the presence of unsaturation in the butadiene component of NBR. Cross-links (sulfuric linkages) are formed between unsaturated sites and macromolecules during vulcanization. As the number of NBR molecules (as weight percentage) increases, the number of active sites for vulcanization increases, too.

Table 9.5 shows the summarized rheometric characteristics of NBR/EPDM elastomeric materials under investigation with and without carbon black and silica.

Table 9.4 Rheometric characteristics of the silica-reinforced nanocomposites based on EPDM and NBR rubbers and EPDM/NBR rubber blends

EPDM/NBR (phr)	M_h (Nm)	M_l (Nm)	t_{S2} (s)	t_{C90} (s)	CRI (s^{-1})
100/0	6.65	2.88	96	1212	0.1
80/20	6.30	4.40	96	456	0.3
60/40	6.30	4.40	126	172	2.2
50/50	7.50	4.70	132	174	2.4
40/60	7.00	3.70	144	181	2.7
20/80	8.90	4.70	156	204	2.1
0/100	7.80	2.48	240	288	2.1

Table 9.5 Rheometric characteristics of NBR, EPDM, and NBR/EPDM rubber blends reinforced with CB and silica

NBR/EPDM (phr)	Filler (phr)		M_h (Nm)	M_l (Nm)	t_{S2} (s)	t_{C90} (s)	CRI (s^{-1})
	CB	Silica					
80/20	50	0	8.36	4.50	48	96	2.08
80/20	60	0	9.93	4.46	126	216	1.11
80/20	70	0	11.07	3.95	103	233	0.76
80/20	80	0	14.96	5.64	102	501	0.25
80/20	90	0	15.19	6.10	89	350	0.38
80/20	100	0	17.20	6.60	60	366	0.32
80/20	35	35	10.78	7.34	138	444	0.32
80/20	0	70	8.13	6.10	12	660	0.2

The minimum torque (M_l) (Table 9.5) can be taken as a measure of the viscosity of the masticated rubber. Whenever there is excessive mastication, a sharp decrease in viscosity is registered. It is observed that the addition of carbon black to rubber blend tends to build torque more rapidly than silica. Regular variations in maximum torque (M_h) and optimum cure time (t_{c90}) were observed for the reinforced blends. NBR/EPDM = 80/20 silica-reinforced rubber blend had higher values for optimum cure time than other NBR/EPDM = 80/20 reinforced rubber blends. Scorch time (t_{s2}) is the time taken to reach a 2-unit increase in torque value above minimum. It is a measure of the premature vulcanization of the material.

The maximum scorch time and maximum scorch safety was observed for the NBR/EPDM blend systems with equivalent amount of CB and silica.

Acidic compounds retard the vulcanization of elastomers. For this reason, precipitated silica, which contains a large number of acidic silanol (Si–OH) groups, is not generally incorporated without an activator in EPDM. The surface of silica is acidic, and therefore, a strong hydrogen bond with polymer groups is formed. NBR rubber has nitrile groups ($-C{\equiv}N$), which are basic, resulting in the hydrogen bonding of the silanol group of silica and a strong silica/NBR interaction. Silica particles also adsorb polar curative molecules on their surfaces, rendering the deactivation of the vulcanization and thus lowering the curing rate.

9.4 Mechanical Properties of Ethylene–Propylene–Diene Rubber Nanoblends

Mechanical properties, such as tensile strength, modulus (%), and elongation at break, were measured with a Zwick 1425 tensile tester according to the ASTM D412-98 standard testing method using a crosshead speed of 500 mm/min and at 25 ± 2 °C. The tensile properties of the blends were examined according to the ASTM D-412 standard testing method. The tear test was conducted, as per the ASTM D-624 standard testing method using a 908 angle test pieces. The experimental conditions and equipment for the tear testing were the same as those of the tensile testing. Five samples from each formulation were tested. The hardness of the samples was measured, as per the ASTM D-2240 standard testing method. For hardness measurements, the sheets having an effective thickness of 6 mm were used. At least five measurements were recorded, and the average values were reported. The rebound resilience measurements were performed using a Schob pendulum, following the ASTM D7121-05 standard testing method. The abrasion resistance was determined as the relative volume loss from the wear test using a Zwick DIN abrader, as per the ASTM D-2228 standard testing method, and the rolling sliding test was employed. In this study, a 0.1-kN load was applied to the rolling sliding cylinder on SiC paper.

Tensile strength is a complex function consisting of the nature and type of cross-links, cross-linking densities, and chemical structure of the used rubber. It is well known that if rubber is deformed by an external force, part of the input energy is stored elastically in the chains and is available (released upon crack growth) as a driving force for fracturing. The remaining energy is dissipated through molecular motions by heat; as such, it is made unavailable to break the chains. At higher cross-linking levels, chain motions become restricted, and the dense network is incapable of dissipating as much energy. This results in a relatively facile brittle fracture at low elongation. The reinforcing effect of filler in rubber depends on the interfacial structure between the polymer and the filler. As the size of the filler reduces, the surface area reduces, and consequently, the interface between the filler and the polymer increases. In the case of nanocomposites, the volume of interface is much larger than that for micro- or macro-composites, and hence, nanocomposite properties are considerably enhanced than micro- or macro-composites [1]. The restrictions imposed by the nanofiller in polymer chain conformations will alter the molecular mobility, relaxation behavior, and thermal transitions of the polymer. It is important to note that fine size alone cannot guarantee good reinforcement as fine particles tend to form agglomerates. The distribution of the nanofiller in the polymer matrix is one of the most important parameters that decide the properties of a nanocomposite.

As the NBR rubber content increases in EPDM/NBR rubber blends, the tensile strength increases to its optimal value and then decreases (Table 9.6).

The tensile strength showed a maximum with the EPDM/NBR = 20/80 rubber blend. If the EPDM content is above 20 phr, the tensile strength decreases. The elongation at break maximum was observed with the same blend. A maximum modulus value was found with the EPDM/NBR = 50/50 blend system. When the NBR rubber content was increased, the hardness values decreased. The abrasion resistance increased with increasing NBR rubber content. The percentage of rebound resilience for the different EPDM/NBR blends afforded average values with low variance between the silica-reinforced nanocomposites based on EPDM and NBR rubbers.

Table 9.6 Mechanical properties of the silica-reinforced nanocomposites based on EPDM and NBR rubbers and EPDM/NBR rubber blends

EPDM/NBR (phr)	Modulus at 200 % elongation (MPa)	Modulus at 300 % elongation (MPa)	Tensile strength (MPa)	Elongation at break (%)	Hardness (Sh°A)	Rebound resilience (%)	Abrasion (mm^3)
100/0	1.86	1.96	3.04	850	74	26	265
80/20	2.84	2.84	2.84	150	75	41	235
60/40	3.34	3.83	4.02	305	75	37	367
50/50	2.94	3.92	5.7	510	74	34	328
40/60	2.65	3.53	7.88	680	73	31	226
20/80	2.11	3.43	11.1	850	71	24	184
0/100	1.96	2.35	16.2	940	69	29	193

The mechanical properties of the silica-reinforced EPDM/NBR = 20/80 rubber blend were improved, and the chemical bonds between phases formed during the networking process. In addition to the networking process, cross-linking between the EPDM and NBR rubbers via sulfide bonds was possible during the co-cross-linking process (Scheme 9.2).

The addition of fillers to polymeric materials leads to the improvement in the mechanical properties of the polymer matrix. The reinforcement effect is directly related to the properties of the interphase and depends on the nature of the specific interactions between polymer and reinforcing fillers [32]. The incorporation of filler into elastomers imparts many interesting and useful properties to the particle-filled composite material. It is well known that the properties mainly depend on the dispersion condition of filler particles and their principal relevant properties: particle size, surface area, aggregate structure, surface activity, and rubber–filler interactions [33]. Carbon black, generally supplied as pellets, has to be dispersed during mixing into smaller entities (such as agglomerates and aggregates). Large particle–particle interactions result in inhomogeneous dispersion and distribution of the filler, processing problems, poor appearance, and inferior properties. Highly sophisticated process technology was the key to enhance the surface activity of these carbon black grades. This progress in process technology resulting in the structuring of carbon black particles was used to generate special carbon blacks for different applications [34]. The effect of added carbon black is directly related to the properties of the interphase and due to the specific interactions between rubber macromolecules and active filler surfaces. It is well known that the properties mainly depend on the dispersion condition of filler particles and their principal relevant properties: particle size, surface area, aggregate structure, surface activity, and rubber–filler interactions [35, 36]. It is obvious that to obtain as many interactions as possible, the polymer active segments must have access to the majority of the carbon black active sites. Since the monounit of CB is the aggregate, the most possible isolated aggregate must be made available to be in contact with the polymer chains and their unsaturation domains. The surface activity, a poorly defined term, but widely used in the filler field, can in a chemical sense be related to different chemical groups on the surface. They are carboxyl, quinone, phenol, and lactone groups (Fig. 9.5).

In a physical sense, variations in surface energy determine the capacity and energy of adsorption. The surface chemistry of carbon blacks has a significant effect

Scheme 9.2 Schematic presentation of the cross-linking of the polymer chains of EPDM and NBR rubbers

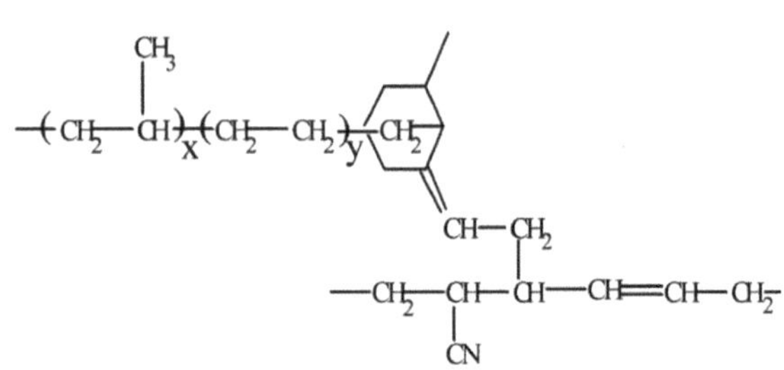

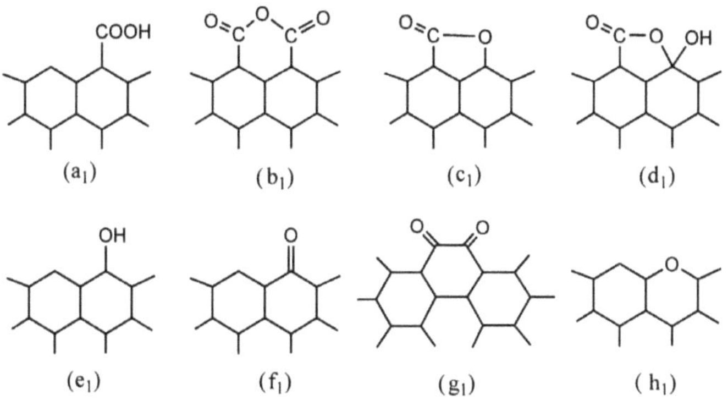

Fig. 9.5 Structures of carbon black surface: (a_1) carboxylic, (b_1) carboxyl anhydride, (c_1) lactonic, (d_1) lactolic, (e_1) phenolic, (f_1) carbonylic, (g_1) quinonic and (h_1) ether groups [37]

only on the vulcanization behavior of filled compounds. Optimal reinforcing power can be achieved only if the filler is well dispersed in the rubber matrix. The chemical or physical interaction between the filler and the rubber is a further important factor in the reinforcing effect [36].

In the case of carbon black, the filler–polymer interaction is mainly of physical nature (physisorption) [38]. Interaction between fillers and rubbers has a significant effect on reinforcement properties of a filled rubber. The chemical and physical properties of both rubbers and filler and the amount of each present in a compound influence this interaction. Rubber–rubber interaction mainly occurs when blends of rubber are used in compounds and are considered to be not as significant as filler–rubber and filler–filler interactions. Filler–rubber interactions are described by the compatibility of the filler with the rubber, while filler–filler interaction is described by the attraction of filler to itself and the ability to form a network. The most important effect of filler–rubber interactions has to do with the occlusion of rubber. The so-called bound rubber is trapped between or within aggregates where it is no longer part of the elastically active rubber matrix. As the amount of carbon increased, the bound rubber content also increases. Carbon black has a highly dispersive component with a stronger filler–rubber interaction and a weaker filler network [39]. Filler–filler interactions are a primary mechanism in reinforcement, especially at high filler loading. These interactions depend on chemical interactions between the filler particle surfaces (filler–filler, filler–rubber), physical interactions (van der Waals forces, hydrogen bonding), morphology of the filler network, and filler volume fraction. Table 9.7 shows the mechanical properties of CB-reinforced NBR/EPDM rubber blends.

As the CB content increases, the tensile strength increases to its optimal value and then decreases (Table 9.7). NBR/EPDM/CB (80/20/70) rubber blend has maximum tensile strength values. When the CB content increases, the elongation at break and rebound resilience values decrease. The hardness values increase with the increase in CB content in rubber composites. The mechanical properties and

Table 9.7 Rheometric characteristics of EPDM/NBR/CB rubber blends

NBR/EPDM (phr)	CB (phr)	Tensile strength (MPa)	Elongation at break (%)	Hardness (Sh°A)	Rebound resilience (%)	Abrasion (mm^3)
80/20	60	15	350	76	25	78
80/20	70	17.41	275	80	23	85
80/20	80	15.4	258	85	21	90
80/20	90	16.68	215	86	20	100
80/20	100	15.94	170	89	19	107

chemical bonds between phases formed during the networking process of the NBR/EPDM/CB (80/20/70) rubber blend composite improved. The cross-linking process of rubber happens through the sulfur bond formation and EPDM and NBR co-cross-linking during the C–C bonds is formed on the border of their phases [1]. Figure 9.6 shows the mechanical properties of NBR/EPDM rubber blends such as hardness, tensile strength, and elongation at break before and after thermal aging at 100 °C for 72 h. The retained tensile strength, elongation at break percentage, and hardness values were then calculated before and after aging (Table 9.8).

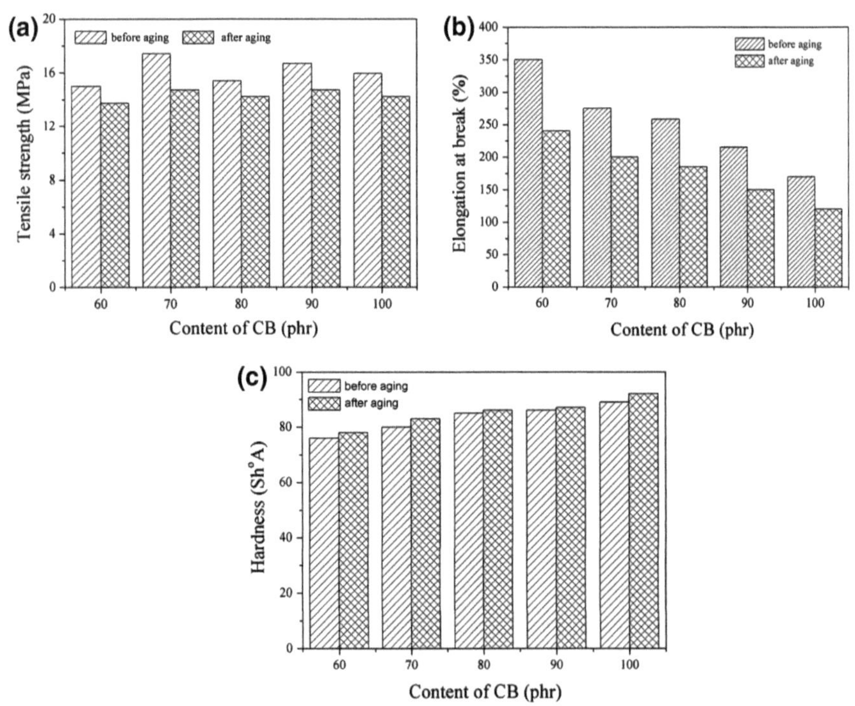

Fig. 9.6 Variation of the **a** tensile strength, **b** elongation at break, and **c** hardness of NBR/EPDM rubber blends with different content of CB filler before and after aging

Table 9.8 Changing of some mechanical properties of EPDM/NBR rubber blends reinforced with different loading CB after aging

NBR/EPDM (phr)	CB (phr)	Tensile strength (%)	Elongation at break (%)	Hardness (Sh°A)
80/20	60	−8.4	−31.4	+2
80/20	70	−4.4	−27.3	+3
80/20	80	−18.3	−28.2	0
80/20	90	−11.7	−30.2	+1
80/20	100	−10.7	−29.4	+3

The hardness of all rubber blend composites increases with increasing CB content, as shown in Fig. 9.6c. This can be attributed to the cross-linking density being increased after thermal aging. The tensile strength is a complex function of the nature and type of cross-links, cross-linking densities, the chemical structure of the used elastomers, and the changes associated with degradation. NBR/EPDM rubber blend with 70 phr of CB has maximum value of tensile strength before and after aging (Fig. 9.6a). Values of elongation at break decrease with increasing CB content. After thermal aging, mechanical properties' values decrease and degradation process can be noticed.

Table 9.9 shows that silica-reinforced rubber blends exhibit higher tensile strength, hardness, and rebound resilience, but lower elongation at break than CB-reinforced rubber blends. Compared to pure NBR and EPDM rubbers, all mechanical properties are improved. At low filler loading, relatively large distances between the filler particles exist. In this case, the rubber chains are attached to one filler particle because of a single attachment or multiple attachments of chain segments of one rubber molecule. With the increase in the concentration of the filler, interparticle attachment predominates, where in rubber molecules it connects two or more filler particles [31].

To investigate the influence of thermal aging on the mechanical properties of the rubbers, the cross-linking reactions were performed in an air-circulating oven operated at 100 °C for 50 h. The retained tensile strength percentage and elongation at break value were calculated (Table 9.9). Tensile properties (tensile strength and elongation at break) and hardness were measured before and after aging.

Based on the results (Table 9.10; Fig. 9.7), it can be noted that for both silica- and carbon black-reinforced NBR/EPDM nanocomposites exhibited smallest changes in mechanical properties after thermal aging. Silica-reinforced nanocomposites exhibited higher changes in all mechanical properties than carbon

Table 9.9 Mechanical properties of EPDM/NBR rubber blends reinforced with CB and silica

NBR/EPDM (phr)	Filler (phr)		Tensile strength (MPa)	Elongation at break (%)	Hardness (Sh°A)	Rebound resilience (%)	Abrasion (mm^3)
	CB	Silica					
80/20	70	0	17.41	275	80	23	85
80/20	0	70	12.4	370	91	42	143

Table 9.10 Changing of some mechanical properties of EPDM/NBR rubber blends reinforced with CB and silica after aging

NBR/EPDM (phr)	Filler (phr)		Tensile strength (%)	Elongation at break (%)	Hardness (Sh°A)
	CB	SiO_2			
80/20	70	0	−4.4	−27.3	+3
80/20	0	70	−10.4	−56.7	+4

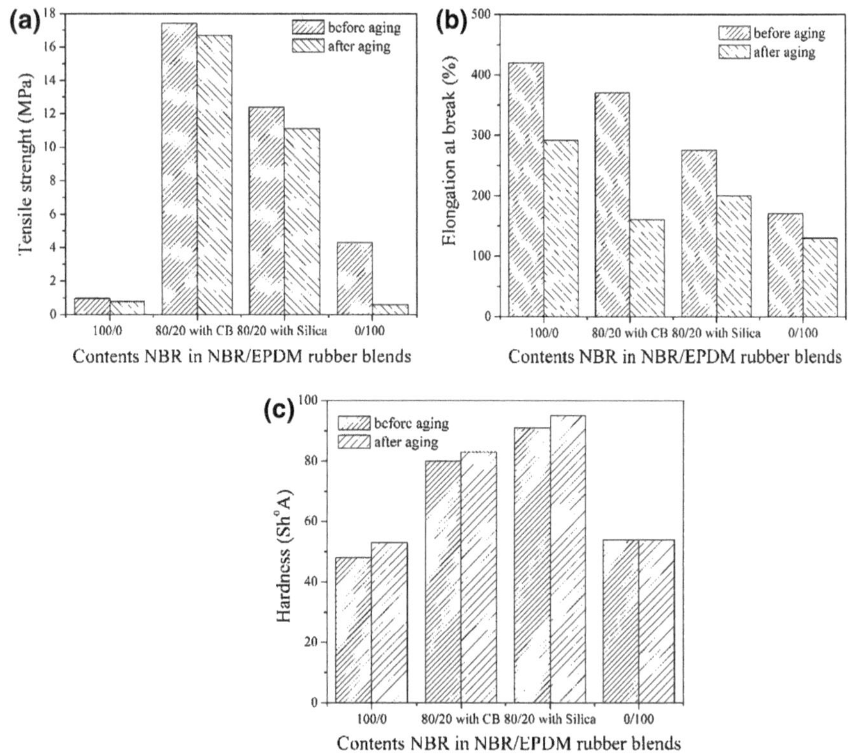

Fig. 9.7 Variation of the **a** tensile strength, **b** elongation at break, and **c** hardness of NBR, EPDM, and rubber blends reinforced with CB and silica before and after aging

black-reinforced nanocomposites. The sulfur vulcanization of the unsaturated rubbers occurs through complicated radical substitution forming mono-, di-, or polysulfide bridges and sulfur intracyclization of the polymer molecules. However, at higher temperatures, the sulfur cross-links are less effective and the physical properties are weaker due to the dissociation of sulfur bonds and rubber chains, which can be attributed to the increased cross-linking density after thermal aging. This can be explained by the sulfur networking addition process of the rubbers and polysulfide cross-linking density reduction process. This polysulfide reacts further to form mono-, di-, and cyclic sulfide bonds during vulcanization via the dissociations, recombination, and rearrangement of the sulfur linkages.

9.5 Thermal Properties of Ethylene–Propylene–Diene Rubber Nanoblends

Generally, the thermal degradation of (polymeric material) polymers follows more than one mechanism. The existence of more than one concurrent chemical reaction is accompanied by the other physical phenomena such as evaporation and ablation introduce further complications for the modeling of degradation kinetics. Kinetic study of thermal degradation provides useful information for the optimization of the successive treatment of polymer materials in order to avoid or at least limit thermal degradation. The analysis of the degradation process becomes more and more important due to an increase in the range of temperatures for engineering applications, recycling of post-consumer plastic waste, as well as the use of polymers as biological implants and matrices for drug delivery, where depolymerization is an inevitable process affecting the lifetime of an article. The degradation of EPDM and its blends has been studied under a variety of conditions including non-isothermal [40] and isothermal degradation [41] under a nitrogen atmosphere, photooxidation [42, 43], radiochemical oxidation [44–46], thermo-oxidation [47, 48], and ultraviolet irradiation [49]. The influence of the factors on the degradation of EPDM, such as metallic oxides [50] and carbon black [51, 52], has also been investigated. It was shown that the structural effects can have influence on the thermal properties and degradation behavior of EPDM [53, 54]. The thermal stability of the NBR and EPDM samples was investigated by non-isothermal thermogravimetric (TG) analysis using a PerkinElmer TGS-2 instrument. The samples with approximately $m \approx$ 10–15 mg were heated at three different linear heating rates of β = 10, 20, and 40 °C min^{-1}, in a dynamic nitrogen atmosphere (the flow rate of 25 cm^3 min^{-1}), in the temperature range from 25 °C up to 600 °C. TGA mass loss curves of EPDM samples, performed at constant heating rates of 10, 20, and 40 °C min^{-1}, are shown in Fig. 9.8.

TG curves of the EPDM samples (Fig. 9.8) show an effect that is reflected in the curve shifting to higher temperatures with an increase of heating rate. The observed heating curves are not completely parallel (Fig. 9.8) and show a smaller degree of asymmetry compared with the TG curves in the case of the NBR samples (Fig. 9.9).

For NBR samples, all DTG curves (Fig. 9.9) are asymmetric, with a single, highly expressed peak. It may be noted that at the heating rate of 10 °C min^{-1}, the one, a less pronounced, shoulder exists at degradation temperature of T = 430.49 °C (Fig. 9.9). At the higher heating rates (20 and 40 °C min^{-1}), we cannot see the shoulders. In the case of EPDM samples, the DTG curves (Fig. 9.8) at all heating rates are also asymmetric, with a single, highly expressed peak. On the presented

DTG curves, there are no shoulders. At all heating rates, the corresponding DTG curves (Fig. 9.8) are much narrower than in the case of DTG curves for the NBR degradation process (Fig. 9.9). In addition, we have observed that the temperatures at which the maximum of the derivative of mass loss occurs are higher (except on the heating rate of 40 °C min^{-1}) for the EPDM samples than those occurring in the case of NBR degradation process (Figs. 9.8 and 9.9; Table 9.11). This suggests a stabilizing effect of the N_2 in the case of EPDM degradation process. All this indicates that the degradation behavior of the NBR and EPDM samples under a nitrogen atmosphere may be different.

From the analysis of the general characteristics of the TG curves, we can conclude that the thermal degradation of the NBR samples, which contain the butadiene units [55], strongly depends on the heating rate. In the case of the thermal degradation of the EPDM samples, this dependence is almost non-existent. Based on the detailed analysis of the behavior of TG and DTG curves at the different heating rates (Fig. 9.8a, b); Table 9.11), we can conclude that the degradation

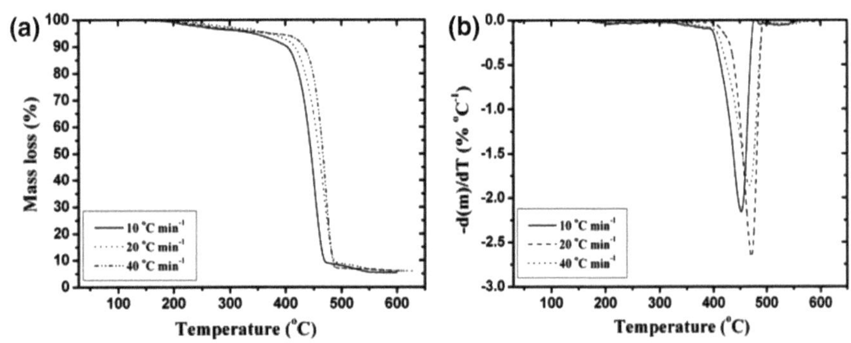

Fig. 9.8 Thermogravimetric (TG) curves (**a**) and derivative thermogravimetric (DTG) curves (**b**) of the EPDM sample degradation, performed at the different heating rates (10, 20, and 40 °C min^{-1})

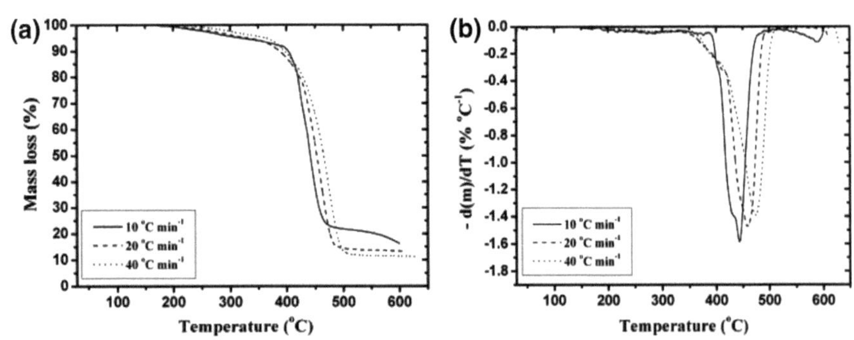

Fig. 9.9 Thermogravimetric (TG) curves (**a**) and derivative thermogravimetric (DTG) curves (**b**) of the NBR sample degradation, performed at the different heating rates (10, 20, and 40 °C min^{-1})

Table 9.11 Temperature at maximum mass loss (T_{max}), mass loss, selected mass loss, total mass loss, and apparent activation energy for NBR, EPDM, and NBR/EPDM/CB composites with 70 phr filler

Sample	Φ (°C/min)	T_m (°C)	Mass loss (%)	$T_{5\%}$ (°C)	$T_{10\%}$ (°C)	$T_{30\%}$ (°C)	Total mass loss (%)	E_a (kJ/mol)[a]
NBR	10	439.5	47.2	293.1	388.2	422.2	79.6	156.0
	20	461.2	63.2	308.8	381.4	432.9	86.2	
	40	473.6	62.5	353.1	392.6	441.2	88.6	
EPDM	10	450.9	63.2	334.3	400.2	431.6	94.6	75.0
	20	471.0	63.0	363.5	430.7	455.3	93.2	
	40	467.0	64.0	360.5	415.0	444.9	93.4	
NBR/EPDM/CB	10	463.4	35.9	276.1	367.9	432.9	51.9	45.1
	20	465.2	39.3	287.9	374.6	430.7	54.1	
	40	472.4	38.3	335.8	388.8	437.4	53.9	

[a]The method for calculating the Eac and Ear is described in our earlier works [25, 52, 57]

processes of EPDM samples cannot be expressed through the simple, one-step kinetic mechanism.

According to the degradation processes, shown in Fig. 9.8, the temperatures of the peaks (T_m) of degradation can be determined. Table 9.11 shows the values of characteristic temperatures (T_m and $T_{5\%}$, $_{10\%}$ and $T_{30\%}$) and also the values of the total mass loss. It is shown in Table 9.11 that the temperatures of the onset and the peak of degradation are both increased with the increase in the heating rate, for all considered samples. Comparing the values of the onset temperatures (T_i), we can conclude that the NBR sample shows the higher thermal stability than the corresponding EPDM sample (Table 9.11). The degradation temperature was found to increase with the increase in the heating rate (β), which corresponds to the time–temperature superposition principle [55].

The apparent activation energy values (E_a) of degradation were determined by the Kissinger method [56] based on the following equation:

$$\ln\left(\Phi/T_m^2\right) = \ln\left(n\mathrm{RAW}_m^{n-1}/E_a\right) - E_a RT_m \quad (9.2)$$

where Φ is the heating rate, T_m is the temperature at the maximum rate of mass loss, n is the apparent reaction order, R is the universal gas constant, A is the pre-exponential factor, and W_m is the weight of the sample at the maximum rate of weight loss. The E_a values were then obtained by the slope of the linear relationship $\ln\left(\Phi/T_m^2\right)$ versus $1/T_m$. We used as T_m values the temperatures of the DTG peaks. We could single out reliable T_m values only for the first DTG peak, because of the broad and irregular shape of the peak at higher temperature. The T_m values of the first degradation stage of the NBR, EPDM, and NBR/EPDM rubber samples with 70 phr CB at the used scanning rates are listed in Table 9.11. E_a for the NBR/EPDM/CB rubber blend composite is lower (45.1 kJ/mol) compared to E_a for the rubber based on NBR (156 kJ/mol) and EPDM (75 kJ/mol), which indicates the easier formation of rubber blend. Also, NBR/EPDM/CB rubber blend composite is a less stable system according to lower temperatures of degradation ($T_{5\%}$, $T_{10\%}$, $T_{30\%}$) compared to the NBR and EPDM individually.

The TG and DTG curves of NBR/EPDM/CB composite with 70 phr of CB filler with a different heating rate are shown in Fig. 9.10a, b.

The information about the matrix-to-filler and filler-to-filler interactions represents a very important part of understanding the properties of composites [58]. Strong interaction between the polymer and the filler particles is essential for achieving good mechanical properties and the thermal stability. These interactions manifest themselves in different ways when investigated by methods of thermal analysis [59]. Since decomposition temperatures of matrix and filler are well separated in inorganic filler/organic rubber composites, the thermogravimetric analysis (TGA) is useful for the determination of the composition of such composites, e.g., rubber/filler weight fractions. Very good examples are quantification of the amount of in situ silica in rubber vulcanizates [60, 61]. TGA affords the most suitable method to quantify the amount of silica based on the big difference in thermal stability between organic rubbers and inorganic fillers. Decomposition processes can be studied in much more detail if the method is coupled with an analytical method for gases (FTIR or MS). Filler-to-matrix and filler-to-filler interactions influence the values of all three important parameters obtained from TGA [62]. Decomposition temperature, rate of mass change, and the apparent activation energy of the decomposition process. The interactions were found to have an opposite effect on the apparent activation energy of the decomposition of the matrix. With an increase of the filler-to-matrix interaction, the apparent activation energy increases due to the physical cross-linking and reduction of the mobility of polymer chains in the interface layer [63]. On the other hand, stronger filler-to-filler interaction accompanied by the appearance of the physical network of filler particles at higher filler content increases the amount of unbound rubber and lowers the apparent activation energy of the decomposition. Some authors reported no change

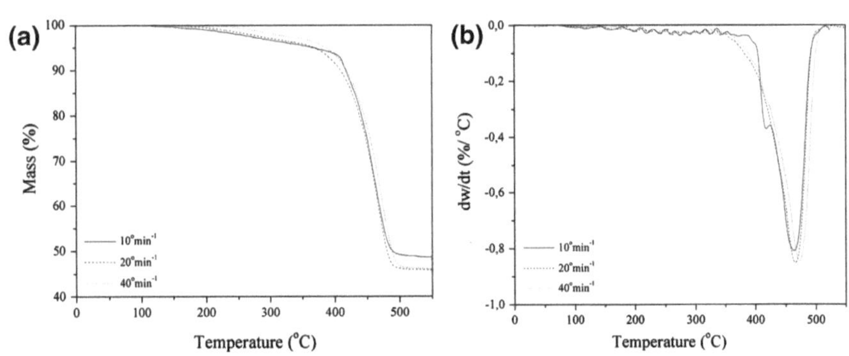

Fig. 9.10 Thermogravimetric (TG) curves (**a**) and derivative thermogravimetric (DTG) curves (**b**) of NBR/EPDM composites with 70 phr of CB at the different heating rates

of the apparent activation energy in the presence of carbon black filler [64]. Thermogravimetric analysis of the natural rubber and synthetic rubbers loaded with carbon black is not influenced by the presence of carbon particles [65]. Addition of silica to a rubber compound offers a number of advantages such as improvement in tear strength, reduction in heat buildup, and increase in the compound adhesion in multicomponent products such as tires. Silica with fine particle size will be effective in improving the interactions. The use of nanosilica as a component of the dry bonding system is expected to improve the thermal stability of composites based on short nylon fiber and elastomers [66].

Figure 9.11 shows typical thermogravimetric (TG) and the derivative thermogravimetric (DTG) (inset) curves for the NBR/EPDM blends reinforced with silica and carbon black/silica fillers in different phr's. All observed thermogravimetric curves are recorded at a constant heating rate of $\beta = 20$ °C $\min^{-1}$.

At a given heating rate (Fig. 9.11), we can see that there is a mass loss of samples by nearly 2.4 %, which can be attributed to the evaporation of the moisture content. The evaporation of water molecules in all the analyzed samples corresponds approximately to the temperature up to $T = 140.00$ °C (Table 9.12). It should be noted that since in the preparation of experimental samples, we used the commercial silica, when it leaves the factory, it has 0.5–3.0 % moisture, which is partially needed for the thickening process. The relatively high evaporation temperature can be explained by the higher temperatures required for breaking the

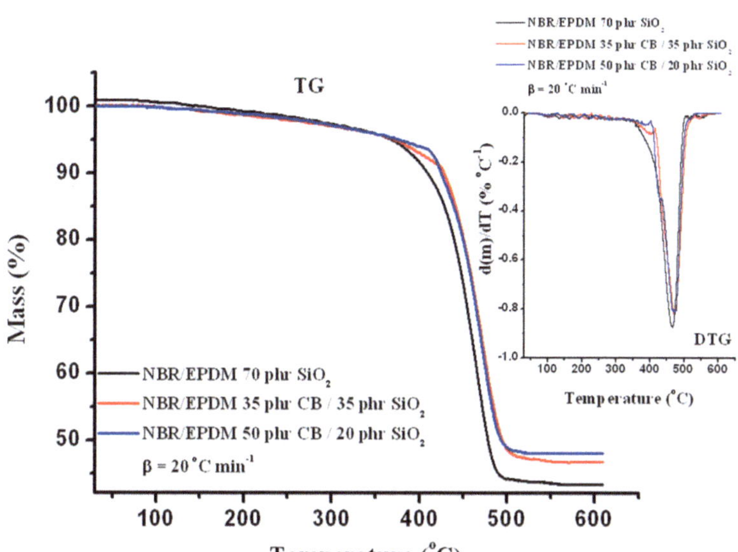

Fig. 9.11 Thermogravimetric (TG) and the derivative thermogravimetric (DTG) curves (presented in the *upper right* corner as an *inset*) of the NBR/EPDM 70 phr SiO_2, NBR/EPDM 35 phr CB/35 phr SiO_2, and NBR/EPDM 50 phr CB/20 phr SiO_2 degradation processes under nitrogen atmosphere, performed at a constant heating rate of $\beta = 20$ °C $\min^{-1}$ (*CB*-carbon black) [67]

Table 9.12 Values of T_i, T_p, and Δm_f for the NBR/EPDM 70 phr SiO_2, NBR/EPDM 35 phr CB/35 phr SiO_2, and NBR/EPDM 50 phr CB/20 phr SiO_2 degradation processes, at the different heating rates (10, 20, and 40 °C min^{-1})

Heating rate, β (°C min^{-1})	NBR/EPDM 70 phr SiO_2		
	T_i (°C)[a]	T_p (°C)[b]	Δm_f (%)
10	113.70	453.28	43.30
20	121.16	463.83	43.35
40	131.75	487.11	43.47
Heating rate, β (°C min^{-1})	NBR/EPDM 35 phr CB/35 phr SiO_2		
	T_i (°C)[a]	T_p (°C)[b]	Δm_f (%)
10	123.59	419.81[c]	46.71
		461.04	
20	127.51	472.86	46.75
40	129.63	491.19	46.81
Heating rate, β (°C min^{-1})	NBR/EPDM 50 phr CB/20 phr SiO_2		
	T_i (°C)[a]	T_p (°C)[b]	Δm_f (%)
10	133.87	426.15[c]	47.96
		466.08	
20	137.95	428.27[c]	48.08
		475.91	
40	142.19	494.11	48.18

[a]From TG curve
[b]From DTG curve
[c]"Shoulder" on the corresponding DTG curve

hydrogen bond formation between neighboring aggregates of silica in the regular network [68]. In addition, because the silica surface is acidic, it forms a strong hydrogen bond with basic materials. It may also be noted that the values of degradation temperature increase (Fig. 9.11) with the degree of loading with carbon black filler. This behavior may be attributed to the increased participation of the interface linking, which is usually accompanied by increased thermal stability. From the observed TG curves, we can see that the main degradation stage occurs in the temperature range of $350.00 \le T \le 510.00$ °C, which is indicated with a clear peak at the corresponding DTG curves (Fig. 9.11). Taking into account that the silica as filler increases the thermal stability of polymer blends and composites [33–36], from the results of peak temperatures (T_p) for the NBR/EPDM 70 phr SiO_2 at $\beta = 20$ °C min^{-1} (for example) (Table 9.12) (in comparison with T_p values for the pure NBR and EPDM rubbers [69]), we can conclude that T_p temperature is shifted to a higher side, i.e., from 459.41 °C (NBR) and 461.62 °C (EPDM) to 463.83 °C (Table 9.12). This indicates the improved thermal stability of the silica-reinforced NBR/EPDM rubber blends. Such a significant improvement in thermal stability of NBR/EPDM/silica rubber compounds can be attributed to the presence of silica mesopore, which offers a great barrier effect to hinder the formation of small molecules during the thermal decomposition and simultaneously resists their movement during desorption from the surface. The introduction of carbon black filler and its increase from 35 to 50 phr compared to the NBR/EPDM rubber compound reinforced with 70 phr of silica lead to an increase in the value of T_p from 472.86 to 475.91 °C (e.g., at β = 20 °C min^{-1}) (Table 9.12). Compared with only silica-reinforced NBR/EPDM rubber blend, this shift in T_p value is much higher and the difference is equal to ΔT_p = 12.08 °C (Table 9.12, β = 20 °C min^{-1}). It can be pointed out that only in the case of NBR/EPDM 35 phr CB/35 phr SiO_2 and NBR/EPDM 50 phr CB/20 phr SiO_2 degradation processes, the "shoulders" appear on the corresponding DTG curves at temperatures below T_p (toward β = 10 and 20 °C min^{-1}) (Fig. 9.11; Table 9.12). The appearance of shoulders at lower temperatures when carbon black is used as filler indicates that the degradation is more complex.

Thermal stability increases significantly in the filler-reinforced overall blends, compared with the pure NBR and EPDM overall rubbers, as can be seen from the values of the residual mass loss (Δm_f), at the considered heating rates (Table 9.12) [69]. This is also confirmed from Fig. 9.11 (TG) that the amount of residue increases with the filler loading. It may be noted that the use of carbon black, as a replacement of silica, has the most significant effect in terms of thermal stability, if one looks at the residual mass and the peak degradation temperatures.

9.6 Degradation Kinetic Analysis of Ethylene–Propylene–Diene Rubber Nanoblends

Generally, the thermal degradation of a polymeric material follows more than one mechanism. The existence of more than one concurrent chemical reaction is accompanied by the other physical phenomena such as evaporation and ablation introduce further complications for the modeling of degradation kinetics. Kinetic study of thermal degradation provides useful information for the optimization of the successive treatment of polymer materials in order to avoid or at least limit thermal degradation. The analysis of the degradation process becomes more and more important due to an increase in the range of temperatures for engineering applications, recycling of post-consumer plastic waste, as well as the use of polymers as biological implants and matrices for drug delivery, where depolymerization is an inevitable process affecting the lifetime of an article.

In order to determine the values of kinetic parameters and the proper function of the reaction mechanism, the different "model-fitting" and "model-free" kinetic methods [70] were used. Based on the kinetic and thermodynamic results, the differences in the degradation mechanisms and thermal stabilities of the observed patterns are discussed.

The rate of degradation reaction can be described in terms of two functions, $k(T)$ and $f(\alpha)$; thus,

$$\frac{d\alpha}{dt} = k(T)f(\alpha) \tag{9.3}$$

where α is the degree of conversion ($\alpha = (m_o - m_t)/(m_o - m_f)$, m_o, m_t, and m_f are the initial, actual, and final mass of the sample in the TGA curves, respectively, t is the reaction time, $k(T)$ is the rate constant, and $f(\alpha)$ is the type of reaction or function of reaction mechanism. By substitution of the Arrhenius equation, $k(T) = A\exp(-E_a/RT)$, the following equation results:

$$\frac{d\alpha}{dt} = A\,\exp\left(-\frac{E_a}{RT}\right)f(\alpha) \tag{9.4}$$

where A is the pre-exponential factor, E_a is the apparent activation energy, R is the gas constant, and T is the absolute temperature in kelvin.

After the introduction of the constant heating rate $\beta = dT/dt$ and rearrangement, one obtains

$$\frac{d\alpha}{f(\alpha)} = \left(\frac{A}{\beta}\right)\exp\left(-\frac{E_a}{RT}\right)dT. \tag{9.5}$$

A subsequent integration of Eq. (9.5) leads to the equation in the form:

$$g(\alpha) = \int_0^{\alpha} \frac{d\alpha}{f(\alpha)} = \frac{A}{\beta}\int_{T_o}^{T} \exp\left(-\frac{E_a}{RT}\right)dT \tag{9.6}$$

where $g(\alpha)$ is the integral form of the reaction mechanism function and T_o is the starting temperature of non-isothermal experiment. Equation (9.6) cannot be expressed by a simple analytical form since its right-hand side corresponds to a series of infinite γ functions. In mathematical practice, Eq. (9.6) can be expressed in the following form [71]:

$$g(\alpha) = \frac{A}{\beta}\int_0^{T} \exp\left(-\frac{E_a}{RT}\right)dT = \frac{AE_a}{\beta R}p(x) \tag{9.7}$$

where is introduced an approximation for the lower limit of the integral on the right-hand side of Eq. (9.6) as $T_o \to 0$, bearing in mind that the value of temperature T_o in practical cases is very small. The variable $x = E_a/RT$ is the reduced apparent activation energy, while $p(x)$ represents the temperature or exponential integral which cannot be exactly calculated [72]. For the different solid-state reaction mechanisms, $f(\alpha)$ and $g(\alpha)$ have different expressions [73]. The most frequently cited kinetic models used in this work are summarized in Table 9.13.

Table 9.13 Algebraic expressions of $f(\alpha)$ and $g(\alpha)$ for the reaction models considered in the present article

№	Symbol	Reaction model	$f(\alpha)$	$g(\alpha)$
1	P1	Power law	$4\alpha^{3/4}$	$\alpha^{1/4}$
2	P2	Power law	$3\alpha^{2/3}$	$\alpha^{1/3}$
3	P3	Power law	$2\alpha^{1/2}$	$\alpha^{1/2}$
4	P4	Power law	$2/3\alpha^{-1/2}$	$\alpha^{3/2}$
5	R1	Zero-order (Polanyi-Wigner equation)	1	α
6	R2	Phase boundary-controlled reaction (contracting area, i.e., bidimensional shape)	$2(1-\alpha)^{1/2}$	$[1-(1-\alpha)^{1/2}]$
7	R3	Phase boundary-controlled reaction (contracting volume, i.e., tridimensional shape)	$3(1-\alpha)^{2/3}$	$[1-(1-\alpha)^{1/3}]$
8	F1	First-order (Mampel)	$(1-\alpha)$	$-\ln(1-\alpha)$
9	F3/2	Three-halves order	$(1-\alpha)^{3/2}$	$2[(1-\alpha)^{-1/2}-1]$
10	F2	Second-order	$(1-\alpha)^2$	$(1-\alpha)^{-1}-1$
11	F3	Third-order	$(1-\alpha)^3$	$(1/2)[(1-\alpha)^{-2}-1]$
12	A3/2	Avrami-Erofe'ev ($n = 1.5$)	$(3/2)(1-\alpha)[-\ln(1-\alpha)]^{1/3}$	$[-\ln(1-\alpha)]^{2/3}$
13	A2	Avrami-Erofe'ev ($n = 2$)	$2(1-\alpha)[-\ln(1-\alpha)]^{1/2}$	$[-\ln(1-\alpha)]^{1/2}$
14	A3	Avrami-Erofe'ev ($n = 3$)	$3(1-\alpha)[-\ln(1-\alpha)]^{2/3}$	$[-\ln(1-\alpha)]^{1/3}$
15	A4	Avrami-Erofe'ev ($n = 4$)	$4(1-\alpha)[-\ln(1-\alpha)]^{3/4}$	$[-\ln(1-\alpha)]^{1/4}$
16	D1	One-dimensional diffusion	$1/2\alpha$	α^2
17	D2	Two-dimensional diffusion (bidimensional particle shape) Valensi equation	$1/[-\ln(1-\alpha)]$	$(1-\alpha)\ln(1-\alpha)+\alpha$
18	D3	Three-dimensional diffusion (tridimensional particle shape) Jander equation	$3(1-\alpha)^{1/3}/2[(1-\alpha)^{-1/3}-1]$	$[1-(1-\alpha)^{1/3}]^2$
19	D4	Three-dimensional diffusion (tridimensional particle shape) Ginstling–Brounstein	$3/2[(1-\alpha)^{-1/3}-1]$	$(1-2\alpha/3)-(1-\alpha)^{2/3}$

In order to determine the kinetic triplet [A, E_a, $f(\alpha)$ or $g(\alpha)$], various methods have been proposed. These methods can, in general, be categorized as (a) isoconversional and (b) model-fitting methods. The isoconversional methods are, in fact, “model-free” methods [74] which assumes that the function $f(\alpha)$ does not change with the variation of heating rate for all values of α. It involves the measuring of the temperatures corresponding to fixed values of α by experiments at the different heating rates β. The isoconversional methods are considered to give accurate values of the activation energy [75]. The pre-exponential factor usually cannot be determined without the assumption of the reaction model ($f(\alpha)$ or $g(\alpha)$). Few of these methods were proposed by Kissinger–Akahira–Sunose (KAS) [56, 76] and Friedman (FR) [77].

The best-known model-fitting methods are the Coats–Redfern (CR) method [78] and the invariant kinetic parameters (IKP) method [79].

9.6.1 Isoconversional (Model-Free) Methods

Kissinger–Akahira–Sunose (KAS) Method

By using the Coats–Redfern approximation for the temperature integral [78] to solve Eq. (9.7) and considering that $2RT/E_a$ is much lower than unity, the Kissinger–Akahira–Sunose (KAS) [56, 76] equation can be written in the form:

$$\ln\left(\frac{\beta}{T^2}\right) = \ln\left[\frac{AR}{g(\alpha)E_a}\right] - \frac{E_a}{RT} \tag{9.8}$$

For each degree of conversion, the linear plot of $\ln(\beta/T^2)$ versus $1/T$ enables E_a and $\ln[AR/g(\alpha)E_a]$ to be determined from the slope and the intercept, respectively. If the reaction model, $g(\alpha)$, is known, the corresponding pre-exponential factor can be calculated for each conversion.

Friedman (FR) Method

The differential isoconversional method suggested by Friedman [77] is based on Eq (9.9):

$$\ln\left[\beta\left(\frac{d\alpha}{dT}\right)\right] = \ln[Af(\alpha)] - \frac{E_a}{RT} \tag{9.9}$$

For a constant α, the plot of $\ln[\beta(d\alpha/dT)]$ versus $1/T$ obtained from curves recorded at several heating rates should be a straight line whose slope gives us the value of E_a. From the intercept of a given straight line ($\ln[Af(\alpha)]$), it is possible to determine the pre-exponential factor for each conversion value, only if the differential form of reaction model $f(\alpha)$ is known.

9.6.2 Model-Fitting Methods

Model-fitting methods involve fitting different kinetic models (Table 9.13) to conversion–temperature curves and simultaneously determining the apparent activation energy E_a and the pre-exponential factor A.

Coats–Redfern Method

The Coats–Redfern (CR) [78] method is based on Eq. (9.8) in the following form:

$$\ln\left[\frac{g(\alpha)}{T^2}\right] = \ln\left(\frac{AR}{\beta E_a}\right) - \frac{E_a}{RT} \tag{9.10}$$

For a given model and heating rate, the linear plot of the left-hand side of Eq. (9.10) versus $1/T$ allowed us to obtain the average apparent activation energy and average pre-exponential factor from the slope and the intercept. Then, we can choose the reaction model with the apparent activation energy similar to that obtained isoconversionally and with a good square of the linear correlation coefficient (R^2).

Invariant Kinetic Parameter (IKP) Method

In this chapter, the method proposed by Budrugeac et al. [79] was used for the purpose of determining the invariant kinetic parameters (IKPs). The method is based on the experimental observation that the same thermo-analytical curve can be described relatively fair by the several different conversion functions. Using an integral [Coats–Redfern (CR)] method (Eq. 9.10), for each heating rate and for each conversion function ($g(\alpha)$), a pair of kinetic parameters (A, E_a) is established. Using the artificial compensation effect that always exists when the reaction model changes, for each heating rate, the compensation parameters, a_v and b_v, are determined in accordance with Eq. (9.12), which represents the artificial compensation effect relation [80]:

$$\ln A_j = a_v + b_v E_{aj} \tag{9.11}$$

where A_j and E_{aj} are the corresponding values of the kinetic parameters which was attributed to the *jth* reaction model (Table 9.13), for a given heating rate β.

The point of intersection of the straight lines of compensations for several heating rates corresponds to the real values of E_a and A, called the invariant kinetic parameters ($E_{a,inv}$ and A_{inv}), as they are independent of the conversion, the model, and the heating rate [79]. Since determining the point of intersection by the graphical method means is uncertain, the invariant kinetic parameters can be defined by the corresponding supercorrelation relation:

$$a_v = \ln A_{\mathrm{inv}} - b_v E_{\mathrm{a,inv}} \tag{9.12}$$

The straight line a_v versus b_v allows us to determine the invariant kinetic parameters ($E_{\mathrm{a,inv}}$ and A_{inv}) from the slope and the intercept. Although the IKP method aims to determine the invariant kinetic parameters independently of the reaction model, comparing them to those obtained using the other methods (such as CR method and isoconversional methods) also allows us to decide which reaction model best describes the investigated degradation process.

9.6.3 Determination of the Most Probable Mechanism Function

Kinetic Method with the Use of the Special Function Z(α)

The values of E_a can be used to find the appropriate kinetic model which best describes the conversion function of the process studied [81]. The most suitable kinetic model can be evaluated with the function $z(\alpha)$, in accordance with Eq. (9.13):

$$z(\alpha) = \pi(x)\left(\frac{\mathrm{d}\alpha}{\mathrm{d}t}\right)\frac{T}{\beta} \tag{9.13}$$

where x is the reduced apparent activation energy ($x = E_a/RT$) and $\pi(x)$ is the expression of the temperature integral, which can be well approximated using the fourth rational approximation proposed by the Senum and Yang [82]. The shape and the maximum $z(\alpha)$ functions for several models are normalized within (0, 1) range. The maxima, α_p^{∞}, of the $z(\alpha)$ function suggests the choice of the most suitable kinetic model characterizing the investigated degradation process. From average E_a value, we can obtain x, and by using x in the Senum–Yang expression for the temperature integral [82], $\pi(x)$ can be obtained. Then, the values of x and $\pi(x)$ are used to calculate $z(\alpha)$ functions, using the Eq. (9.13).

Integral "Master" Plot Method

Using the reference point at $\alpha = 0.50$, the following integral "Master" equation can be derived from Eq. (9.7):

$$\frac{g(\alpha)}{g(0.50)} = \frac{p(x)}{p(x_{0.50})} \tag{9.14}$$

where $p(x_{0.50})$ is the temperature integral at the value of $\alpha = 0.50$. In this paper, the fourth rational approximation of Senum and Yang [82] is used for $p(x)$. In order to determine the reaction model, it is necessary that the conversion function $f(\alpha)$ (or $g(\alpha)$), which is the characteristic of the investigated degradation process, should be included in the analyzed set of analytical functions (Table 9.13).

Determination of the Pre-Exponential Factor (A)

The pre-exponential factor (A) can be calculated from the Eq. (9.7) after an appropriate rearranging, only if we know the analytical form of the $g(\alpha)$ function. In this case, we obtain [79]:

$$\ln\left(\frac{\beta R}{E_a}\right) - \ln[p(x)] = \ln A - \ln[g(\alpha)] \qquad (9.15)$$

where β is the observed value of the heating rate, while E_a represents the value of the apparent activation energy, which is calculated from the differential (Friedman) isoconversional method. If the reaction model (function $g(\alpha)$) is properly defined and if the value of E_a is accurately determined, then at all the considered heating rates, the experimental points will lie on a single straight line. For each investigated system, from the intercept of the obtained straight line, ln A can be calculated and then a value of A, which for us was the ultimate goal.

TGA experiments have been performed in dynamic mode, in order to obtain the variation of the sample mass as a function of temperature, required for consideration of thermal stability and influence of fillers on the kinetic behaviors of the investigated polymer blends. TGA and DTA curves of EPDM and NBR samples, performed at constant heating rates of 10, 20, and 40 °C min^{-1}, are shown in Figs. 9.8 and 9.9, respectively. Figures 9.11 and 9.12 show typical thermogravimetric (TG) and the derivative thermogravimetric (DTG) (inset) curves for the NBR/EPDM blends reinforced with silica and carbon black/silica fillers in different phr's and NBR/EPDM rubber blend reinforced with 50 phr CB. All observed thermo-analytical curves are performed at a constant heating rate of $\beta = 20$ °C min^{-1}.

Thermogravimetric (TG) curves of the NBR samples (Fig. 9.9) are shifted to higher degradation temperatures, which is typical for thermally activated processes. Also, heating curves are not parallel and show the asymmetry. However, it is evident that there is a dependence of the residual mass loss (Δm_f) on the heating rate (β). It is shown in Fig. 9.9 that the greatest mass loss is realized on the value of the heating rate of 40 °C min^{-1}.

TG curves of the EPDM samples (Fig. 9.8) also show an effect that is reflected in the curve shifting to higher temperatures with increasing heating rate. The observed heating curves are not parallel (Fig. 9.8) and show a smaller degree of asymmetry compared with the TG curves in the case of the NBR samples (Fig. 9.9). On the other hand, there is a very little expressed dependence of Δm_f on β,

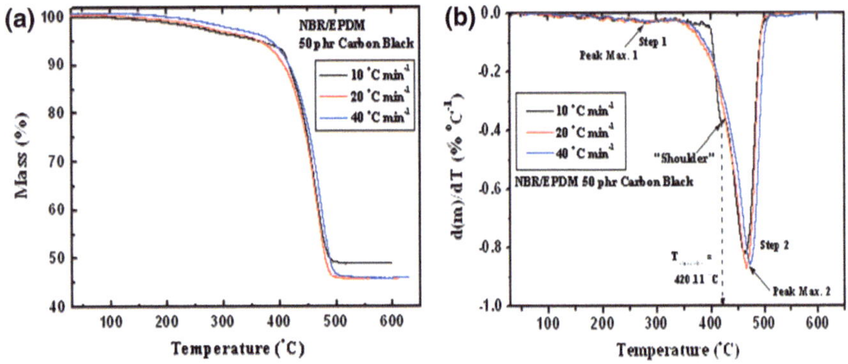

Fig. 9.12 Thermogravimetric (TG) (**a**) and the derivative thermogravimetric (DTG) (**b**) curves of the NBR/EPDM rubber blend reinforced with 50 phr CB degradation processes under nitrogen atmosphere, performed at a constant heating rate of β = 20 °C min^{-1} (*CB*-carbon black)

compared with the same type of dependency in the case of the NBR heating curves (Fig. 9.9). From the analysis of the general characteristics of the TG curves, we can conclude that the thermal degradation of the NBR samples, which contain the butadiene units [83], strongly depends on the heating rate. In the case of the thermal degradation of the EPDM samples, this dependence is almost non-existent. Comparing the thermogravimetric (TG) curves of the pure polymer rubbers (NBR and EPDM) [Figs. 9.8 and 9.9 (β = 20 °C min^{-1})] and TG curves (Fig. 9.12) of CB-reinforced NBR/EPDM blends, it is indicated that shape of TG curves is affected by rather large loading of the carbon black filler. At all heating rates, TG curves for the NBR/EPDM blend reinforced with 50 phr of carbon black filler (Fig. 9.12) exhibit less sharpness (in the slope) compared with the pure EPDM and NBR (Figs. 9.8 and 9.9). Also, the presence of unreacted compounding ingredients in the vulcanizates and weak links in the skeleton of the macromolecules can cause early degradation, in comparison with raw rubbers, individually [84]. It can be pointed out that the presence of the carbon black filler in the NBR/EPDM blend decreases the maximum rate of the main degradation step, compared to the virgin rubbers [85].

Figure 9.13 shows the results of isoconversional analysis using Eqs. (9.8) and (9.9), for degradation processes of the NBR, EPDM, and NBR/EPDM rubber blends reinforced with 70 phr of silica, NBR/EPDM rubber blend reinforced with carbon black/silica (35/35), NBR/EPDM rubber blend reinforced with carbon black/silica (50/20), and NBR/EPDM rubber blend reinforced with 50 phr carbon black, respectively.

The apparent activation energies of degradation of the NBR sample from α = 0.05 (=5 %) to α = 0.95 (=95 %) conversions are shown in Fig. 9.13a. It is shown in Fig. 9.13a that there is some difference in the values of E_a calculated by using the Kissinger–Akahira–Sunose (KAS) (designated by E_a^{KAS}) and the Friedman (FR) (designated by E_a^{FR}) method. It is shown that when the values of the

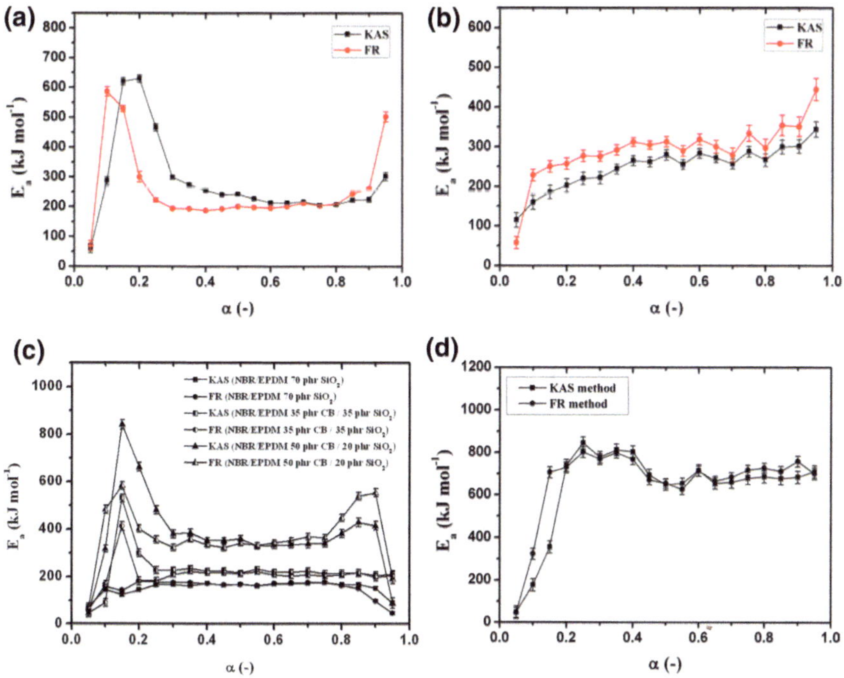

Fig. 9.13 The apparent activation energies (E_a) versus the degree of conversion (α), obtained by the Kissinger–Akahira–Sunose (KAS) and the Friedman (FR) isoconversional methods, for the **a** NBR, **b** EPDM, and **c** NBR/EPDM rubber blend reinforced with 70 phr of silica, NBR/EPDM rubber blend reinforced with carbon black/silica (35/35), NBR/EPDM rubber blend reinforced with carbon black/silica (50/20), and **d** NBR/EPDM rubber blend reinforced with 50 phr carbon black degradation processes

apparent activation energy (E_a) depend on the degree of conversion (α), we can infer that the $E_a^{KAS} \neq E_a^{FR}$ [86]. It should be noted that the existence of significant differences between E_a^{FR} and E_a calculated by using the integral isoconversional method (KAS method) (Fig. 9.13a) can be attributed to the way of derivation of a given relation, where the observed integral method was based.

Namely, the difference of the calculated E_a by two methods can be explained by a systematic error due to improper integration [86]. The FR method employs instantaneous rate values being, therefore, it becomes sensitive to the experimental data. On the other hand, the relation of KAS method is derived assuming a constant E_a value, introducing systematic error in the estimation of E_a in the case that E_a varies with α.

It is shown in Fig. 9.13a that at the lower conversions (up to $\alpha = 0.10$) (looking the values of E_a obtained using FR method), we have a sharp increase in the E_a value from $E_a = 70.7$ kJ mol^{-1} to $E_a = 587.1$ kJ mol^{-1}. After $\alpha = 0.10$, in the conversion range of $0.15 \leq \alpha \leq 0.25$, we can clearly see that there exists a significant decrease in the E_a value (Fig. 9.13a). After that (in the conversion range of

$0.30 \le \alpha \le 0.80$), E_a is almost stable, and the value of E_a can be taken as a constant. It was found that the average value of E_a in the range of $0.30 \le \alpha \le 0.80$ is equal to the $\langle E_a \rangle = 199.6$ kJ mol^{-1}. In addition, for $\alpha > 0.80$, we can detect a slight increase in the E_a value (Fig. 9.13a). The shape of dependence $E_a = E_a(\alpha)$ in the case of the NBR samples (Fig. 9.13a) indicates a change in the rate-limiting step of the degradation kinetics [87], caused by a shift of the rate-limiting step from initiation to the degradation initiated by the random scission. This dependence E_a on α indicates a complex process with the participation of at least of two mechanisms, which is a rather typical phenomenon for many polymer rubbers [88].

The first mechanism corresponds to the part where the small mass loss appears (in the temperature range of $\Delta T_I = 210$–375 °C on TG curves (Fig. 9.9), which corresponds to conversion level up to $\alpha = 0.10$ (=10 %) (Fig. 9.13a), while the second part where the substantial mass loss takes place is due to the main degradation mechanism ($\Delta T_{II} = 400$–490 °C on TG curves (Fig. 9.9), which corresponds to conversion levels after $\alpha = 0.15$ (=15 %) (Fig. 9.13a).

From the obtained values of E_a and the shape of dependency E_a on α (Fig. 9.13a), we can derive the following items: (a) The first step is due to the volatile depolymerization (up to $\alpha = 0.10$) and (b) the second step can be attributed to the decomposition of cyclized and the cross-linked butadiene (BD) residue, together with the decomposition of other degradation products (arising from the degradation of acrylonitrile unit, releasing ammonia, which is rich in energy), resulting after the first step (after conversions $\alpha = 0.10$) (Fig. 9.13a) [89].

It can be pointed out that the value of 587.1 kJ mol^{-1} at $\alpha = 0.10$ (Fig. 9.13a) corresponds to bond breaking of the carbon–carbon double bond (bond dissociation energy for –C=C– is in the range 580–610 kJ mol^{-1} that is about two times stronger than a –C–C– single bond) in the butadiene unit [90, 91]. The thermal stability of the butadiene units in the NBR structure follows the order: *trans* 1,4 BD < *cis* 1,4 BD < 1,2 BD [91]. Zhou and Mark found that *trans* 1,4 BD decomposes at 475 °C with little devolatilization at 330 °C under nitrogen atmosphere at a heating rate of 20 °C min^{-1} [92]. For our system, the value of peak temperature (T_m) at a heating rate of 20 °C min^{-1} is equal to 461.2 °C (Table 9.11), where this temperature is slightly lower than the temperature obtained by Zhou and Mark [92].

The apparent activation energies of degradation of the EPDM sample from $\alpha = 0.05$ (=5 %) to $\alpha = 0.95$ (=95 %) conversions are shown in Fig. 9.13b. Figure 9.13b shows that the shape of dependence $E_a = E_a(\alpha)$ for the EPDM degradation is different from the $E_a = E_a(\alpha)$ dependency in the case of the NBR sample degradation (Fig. 9.13a). For the EPDM degradation process, the apparent activation energy (E_a) increases moderately from $E_a = 115.2$ kJ mol^{-1} to $E_a = 222.5$ kJ mol^{-1} in the conversion range of $0.05 \le \alpha \le 0.30$ (for the KAS method) and from $E_a = 229.5$ kJ mol^{-1} to $E_a = 276.3$ kJ mol^{-1} in the conversion range of $0.10 \le \alpha \le 0.30$ (for the FR method) (Fig. 9.13b). As in the case of NBR degradation, for the EPDM samples, we have also that the $E_a^{KAS} \ne E_a^{FR}$. It was found that in the conversion range of $0.35 \le \alpha \le 0.80$ ($\Delta T = 430$–470 °C), the value of the apparent activation energy (E_a) can be considered as a constant, within the limits of experimental error. For a given conversion range, the average values of E_a estimated from the KAS and FR methods are as follows: $<E_a>^{KAS} = 267.7$ kJ mol^{-1} and $<E_a>^{FR} = 304.9$ kJ mol^{-1} (Fig. 9.13b). For $\alpha > 0.80$, we can see that there is some increase in the value of the apparent activation energy (E_a).

As has been reported by Peterson et al. [88], the observed variation in the apparent activation energy can be attributed to the degradation kinetics being governed by different processes at the initial and final stages and the lower value of the apparent activation energy being associated with the initial process that occurs at the weak links. In this way, polyethylene units can act as the weak links. As these weak links are consumed, the limiting step of degradation shifts toward the degradation initiated by the random chain scission, where it is progressively taking place along the chain until the fragments are small enough to volatilize. This type of degradation requires higher apparent activation energy. It was shown that the variation magnitude of the E_a values is controlled by both the ethylene content and the microstructural changes [93, 94].

The obtained average value of E_a estimated from the FR isoconversional method ($< E_a>^{FR} = 304.9$ kJ mol^{-1}) for $0.35 \leq \alpha \leq 0.80$ (Fig. 9.13b) is somewhat lower, than the corresponding value of E_a ($E_a = 330$ kJ mol^{-1}) obtained by Gamlin et al. [93], which was calculated using the same kinetic method. The reaction model for the NBR and EPDM degradation processes can be determined only in the conversion range, where the apparent activation energy (E_a) shows a constant value (Fig. 9.13a and b). It is shown in Fig. 9.13c that there is some difference in the E_a values calculated by using the Kissinger–Akahira–Sunose (KAS) (designated by E_a^{KAS}) and the Friedman (FR) (designated by E_a^{FR}) method. The observed difference between the E_a^{KAS} and the E_a^{FR} values increases with increasing CB-loaded blend. On the other hand, for all three systems, the largest difference between E_a^{KAS} and E_a^{FR} values can be observed at low and high values of α.

For both methods used for all test systems, we have the appearance of the increased values of the apparent activation energy (E_a), at lower values of α (Fig. 9.11). In the case of NBR/EPDM rubber blend reinforced with 70 phr of silica, the NBR has basic nitrile groups (–CN), and a hydrogen bond with silica can be made, for which the filler dispersion is improved. For the FR method, in this case, E_a value increases up to 158.7 kJ mol^{-1}. A rather high value of E_a is a consequence of the energy needed to break these hydrogen bonds, which establishes the primary reaction groups of NBR rubber and the silanol groups of silica. The contribution of hydrogen bonds cannot be neglected. These bonds can be effective in hindering the motion of the side groups, whose elimination begins at the lower temperatures. In addition, it should still be noted that because the silica has a hydrophilic behavior, in this case it shows a lower compatibility with the EPDM. It is well known from the literature [95] that some solid acids, such as amorphous alumina, silica–alumina, and clays, which have Brønsted or Lewis strong acid centers, can promote the polymer chain cracking. This fact could be one of the reasons for the appearance of the lower value of E_a for the degradation process of NBR/EPDM rubber blend reinforced with 70 phr of silica, compared with the values of E_a for the degradation processes, which are higher in the case of NBR/EPDM rubber blend reinforced with carbon black/silica (35/35) and NBR/EPDM rubber blend reinforced with carbon black/silica (50/20) (Fig. 9.13c). Also, since the shapes of the TG curves shown in Fig. 9.11 are different, it could be expected that the values of E_a and A in the rate equation (Eq. 9.7) characterizing the degradation kinetics of the studied systems would be different (Fig. 9.13c).

For the other two investigated systems (NBR/EPDM rubber blend reinforced with carbon black/silica (35/35) and NBR/EPDM rubber blend reinforced with carbon black/silica (50/20)), degradation process at lower α values (up to $\alpha = 0.15$) occurs with even greater increase in the apparent activation energy, compared with a jump of E_a values in the case of NBR/EPDM rubber blend reinforced with 70 phr of silica degradation, in approximately the same α range (i.e., to $\alpha = 0.10$) (Fig. 9.13c). If we look E_a values calculated by the FR method, for NBR/EPDM rubber blend reinforced with carbon black/silica (35/35) degradation, in the α range of $0.05 \leq \alpha \leq 0.15$, the value of E_a increased up to 412.7 kJ mol^{-1}, while for NBR/EPDM rubber blend reinforced with carbon black/silica (50/20), in the same α range, the value of E_a increased up to 583.4 kJ mol^{-1} (Fig. 9.13c).

Such behavior of the studied systems, in this stage of the process, indicates that the degradation process requires much higher values of the apparent activation energy, which may include a series of parallel reactions that take place at the same time, and therefore the presence of more complex nature of the reactions, than is the case with silica-reinforced polymer blends. The high E_a values are observed only in the initial stages of degradation in the case of NBR rubber, while in the case of degradation of EPDM rubber, such high values of E_a were not detected [85]. Nevertheless, since the small amount of EPDM in the mixture of NBR/EPDM is presented, it should be noted that in accordance with Kurian et al. [96], the EPDM structure comprises rigid domains and flexible regions. The rigid domains, related to polyethylene blocks and cross-linked dienes, probably inhibit the decomposition, increasing the E_a value. In addition, the presence of fillers can reduce the diffusion rate of the decomposed compounds, also increasing the E_a value.

For the NBR/EPDM rubber blend reinforced with carbon black/silica (35/35) and NBR/EPDM rubber blend reinforced with carbon black/silica (50/20) degradation processes, we have α ranges where the value of E_a can be approximately taken as the constant. These ranges of α value, where E_a does not vary greatly with conversion, are as follows: $0.25 \leq \alpha \leq 0.95$ and $0.25 \leq \alpha \leq 0.75$ for NBR/EPDM rubber blend reinforced with carbon black/silica (35/35) and NBR/EPDM rubber blend reinforced with carbon black/silica (50/20) degradations, respectively. For the considered systems, the average value of E_a calculated by KAS and FR methods is as follows: $\langle E_a \rangle^{KAS}$ = 220.7 kJ mol^{-1}; $\langle E_a \rangle^{FR}$ = 208.8 kJ mol^{-1} (NBR/EPDM rubber blend reinforced with carbon black/silica (35/35)); $\langle E_a \rangle^{KAS}$ = 353.6 kJ mol^{-1} (only for $0.30 \leq \alpha \leq 0.80$); and $\langle E_a \rangle^{FR}$ = 345.4 kJ mol^{-1} (NBR/EPDM rubber blend reinforced with carbon black/silica (50/20)) (Fig. 9.13c). Just in the case of NBR/EPDM rubber blend reinforced with carbon black/silica (50/20) degradation, we have increased the behavior of the E_a value, after $\alpha > 0.75$ (Fig. 9.13c).

On the other hand, the total dependence of E_a on α for the NBR/EPDM rubber blend reinforced with 70 phr of silica degradation is somewhat different than in the previous two cases. In this case, after $\alpha = 0.10$, the value of E_a decreases up to $\alpha = 0.15$ and then increases again to the value of $\alpha = 0.25$, which is contrary to the behavior of E_a with α, for the NBR/EPDM rubber blend reinforced with carbon black/silica (35/35) and NBR/EPDM rubber blend reinforced with carbon black/silica (50/20) degradation processes (Fig. 9.13c). For $\alpha > 0.25$, we have a α range where the value of E_a can be taken to remain constant (the average values of E_a calculated by the KAS and FR methods for the NBR/EPDM rubber blend reinforced with 70 phr of silica degradation are as follows: $\langle E_a \rangle^{KAS} = 168.3$ kJ mol^{-1} ($0.25 \leq \alpha \leq 0.85$) and $\langle E_a \rangle^{FR} = 172.7$ kJ mol^{-1} ($0.20 \leq \alpha \leq 0.80$)).

Generally speaking, if we look all the presented dependence of E_a on α in Fig. 9.13c, we note that the apparent activation energy increases with the char filler content. The higher the apparent activation energy of the degradation of the reinforced polymer blend, the better will be the thermal stability of the investigated system. The apparent activation energy for the main degradation stage of the investigated rubber compounds increases from 199.6 kJ mol^{-1} (FR method) [85] to 208.8 kJ mol^{-1} by the addition of 35 phr CB; it further increases to 345.4 kJ mol^{-1} in the presence of 50 phr CB. It can be pointed out that the blend, which contains the commercial silica as in our case, has an apparent activation energy that is less than that of nanosilica blend [97, 98].

It is shown in Fig. 9.13d that there is some difference in the E_a values calculated by using the Kissinger–Akahira–Sunose (KAS) (designated by E_a KAS) and the Friedman (FR) (designated by E_a FR) method. It is shown that when the values of the apparent activation energy (E_a) depend on the degree of conversion (α), we can infer that the $E_a^{KAS} \neq E_a^{FR}$ [86]. It should be noted that the existence of significant differences between E_a^{FR} and E_a calculated by using the integral isoconversional method (such as KAS method) (Fig. 9.13d) can be attributed to the way of derivation of a given relation, where the observed integral method was based [86]. Namely, the difference of the calculated E_a by two methods can be explained by a systematic error due to improper integration [86]. The FR method employs instantaneous rate values being, therefore, it becomes sensitive to the experimental data.

On the other hand, the relation of KAS method is derived assuming a constant E_a value, introducing systematic error in the estimation of E_a in the case that E_a varies with a. It is shown in Fig. 9.13d that the apparent activation energy increases (includes both isoconversional methods) up to the values of $E_a = 845.2$ kJ mol^{-1} (until $\alpha = 0.25$ for KAS) and 801.8 kJ mol^{-1} (till $\alpha = 0.25$ for FR). For both considered methods, we can see that the value of E_a slightly varied in a range of $0.30 \leq \alpha \leq 0.40$, but still remains fairly high. After $\alpha > 0.40$, then the value of E_a decreases up to 693.9 kJ mol^{-1} (at $\alpha = 0.45$ for the KAS method) and 670.2 kJ mol^{-1} (at $\alpha = 0.45$ for the FR method). In the conversion range of $0.45 \leq \alpha \leq 0.95$, the E_a value is almost stable, and the value of E_a can be taken as a constant. It was found that the average values of E_a in a range of $0.45 \leq \alpha \leq 0.95$ calculated by the KAS and FR methods are equal to $\langle E_a \rangle^{KAS} = 678.0$ kJ mol^{-1} and $\langle E_a \rangle^{FR} = 692.8$

kJ mol^{-1}, respectively. The resulting very high value of the apparent activation energy (E_a) indicates a very strong filler-to-matrix interaction, where probably there is an abrupt decrease in the mobility of polymer chains.

If E_a does not vary significantly with conversion, it is possible to describe the whole reaction interval by a single kinetic model even if the real mechanism is complex. Taking into account the conversion intervals which are noted above, for the each investigated system, the CR method was applied. For all the considered systems, the CR method demonstrated the presence of several statistically equivalent reaction models at each of the observed heating rates. These results prevent us to determine with high reliability the values of the kinetic parameters (A and E_a) extracted from a single thermogravimetric (TG) curve.

As a result of the application of the CR method which involves the mechanisms related to the chemical process (or mechanism non-invoking equations), the acceleratory rate equations, the sigmoidal rate equations or random nucleation and subsequent growth, deceleratory rate equations which include phase boundary reactions and equations based on the diffusion mechanism [99]) in order to calculate the kinetic parameters, the artificial compensation effect (*art*-CE) [80] was found. An artificial compensation effect usually occurs as a result of force-fitting the thermo-analytical data to the different kinetic models. The *art*-CE plots for the NBR/EPDM rubber blend reinforced with 70 phr of silica (in the conversion range of $0.20 \leq \alpha \leq 0.80$), NBR/EPDM rubber blend reinforced with carbon black/silica (35/35) (in the conversion range of $0.25 \leq \alpha \leq 0.95$), and NBR/EPDM rubber blend reinforced with carbon black/silica (50/20) (in the conversion range of $0.25 \leq \alpha \leq 0.75$) degradation processes at the different heating rates (β = 10, 20, and 40 °C min^{-1}) are shown in Fig. 9.14.

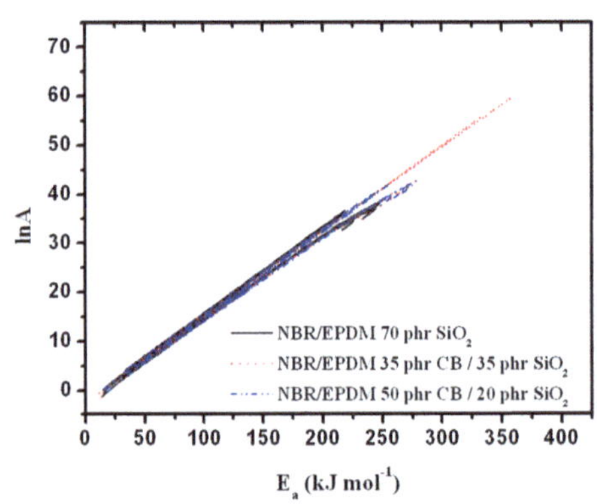

Fig. 9.14 The artificial compensation effect plots for the NBR/EPDM rubber blend reinforced with 70 phr of silica (in the conversion range of $0.20 \leq \alpha \leq 0.80$) (—), NBR/EPDM rubber blend reinforced with carbon black/silica (35/35) (in the conversion range of $0.25 \leq \alpha \leq 0.95$) (⋯⋯), and NBR/EPDM rubber blend reinforced with carbon black/silica (50/20) (in the conversion range of $0.25 \leq \alpha \leq 0.75$) (··-··-) degradation processes, at the different heating rates (β = 10, 20, and 40 °C min^{-1})

In order to determine the reaction models and the kinetic parameters (A, E_a) of the investigated degradation processes, we apply the CR method that uses only one TG curve at a given heating rate, which belongs to the group of model-fitting methods. Tables 9.14 and 9.15 show the results of the direct application of the CR method on thermo-analytical data for the degradation processes of the NBR and EPDM samples, at the different heating rates.

We can see from the results presented in Tables 9.14 and 9.15 that the values of kinetic parameters (ln A, E_a) strongly depend on the reaction model, with the existence of several reaction models that are statistically equivalent. For the NBR degradation, the statistically equivalent reaction models are as follows: R2, R3, F1, A3/2, A2, A3, A4, and D3 (10 °C min^{-1}); R2, R3, F1, A3/2, A2, D3, and D4 (20 °C min^{-1}); and P2, P3, P4, R1, D1, and D2 (40 °C min^{-1}). In the case of the EPDM degradation, the statistically equivalent reaction models are as follows: R2, R3, F1, D2, D3, and D4 (at all considered heating rates) (in Tables 9.14 and 9.15, the statistically equivalent reaction models are designated by the *bold* marked adj. R-square (R^2)).

On the other hand, there are several reaction models whose E_a values are close to the average value of E_a calculated by the FR isoconversional method ($\langle E_a \rangle^{NBR}$ = 199.6 kJ mol^{-1} and $\langle E_a \rangle^{EPDM}$ = 304.9 kJ mol^{-1})), in the considered conversion ranges (Tables 9.14 and 9.15).

Based on the obtained results using the CR method, it is impossible to reliably determine the values of kinetic parameters and the analytical form of the reaction model, for both investigated degradation processes. Therefore, it is necessary to access the further kinetic analysis to obtain the exact form of the kinetic triplets. Using the results obtained from the application of the CR method observed at each heating rate, the appropriate compensation parameters (a_v and b_v) and the values of isokinetic temperature ($T_{iso} = 1/Rb_v$) were derived.

Table 9.16 shows the values of the compensation parameters (a_v and b_v) (calculated using Eq. (9.11)) and also the values of isokinetic temperature (T_{iso}), for the NBR, EPDM, and NBR/EPDM rubber blends reinforced with 70 phr of silica, NBR/EPDM rubber blend reinforced with carbon black/silica (35/35), NBR/EPDM rubber blend reinforced with carbon black/silica (50/20) obtained from the *art*-CE plots (Fig. 9.14), and NBR/EPDM rubber blend reinforced with 50 phr carbon black degradation processes, at the different heating rates (β).

When studying the kinetics of reactions of the same type using the correct $g(\alpha)$ function, the straight lines characterizing the kinetics of these reactions usually cross at a point with coordinates lnk_{iso} (ln$k_{iso} = a_v$ (k_{iso} is the isokinetic rate constant)) and $1/T_{iso}$. This point corresponds to the so-called isokinetic temperature T_{iso}, at which the rate constants of the reactions have the same value—k_{iso}. This means that at temperatures higher than T_{iso}, the reactions occurring at higher E_a would have the higher rate viz. higher rate constants than those with lower apparent activation energies. For both processes, values of T_{iso} (Table 9.16) lie in the range of the experimental temperatures, which means that only one of the separated statistically equivalent models (Tables 9.14 and 9.15) is the fact the reaction model

Table 9.14 The kinetic parameters (ln A, E_a) calculated using the Coats–Redfern (CR) method at the different heating rates (β = 10, 20, and 40 °C min^{-1}) for the non-isothermal degradation of NBR, in the conversion range of $0.30 \le \alpha \le 0.80$

Symbol	Heating rate, β								
	10 °C min^{-1}			20 °C min^{-1}			40 °C min^{-1}		
	ln A	E_a (kJ mol^{-1})	R^2	ln A	E_a (kJ mol^{-1})	R^2	ln A	E_a (kJ mol^{-1})	R^2
P1	0.85	23.1	0.98441	0.80	20.1	0.98923	0.46	15.8	0.99875
P2	3.18	34.8	0.98782	2.95	30.8	0.99197	2.39	25.1	**0.99916**
P3	7.53	58.1	0.99025	6.93	52.2	0.99381	5.89	43.7	**0.99940**
P4	31.76	197.8	0.99252	28.89	180.7	0.99545	24.81	155.6	**0.99959**
R1	19.82	127.9	0.99203	18.09	116.4	0.99510	15.53	99.6	**0.99955**
R2	25.33	162.1	**0.99778**	22.85	146.7	**0.99953**	19.43	125.2	0.99837
R3	27.24	174.9	**0.99856**	24.47	180.7	**0.99986**	20.72	134.7	0.99705
F1	33.33	202.8	**0.99877**	29.93	182.6	**0.99912**	25.46	155.2	0.99334
F3/2	41.76	250.0	0.99647	37.27	224.2	0.99538	31.57	190.0	0.98588
F2	51.29	303.8	0.99227	45.55	271.3	0.98976	38.42	229.3	0.97732
F3	73.26	428.5	0.98231	64.59	380.7	0.97731	54.15	320.1	0.96080
A3/2	20.88	131.2	**0.99870**	18.79	117.7	**0.99907**	15.97	99.4	0.99284
A2	14.55	95.5	**0.99862**	13.12	85.3	**0.99902**	11.12	71.5	0.99228
A3	8.07	59.7	**0.99844**	7.29	52.8	0.99889	6.10	43.6	0.99091
A4	4.71	41.8	**0.99822**	4.25	36.6	0.99873	3.46	29.7	0.98912
D1	43.57	267.7	0.99275	39.55	244.9	0.99561	33.94	211.5	**0.99961**
D2	50.33	309.5	0.99646	45.41	282.0	0.99859	38.75	242.8	**0.99941**
D3	58.08	361.7	**0.99866**	51.98	328.1	**0.99987**	43.99	281.5	0.99727
D4	51.89	326.8	0.99750	46.57	297.2	**0.99931**	39.48	255.6	0.99889

Table 9.15 The kinetic parameters (ln A, E_a) calculated using the Coats–Redfern (CR) method at the different heating rates (β = 10, 20, and 40 °C min^{-1}) for the non-isothermal degradation of EPDM, in the conversion range of $0.35 \leq \alpha \leq 0.80$

Symbol	Heating rate, β								
	10 °C min^{-1}			20 °C min^{-1}			40 °C min^{-1}		
	ln A	E_a (kJ mol^{-1})	R^2	ln A	E_a (kJ mol^{-1})	R^2	ln A	E_a (kJ mol^{-1})	R^2
P1	1.88	28.4	0.99546	1.66	24.5	0.99429	5.58	41.5	0.99560
P2	4.47	41.9	0.99632	4.02	36.8	0.99554	8.80	59.4	0.99621
P3	9.37	68.8	0.99697	8.45	61.3	0.99644	15.00	95.3	0.99671
P4	36.98	230.4	0.99760	33.15	208.3	0.99728	50.55	310.5	0.99723
R1	23.34	149.6	0.99746	20.98	134.8	0.99710	32.94	202.9	0.99711
R2	30.19	192.1	**0.99983**	26.96	172.9	**0.99993**	41.66	258.2	**0.99987**
R3	32.62	208.1	**0.99967**	29.05	187.2	**0.99987**	44.77	278.9	**0.99990**
F1	39.85	243.0	**0.99821**	35.56	218.4	**0.99868**	53.50	324.1	**0.99889**
F3/2	50.26	302.7	0.99397	44.73	271.6	0.99449	66.44	401.0	0.99534
F2	62.09	370.7	0.98842	55.12	332.3	0.98899	81.11	488.6	0.99044
F3	89.50	529.3	0.97712	79.17	473.5	0.97764	115.10	692.2	0.98005
A3/2	25.28	158.0	0.99813	22.60	141.5	0.99850	34.93	212.0	0.99884
A2	17.89	115.5	0.99804	16.02	103.1	0.99843	25.54	155.9	0.99880
A3	10.37	73.0	0.99784	9.29	64.7	0.99825	16.02	99.8	0.99871
A4	6.49	51.8	0.99761	5.80	45.4	0.99803	11.15	71.8	0.99861
D1	50.48	311.2	0.99767	45.20	281.8	0.99736	68.03	418.1	0.99729
D2	58.84	363.0	**0.99948**	52.51	328.2	**0.99943**	78.70	485.6	**0.99933**
D3	68.68	428.2	**0.99968**	61.01	386.6	**0.99988**	91.37	570.2	**0.99990**
D4	61.08	384.5	**0.99979**	54.31	347.5	**0.99983**	81.87	513.5	**0.99975**

Table 9.16 The values of the compensation parameters (a_v and b_v) and the isokinetic temperature (T_{iso}), for the NBR, EPDM, and NBR/EPDM rubber blends reinforced with 70 phr of silica, NBR/EPDM rubber blend reinforced with carbon black/silica (35/35), NBR/EPDM rubber blend reinforced with carbon black/silica (50/20), and NBR/EPDM rubber blend reinforced with 50 phr carbon black degradation processes, at the different heating rates (β = 10, 20, and 40 °C min^{-1})

NBR sample					*EPDM sample*				
β (°C min^{-1})	a_v (min^{-1})	b_v (mol kJ^{-1})	T_{iso} (°C)	R^2	β (°C min^{-1})	a_v (min^{-1})	b_v (mol kJ^{-1})	T_{iso} (°C)	R^2
10	−2.46230	0.17264	423.54	0.99782	10	−2.33113	0.17033	432.99	0.99846
20	−1.90242	0.17000	434.36	0.99721	20	−1.79287	0.16743	445.22	0.99803
40	−1.42111	0.16801	442.74	0.99608	40	−0.66822	0.16476	456.87	0.99905
NBR/EPDM 70 phr SiO_2					*NBR/EPDM 35 phr CB/35 phr SiO_2*				
β (°C min^{-1})	a_v (min^{-1})	b_v (mol kJ^{-1})	T_{iso} (°C)	R^2	β (°C min^{-1})	a_v (min^{-1})	b_v (mol kJ^{-1})	T_{iso} (°C)	R^2
10	−2.81383	0.17232	424.84	0.99410	10	−3.04206	0.17255	423.91	0.99673
20	−2.03924	0.16915	437.92	0.99490	20	−2.33343	0.16932	437.21	0.99708
40	−1.44820	0.16457	457.71	0.99444	40	−1.82605	0.16728	445.87	0.99598
NBR/EPDM 50 phr CB/20 phr SiO_2					*NBR/EPDM 50 phr CB*				
β (°C min^{-1})	a_v (min^{-1})	b_v (mol kJ^{-1})	T_{iso} (°C)	R^2	β (°C min^{-1})	a_v (min^{-1})	b_v (mol kJ^{-1})	T_{iso} (°C)	R^2
10	−2.63556	0.16830	441.51	0.99605	10	−2.75415	0.17099	430.27	0.99862
20	−1.98778	0.16648	449.32	0.99575	20	−2.13353	0.17015	433.74	0.99847
40	−1.38766	0.16485	456.47	0.99512	40	−1.66255	0.16937	437.00	0.99855

can fully describe the observed process in the given range of α, where E_a show no significant change.

Table 9.16 shows the values of the compensation parameters (a_v and b_v) (calculated using linear equation in the form $\ln A = a + bE_a$, for each observed reaction model [99]) and also the values of isokinetic temperature (T_{iso}), for the NBR/EPDM rubber blend reinforced with 70 phr of silica, NBR/EPDM rubber blend reinforced with carbon black/silica (35/35), and NBR/EPDM rubber blend reinforced with carbon black/silica (50/20) degradation processes, at the different heating rates, β. At the temperature T_{iso}, all the compounds studied are characterized by the same rate constant, equal to k_{iso} (isokinetic rate constant). At temperatures higher than T_{iso}, the reactions occurring at the higher values of E_a would have a higher rate (a higher rate constant) than those with the lower apparent activation energies. It is shown in Table 9.16 that the largest difference between the T_{iso} values is presented in the case of NBR/EPDM rubber blend reinforced with 70 phr of silica and NBR/EPDM rubber blend reinforced with carbon black/silica (50/20) degradations, at $\beta = 10$ and 20 °C min^{-1}. At the highest heating rate (40 °C min^{-1}), we can detect the close values of T_{iso} (Table 9.16).

For all processes, values of T_{iso} (Table 9.16) lie in the range of the experimental temperatures, which means that only one of the statistically equivalent models is the fact the reaction model can fully describe the observed degradation process, in the given range of α where E_a does not show significant variation with α. Using the supercorrelation relation (Eq. (9.12)), the corresponding values of the invariant kinetic parameters ($E_{a,inv}$ and A_{inv}) for both degradation processes were calculated. For the NBR degradation, the following values of $E_{a,inv}$ and A_{inv} were found: $E_{a,inv}^{NBR} = 224.2$ kJ mol^{-1} and $A_{inv}^{NBR} = 5.43 \times 10^{15}$ min^{-1}. On the other hand, for the EPDM degradation, we have the following values of the invariant kinetic parameters ($E_{a,inv}$ and A_{inv}): $E_{a,inv}^{EPDM} = 296.9$ kJ mol^{-1} and $A_{inv}^{EPDM} = 8.07 \times 10^{20}$ min^{-1}. If we compare the value of $E_{a,inv}^{NBR}$ (224.2 kJ mol^{-1}) with the average values of E_a which were obtained using the KAS and the FR methods, we can conclude that the $E_{a,inv}^{NBR}$ value is approximately between the values of $\langle E_a \rangle^{FR} = 199.6$ kJ mol^{-1} and $\langle E_a \rangle^{KAS} = 236.3$ kJ mol^{-1} (($0.30 \le \alpha \le 0.80$) Fig. 9.13a). For the EPDM degradation, the $E_{a,inv}^{EPDM}$ value (296.9 kJ mol^{-1}) is in good agreement with the average value of E_a, calculated using the FR method ($\langle E_a \rangle^{FR} = 304.9$ kJ mol^{-1}) (($0.35 \le \alpha \le 0.80$) Fig. 9.13b). For NBR/EPDM with 50 phr CB, we have found the following values of the invariant kinetic parameters: $E_{a,inv} = 674.76 \pm 1.8$ kJ mol^{-1} and $A_{inv} = 8.13 \times 10^{48}$ min^{-1}. A very high value in the magnitude of the pre-exponential factor can indicate that some form of the activated complex is created during the degradation, so it is probably highly restricted in the rotation, compared to the initial reagent. It can be seen that the invariant apparent activation energy value obtained by the IKP method is comparable with the value of E_a calculated by the KAS isoconversional method ($\langle E_a \rangle^{KAS} = 678.0$ kJ mol^{-1}). It is shown in Table 9.16 that the value of T_{iso} increases with increasing heating rate for sample with 50 phr CB.

From the $\langle E_a \rangle^{FR}$ values, we can obtain the corresponding $z(\alpha)$ functions at the different heating rates, for all considered systems. Figure 9.15a–c shows the variation of $z(\alpha)$ values with conversion (α) at the different heating rates, for the NBR,

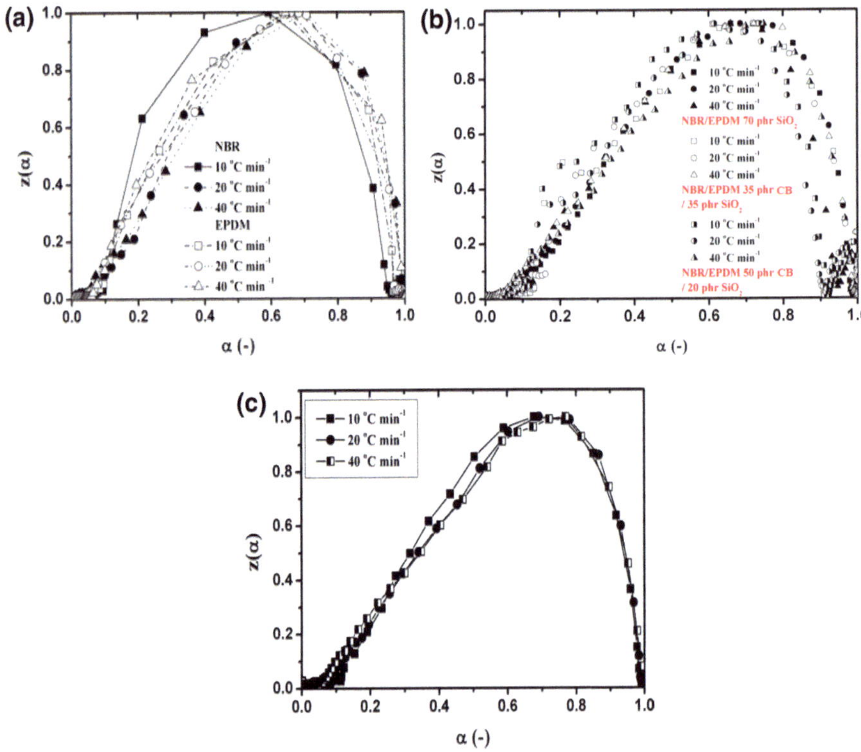

Fig. 9.15 **a** Variation of the z(α) functions versus α for the NBR and EPDM; **b** NBR/EPDM rubber blend reinforced with 70 phr of silica, NBR/EPDM rubber blend reinforced with carbon black/silica (35/35), and NBR/EPDM rubber blend reinforced with carbon black/silica (50/20); and **c** carbon black-loaded NBR/EPDM blend degradation processes at the different heating rates (β = 10, 20, and 40 °C min^{-1})

EPDM, and NBR/EPDM rubber blends reinforced with 70 phr of silica, NBR/EPDM rubber blend reinforced with carbon black/silica (35/35), and NBR/EPDM rubber blend reinforced with carbon black/silica (50/20) degradation processes, respectively. The values of $z(\alpha)$ functions are normalized within (0,1). In addition, the values of α_p^∞ can be obtained from Fig. 9.15a–c. Figure 9.15b shows that $z(\alpha)$ curves for all degradation processes are not invariant with respect to heating rate. The $z(\alpha)$ curves for the degradation of all samples show much greater sensitivity at higher values of $\alpha(\alpha > 0.750)$, when changing the heating rate. For all cases, each $z(\alpha)$ curve exhibits a clear maximum which is consistent with the fact that the $z(\alpha)$ function has a maximum at α_p^∞ for all reaction models.

Figure 9.15c shows that the $z(\alpha)$ curves for the degradation process are not invariant with respect to heating rate, β. The $z(\alpha)$ curves show much greater sensitivity at the lower values of conversions $\alpha(\alpha < 0.650)$, when changing the heating rate. At the higher values of α ($\alpha > 0.650$), the $z(\alpha)$ curves are the almost independent of the heating rate. For all the heating rates (10, 20, and 40 °C min^{-1}), each of the $z(\alpha)$ curves exhibits a clear maximum which is consistent with the fact that the corresponding $z(\alpha)$ function has a maximum at α_p^{∞} for all reaction models.

The determined values of α_p and α_p^{∞} for the NBR, EPDM, NBR/EPDM rubber blends reinforced with 70 phr of silica, NBR/EPDM rubber blend reinforced with carbon black/silica (35/35), NBR/EPDM rubber blend reinforced with carbon black/silica (50/20) and NBR/EPDM rubber blend reinforced with 50 phr carbon black degradation processes at the different heating rates, are listed in Table 9.17.

For both degradation processes for NBR and EPDM rubbers, each individual function $z(\alpha)$ is not invariant with respect to heating rate. The $z(\alpha)$ curves for the degradation of EPDM samples (Fig. 9.15a) show less sensitivity when changing the heating rate, compared with the corresponding $z(\alpha)$ curves in the case of NBR degradation (Fig. 9.15a). In both cases, each $z(\alpha)$ curve also exhibits a clear maximum which is consistent with the fact that the $z(\alpha)$ function has a maximum at α_p^{∞} for all reaction models.

It is clearly shown from α_p^{∞} values (i.e., not equal to 0.632, 0.704, 0.776, or 0.834 (Table 9.17)) that the thermal degradation reaction model for the NBR and EPDM samples can be described using the reaction order models (F*n*) (also, for the NBR degradation process, it should be taken into account the value of $\alpha_p^{\infty} = 0.707$ (at 40 °C min^{-1}) (Table 9.17) which is close to $\alpha_p^{\infty} = 0.704$ and where it belongs to the D3 reaction model). In this case, the application of the parameter α_p^{∞} needs to be discussed. According to this parameter, the model can be characterized, and

Table 9.17 The values of α_p and α_p^{∞} for the NBR, EPDM, and NBR/EPDM rubber blends reinforced with 70 phr of silica, NBR/EPDM rubber blend reinforced with carbon black/silica (35/35), NBR/EPDM rubber blend reinforced with carbon black/silica (50/20), and NBR/EPDM rubber blend reinforced with 50 phr carbon black degradation processes at the different heating rates (β = 10, 20, and 40 °C min^{-1})

NBR degradation			*EPDM degradation*		
β (°C min^{-1})	α_p^{∞}	α_p^{a}	β (°C min^{-1})	α_p^{∞}	α_p^{a}
10	0.593	0.593	10	0.664	0.664
20	0.686	0.686	20	0.662	0.662
40	0.707	0.707	40	0.645	0.645
NBR/EPDM 70 phr SiO_2			*NBR/EPDM 35 phr CB/35 phr SiO_2*		
β (°C min^{-1})	α_p^{∞}	α_p^{a}	β (°C min^{-1})	α_p^{∞}	α_p^{a}
10	0.659	0.659	10	0.664	0.664
20	0.685	0.685	20	0.662	0.662
40	0.659	0.659	40	0.645	0.645
NBR/EPDM 50 phr CB/20 phr SiO_2			*NBR/EPDM 50 phr CB*		
β (°C min^{-1})	α_p^{∞}	α_p^{a}	β (°C min^{-1})	α_p^{∞}	α_p^{a}
10	0.736	0.736	10	0.678	0.678
20	0.615	0.615	20	0.692	0.692
40	0.748	0.748	40	0.770	0.770

[a] α_p is the conversion value corresponding to the DTG curve peak

consequently, the reaction order n can be calculated, if it is an Fn ($n > 1$) model, based on the equation [100]:

$$\alpha_p^{\infty} = 1 - n^{\frac{1}{(1-n)}}. \tag{9.16}$$

Using this equation, n can be calculated in an iterative manner. For the NBR degradation process, the following values of n were found: 1.23(β = 10 °C min^{-1}), 0.74(β = 20 °C min^{-1}), and 0.65(β = 40 °C min^{-1}). On the other hand, for the EPDM degradation process, we calculate the following values of n: 0.84(10 °C min^{-1}), 0.85(20 °C min^{-1}), and 0.93(40 °C min^{-1}). However, these values (except for value of n at 10 °C min^{-1} for the NBR degradation) are obviously inappropriate due to failing to satisfy the condition $n > 1$. The reason for the unsatisfactory result might be due to the existence of large variations of the $z(\alpha)$ curves with heating rate, β. In the case, when there is no dependence on the heating rate (or when all $z(\alpha)$ curves are practically invariant with respect to the temperature), $z(\alpha)$ curves are typically smoothed and mutually consistent.

Since the reaction model used assumes that only one pseudo-component with only one mechanism occurred during the degradation process, the discrepancies between theoretical and experimental $z(\alpha)$–α curves are unavoidable. Therefore, in the case of the NBR and EPDM degradation processes, one cannot expect to get appropriate model parameter (reaction order, n), if the calculation is only based on the theoretical equation (Eq. 9.16).

In order to obtain the correct function of reaction mechanism that will properly describe the investigated degradation processes, the kinetic analysis should be continued. Since, the α_p^{∞} values are not approximately equal to 0.632, 0.704 or 0.776, than the thermal degradation reaction model for the NBR/EPDM rubber blend reinforced with 70 phr of silica (Table 9.17), can be described using the reaction order models (Fn) [81]. According to the α_p^{∞} parameter, the model can be characterized, and consequently, the reaction order n can be calculated, if it is an Fn ($n > 1$) model, based on the Eq. (9.16) [100]. Using this equation, n can be calculated in an iterative manner. For the NBR/EPDM rubber blend reinforced with 70 phr of silica degradation process, the following values of n were found as follows: 0.86(10 °C min^{-1}), 0.75(20 °C min^{-1}), and 0.86(40 °C min^{-1}). However, these values are obviously inappropriate due to the failing to satisfy the condition $n > 1$. The reason for the unsatisfactory result might be due to the existence of fairly large variations of the $z(\alpha)$ curves with heating rate, β, especially at higher α (Fig. 9.15b). In the case, when there is no dependence on the heating rate (or when all $z(\alpha)$ curves are practically invariant with respect to the temperature), the $z(\alpha)$ curves are typically smoothed and mutually consistent.

Since the reaction model used assumes that only one pseudo-component with only one mechanism occurred during the degradation process, the discrepancies between theoretical and experimental $z(\alpha)$–α curves are unavoidable. Therefore, in the case of the NBR/EPDM rubber blend reinforced with 70 phr of silica degradation process, one cannot expect to get appropriate model parameter (the reaction order, n), if the calculation is only based on the theoretical equation [100].

On the other hand, for the NBR/EPDM rubber blend reinforced with carbon black/silica (35/35) and NBR/EPDM rubber blend reinforced with carbon black/silica (50/20) degradation processes, the values of α_p^∞ lie in the range of $0.719 \le \alpha_p^\infty \le 0.733$ and $0.615 \le \alpha_p^\infty \le 0.748$, respectively. If we consider α_p^∞ values obtained for the NBR/EPDM rubber blend reinforced with carbon black/silica (35/35) degradation, as a conclusion clearly arises, that qua mechanisms of the process we can extract the groups of the kinetic models which are based on the phase boundary reactions (Rn(R2 and R3)) and the diffusion-controlled reactions (Dn(D2 and D3)) [81, 99]. In addition, for the NBR/EPDM rubber blend reinforced with carbon black/silica (50/20) degradation process, the values of α_p^∞ lie in the range of $0.615 \le \alpha_p^\infty \le 0.748$, where these values can be assigned also to the phase boundary reactions (Rn(R2 and R3)) and the diffusion-controlled reactions (Dn(D2 and D3)) (except for the heating rate of 20 °C min^{-1}).

Because, the α_p^∞ values are equal to 0.678, 0.692, and 0.770, than the degradation process of the 50 phr carbon black-loaded NBR/EPDM (Table 9.17), can be described using the diffusion kinetic models (Dn) [81]. According to the values of α_p^∞ parameter, the investigated process can be best described by the three-dimensional diffusion reaction model, with spherical (D3) (at $\beta = 10$ and 20 ° C min^{-1}) (Jander equation) and with cylindrical (D4) (at β = 40 °C min^{-1}) (Ginstling–Brounstein equation) symmetry [99].

Figure 9.16 shows the experimental "Master" plots for the NBR and EPDM degradation processes, at the different heating rates, together with the theoretical ones for P4, F1, F3/2, D2, D3 (NBR) and R2, R3, F1, F3/2, F2, D3, D4 (EPDM) reaction models.

Figure 9.17 shows the experimental "Master" plots for the NBR/EPDM rubber blend reinforced with 70 phr of silica, NBR/EPDM rubber blend reinforced with

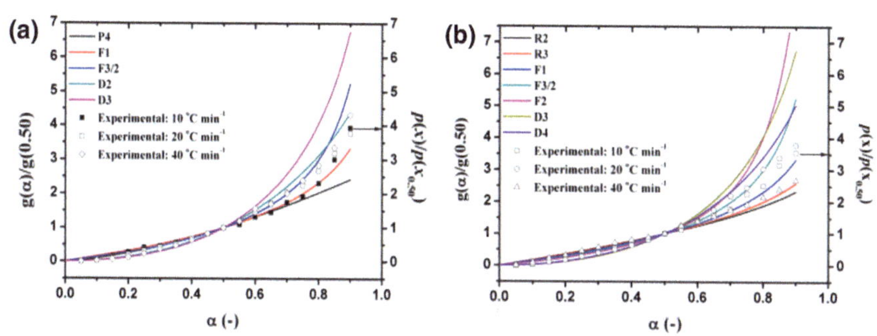

Fig. 9.16 The experimental "Master" plots for the **a** NBR degradation process, at the different heating rates β, together with the theoretical ones for P4, F1, F3/2, D2 and D3 models and **b** EPDM degradation process, at the different heating rates β, together with the theoretical ones for R2, R3, F1, F3/2, F2, D3, and D4 models

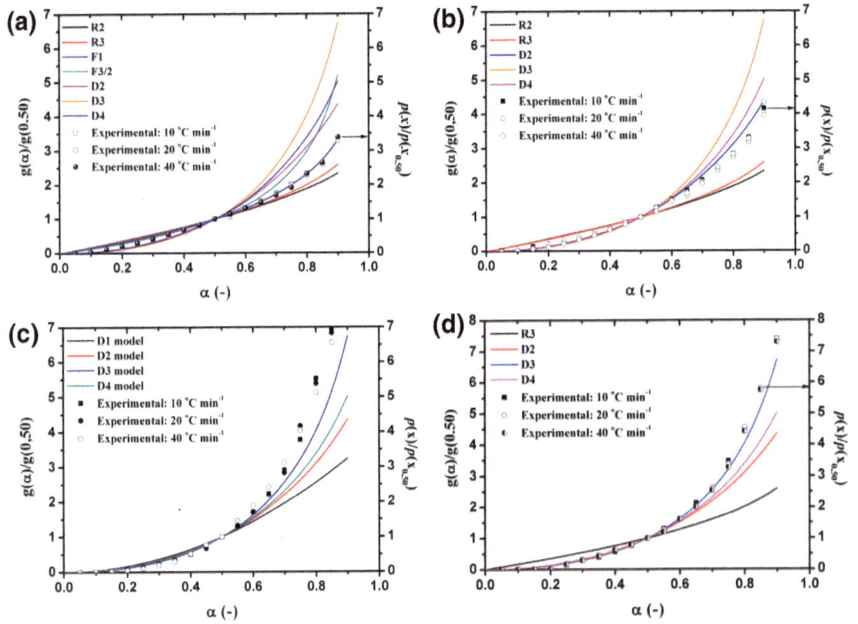

Fig. 9.17 The experimental "Master" plots for the **a** NBR/EPDM rubber blend reinforced with 70 phr of silica, **b** NBR/EPDM rubber blend reinforced with CB/silica (35/35), **c** NBR/EPDM rubber blend reinforced with CB/silica (50/20), and **d** NBR/EPDM rubber blend reinforced with 50 phr CB degradation processes, at the different heating rates

carbon black/silica (35/35), NBR/EPDM rubber blend reinforced with carbon black/silica (50/20) and NBR/EPDM rubber blend reinforced with 50 phr carbon black degradation processes, at the different heating rates, together with the theoretical ones for R2, R3, F1, F3/2, D2, D3, D4 (NBR/EPDM rubber blend reinforced with 70 phr of silica), R2, R3, D2, D3, D4 (NBR/EPDM rubber blend reinforced with carbon black/silica (35/35)), D1, D2, D3 and D4 (NBR/EPDM rubber blend reinforced with carbon black/silica (50/20)), and R3, D2, D3 and D4 (NBR/EPDM rubber blend reinforced with 60 phr carbon black) reaction models.

It can be seen that for the NBR degradation (Fig. 9.16a), the experimental points (for all heating rates) at lower conversions ($\alpha < 0.50$) generally lie on the "Master" curve, which corresponds to the D2 reaction model. On the other hand, at higher conversions ($\alpha > 0.50$), the experimental points lie between the "Master" curves, which correspond to the F1 and F3/2 reaction models. It can be pointed out that at the higher values of heating rates (20 and 40 °C min^{-1}), the experimental "Master" curves tend to lie more on the theoretical "Master" curve, which corresponds to the F3/2 reaction model. In addition, for the EPDM degradation (Fig. 9.16b), the experimental points (for all heating rates) in the whole observed conversion range are grouped around the F1 and F3/2 reaction models.

Based on the obtained results, we cannot decide which of the separated reaction models (F1, F3/2, and D2 for the NBR and F1 and F3/2 for the EPDM degradation process) describe the most realistic investigated processes. It can be seen that for the NBR/EPDM rubber blend reinforced with 70 phr of silica degradation (Fig. 9.17a), the experimental points (for all heating rates) at lower conversions ($\alpha < 0.50$) go closer to the F3/2 reaction model, while these points at higher conversions ($\alpha > 0.50$) lie more on the "Master" curve, which corresponds to the F1 reaction model. In addition, for the NBR/EPDM rubber blend reinforced with CB/silica (35/35) degradation (Fig. 9.17b), the experimental points (for all heating rates) in the whole observed conversion range are very close to the theoretical "Master" plot, which corresponds to the D2 reaction model. On the other hand, for the NBR/EPDM rubber blend reinforced with CB/silica (50/20) degradation process, the experimental points at lower conversions ($\alpha < 0.55$) lie much closer to the theoretical curve which corresponds to the D3 reaction model, whereas at higher values of α ($\alpha > 0.55$), there was a temporary deviation from the same theoretical curve (Fig. 9.17c).

Figure 9.17d shows the experimental "Master" plots for the investigated degradation process at the different heating rates, together with theoretical reaction models R2 (phase boundary reaction—contracting sphere), D2 (two-dimensional diffusion—Valensi equation), D3 (three-dimensional diffusion—Jander equation), and D4 (three-dimensional diffusion—Ginstling–Brounstein equation). It can be seen that for the 50 phr loaded NBR/EPDM blend degradation (Fig. 9.17d), the experimental points (symbols) (for all heating rates) are on and very close to the theoretical "Master" plot, for the three-dimensional diffusion-controlled process (D3), which is described by the Jander equation ($f(\alpha) = (3/2)\cdot(1-\alpha)^{2/3}\cdot\left[1-(1-\alpha)^{1/3}\right]^{-1}$). From the results displayed in Fig. 9.17d, we can clearly see that the D3 reaction model represents the best kinetic model for describing the degradation process, under the nitrogen atmosphere.

To check the validity of the above-calculated kinetic parameters (A, E_a) and assumed reaction model (D3), we performed the corresponding numerical calculations using the Eq. (9.7) with the $f(\alpha)$ function, which corresponds to the three-dimensional diffusion-controlled process, in the conversion range of 0.45 ≤ ≤ 0.95, at heating rates of 10, 20, and 40 °C min^{-1}. For the investigated degradation process, we used the values of kinetic parameters obtained by applying the IKP method ($E_a = 674.7$ kJ mol^{-1}; $A = 8.133 \times 10^{48}$ min^{-1}).

From these results, it is very difficult to determine the reaction mechanism that best describes the investigated degradation processes. Namely, the discrepancies that are observed in the use of the last two methods are caused due to the variability of the apparent activation energy and by the strong dependence of these methods on the apparent activation energy values. However, if a good global agreement in the entire reaction process is reached, one can expect a reasonable model. Therefore, the model obtained must be evaluated to quantify the goodness of fit. This consequently implies the reliability of the obtained kinetic parameters, as well as the function of reaction mechanism. Goodness of fit is necessary, but not sufficient, for the evaluation of a thermal model because it cannot evaluate latent force-fitting caused by the compensation effect. Compensation effect is caused by the exponent format of the Arrhenius equation and refers to the fact that the co-variability of the parameters E_a and A makes the experimental data to fit the several different models well with different $f(\alpha)$. In order to overcome this problem, we apply the composite integral method I [see Eq. (9.15)] that allows us to reliably determine the required reaction model. It is worth noting that the key point in this method is getting a single straight line, using data from the different heating rates. It can be pointed out that sometimes an inappropriate kinetic model can also lead to perfect straight lines, but those lines are parallel to each other instead of superposing on to a one single curve. Table 9.18 shows the results of regression analysis and the values of kinetic parameters (A and E_a) calculated from the composite integral method I (Eq. 9.15), for all investigated degradation processes.

It can be seen that for the F3/2 reaction model in the case of NBR/EPDM rubber blend reinforced with 70 phr of silica degradation, calculated E_a value (Table 9.18) is relatively close to the average value of E_a estimated from the integral isoconversional method (KAS $\langle E_a \rangle^{KAS} = 168.3$ kJ mol^{-1} for $0.25 \leq \alpha \leq 0.85$). In the case of NBR/EPDM rubber blend reinforced with CB/silica (35/35) degradation, calculated value of E_a for D2 reaction model (Table 9.18) is closest to the average

Table 9.18 Values of the kinetic parameters (A and E_a) evaluated for different reaction models, in the case of NBR/EPDM rubber blend reinforced with 70 phr of silica, NBR/EPDM rubber blend reinforced with CB/silica (35/35), and NBR/EPDM rubber blend reinforced with CB/silica (50/20) degradation processes, using the composite integral method I

NBR/EPDM 70 phr SiO_2 degradation				
Model	E_a (kJ mol^{-1})	ln A	A (min^{-1})	R^2
F1	143.9	23.04	1.01×10^{10}	0.97078
F3/2	159.9	25.86	1.70×10^{11}	0.97883
D2	197.5	30.61	1.97×10^{13}	0.97417
D3	213.8	32.11	8.81×10^{13}	0.95741
NBR/EPDM 35 phr CB/35 phr SiO_2 degradation				
Model	E_a (kJ mol^{-1})	ln$A$	A (min^{-1})	R^2
R2	145.8	22.24	4.56×10^{9}	0.90570
D2	224.6	34.93	1.48×10^{15}	0.99587
D3	259.8	39.63	1.63×10^{17}	0.98959
D4	236.1	35.45	2.49×10^{15}	0.99569
NBR/EPDM 50 phr CB/20 phr SiO_2 degradation				
Model	E_a (kJ mol^{-1})	ln$A$	A (min^{-1})	R^2
D1	249.4	39.51	1.44×10^{17}	0.93798
D2	270.7	42.60	3.17×10^{18}	0.96309
D3	296.1	45.58	6.24×10^{19}	0.98251
D4	279.1	42.58	3.11×10^{18}	0.97033

value of E_a estimated from the KAS method (KAS $<E_a>^{KAS} = 220.7$ kJ mol^{-1} for $0.25 \leq \alpha \leq 0.95$). On the other hand, for the NBR/EPDM rubber blend reinforced with CB/silica (50/20) degradation, calculated value of E_a is the highest for the D3 reaction model (Table 9.18), but this value compared to apparent activation energies, which are calculated for the other models, however, deviates from the $<E_a>^{KAS}$ (KAS: $<E_a>^{KAS} = 353.6$ kJ mol^{-1} for $0.30 \leq \alpha \leq 0.80$).

Based on the obtained results, we cannot decide which of the two separated reaction models (*F*1 and *F*3/2 for the NBR/EPDM rubber blend reinforced with 70 phr of silica, D2 and D4 for the NBR/EPDM rubber blend reinforced with carbon black/silica (35/35), and D3 and D4 for the NBR/EPDM rubber blend reinforced with CB/silica (50/20) degradations, respectively) (Table 9.18) together with evaluated kinetic parameters can best describe the given process.

It can be pointed out that the apparent activation energies calculated for the above-mentioned reaction models (Table 9.18) are quite comparable with the values of E_a estimated from the KAS isoconversional method ($e \leq 5.00$ % [85]). The deviation occurs only in the case of the NBR/EPDM rubber blend reinforced with CB/silica (50/20) degradation, but the values of F and S^2 clearly indicate that the reaction model D3 is the exactly best model which can describe the investigated degradation. Table 9.19 shows the values of the pre-exponential factor (A) calculated by the Eq. (9.15). For all the observed degradation processes, the obtained straight lines have a linear correlation coefficient, which is higher than 0.99500. The average values of E_a calculated using Friedman's method were used in all three cases.

It may be noted that there is a good agreement between the values of A calculated by Eq. (9.15) and the values of A which are calculated using the composite integral method I (Table 9.19; for the F3/2 and D2 models) for the NBR/EPDM rubber blend reinforced with 70 phr of silica and NBR/EPDM rubber blend reinforced with CB/silica (35/35) degradations. However, we can see that the composite integral method I exhibits high sensitivity and becomes less reliable for the testing of degradation process, for the polymer blend reinforced with the highest content of CB. On the other hand, it can be observed that the parameter A greatly increases when the polymer blend is filled with the highest content of CB. This increase is approximately of twelve orders of magnitude for the blend filled with up to 50 phr of CB.

To check the validity of the above-calculated kinetic parameters (A, E_a) and assumed reaction models, we performed the corresponding numerical calculations

Table 9.19 Values of the pre-exponential factor (A) calculated using the Eq. (9.15), for NBR/EPDM rubber blend reinforced with 70 phr of silica, NBR/EPDM rubber blend reinforced with CB/silica (35/35), and NBR/EPDM rubber blend reinforced with CB/silica (50/20) degradation processes

NBR/EPDM 70 phr SiO_2 degradation			NBR/EPDM 35 phr CB/35 phr SiO_2 degradation			NBR/EPDM 50 phr CB/20 phr SiO_2 degradation		
Model F3/2			Model D2			Model D3		
ln *A*	*A* (min^{-1})	R^2	ln *A*	*A* (min^{-1})	R^2	ln *A*	*A* (min^{-1})	R^2
27.41	8.02×10^{11}	0.99725	34.99	1.57×10^{15}	0.99503	53.08	1.13×10^{23}	0.99596

using the Eq. (9.7), with $g(\alpha)$ functions, which correspond to the F3/2, D2, and D3 [99]. For all investigated degradation processes, we used the average values of E_a calculated by applying the Friedman method, in the conversion ranges, where the E_a values show no significant variation. For the values of the pre-exponential factor (A), we took the results shown in Table 9.19.

Figure 9.18 shows the comparison between experimental (color lines) and calculated (symbols) conversion ($\alpha - T$) curves, at the heating rate of 20 °C min^{-1}, for the NBR/EPDM rubber blend reinforced with 70 phr of silica, NBR/EPDM rubber blend reinforced with CB/silica (35/35), and NBR/EPDM rubber blend reinforced with CB/silica (50/20) degradation processes, respectively.

From Fig. 9.18, we can see that there is a very good agreement between experimentally obtained $\alpha - T$ curves and numerically calculated conversion curves for the investigated degradation processes, using kinetic triplets, which include the apparent activation energy (E_a) calculated by Friedman's method, the values of A calculated using Eq. (9.15) and reaction models F3/2, D2, and D3. From the obtained results, it follows that with the addition of silica filler (SiO_2) to the polymer blend (NBR/EPDM), the non-isothermal degradation mechanism tends to be more complex, and the reaction order (n) for the NBR/EPDM rubber blend reinforced with 70 phr of silica is, therefore, higher than unity ($n > 1$; there is a deviation from the first-order kinetics). This phenomenon can be clearly seen in Fig. 9.19, where we have the dependence presented in the form ln(dα/dt) *vs* ln($1 - \alpha$) (dα/dt is the rate of process), for the NBR/EPDM rubber blend reinforced with 70 phr of silica degradation.

If the dependence of ln(dα/dt) *vs* ln($1 - \alpha$) is the nonlinear ("curved"), in this case there is a deviation from the simple first-order kinetics [101]. If the considered curve at a given heating rate is the concave upward, then the examined process can be considered to take place according to the kinetics of nth order, with $n > 1$ [101]. Figure 9.19 clearly shows that the degradation process of NBR/EPDM rubber blend

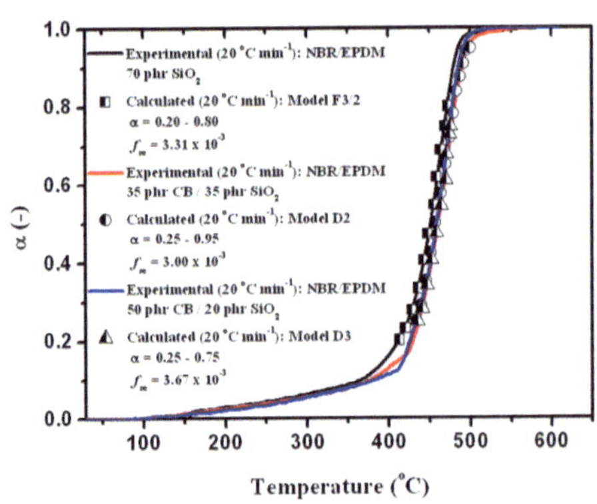

Fig. 9.18 The comparison between experimental (*color lines*) and calculated (*symbols*) conversion (α–T) curves, at the heating rate of 20 °C min^{-1}, for the NBR/EPDM rubber blend reinforced with 70 phr of silica ($\alpha = 0.20 - 0.80$ (F3/2)), NBR/EPDM rubber blend reinforced with CB/silica (35/35) ($\alpha = 0.25 - 0.95$ (D2)), and NBR/EPDM rubber blend reinforced with CB/silica (50/20) ($\alpha = 0.25 - 0.75$ (D3)) degradation processes

reinforced with 70 phr of silica takes place over the kinetics of reaction order higher than unity, more accurately described with reaction model, F3/2. In addition, it can be pointed out that the chain scission reactions are also intensively involved when SiO_2 particles are introduced into polymer host.

For the NBR/EPDM rubber blend reinforced with CB/silica (35/35) and NBR/EPDM rubber blend reinforced with CB/silica (50/20) degradation processes, the equations of Valensi and Jander [99] referring to the diffusion-controlled reactions were established as the most appropriate kinetic models. In the case of NBR/EPDM rubber blend reinforced with CB/silica (35/35) and NBR/EPDM rubber blend reinforced with CB/silica (50/20), the degradation mechanism can primarily be presented by a diffusion model with a plane and spherical symmetry (D2; D3), which would be in accordance with the formation of a physical barrier. In this case, for the char-loaded polymer blends, we have a self-protective behavior leading to the formation of a protective barrier [102, 103]. This effect is most pronounced in the case of NBR/EPDM rubber blend reinforced with CB/silica (50/20) degradation. In addition, with increasing char content, the various effects, such as the branching effect, nucleation, or surface effects, are reduced, so the diffusion of decomposition products from the bulk polymer into the gas phase is accelerated. This is not the case for the degradation of NBR/EPDM rubber blend reinforced with 70 phr of silica, where the diffusion is probably delayed.

In such a case, it is necessary to find which of the observed reaction models give the lowest value of RSS (residual sum of squares). Meanwhile, the lowest value of RSS may not differ significantly from the second smallest. One is advised to use the *F*-test [104] to check whether the difference in the values of RSS is significant. Note that RSS can be readily converted to the variance, S^2, as follows [105]:

$$S^2 = \frac{RSS}{N - p} \tag{9.17}$$

where N is the total number of experimental points used in the calculation and p is the total number of kinetic parameters determined as a result of the calculation. The

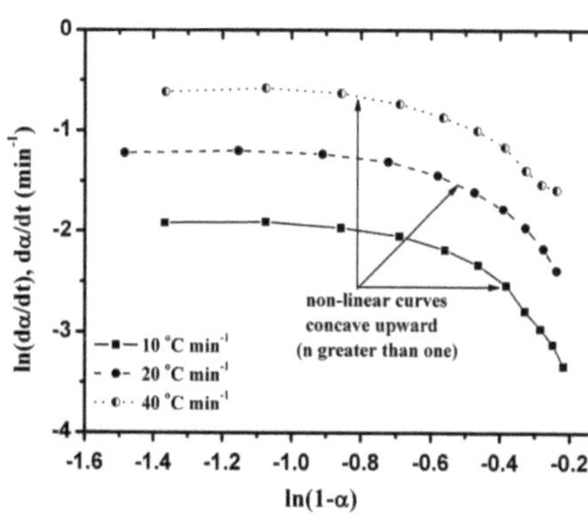

Fig. 9.19 Dependence of ln (dα/dt) *versus* ln(1 − α) in the case of the NBR/EPDM rubber blend reinforced with 70 phr of silica degradation process

significance of the difference in two variances is readily checked by using the regular *F*-test.

Table 9.20 shows the S^2 and *F*-test values for the considered reaction models, in the case of the NBR, EPDM, and NBR/EPDM rubber blends reinforced with 70 phr of silica, NBR/EPDM rubber blend reinforced with carbon black/silica (35/35), and NBR/EPDM rubber blend reinforced with carbon black/silica (50/20) degradation processes, respectively.

From the results displayed in Table 9.20, we can clearly see that the F3/2 (one-and-a-half-order reaction) and F1 (first-order reaction) represent the best kinetic models for describing the NBR and EPDM degradation processes, respectively. From the results displayed in Table 9.20, we can clearly see that the F3/2 (one-and-a-half-order reaction) represents the best kinetic model for describing the

Table 9.20 The S^2 and *F*-test values for considered reaction models, in the case of the NBR, EPDM, and NBR/EPDM rubber blends reinforced with 70 phr of silica, NBR/EPDM rubber blend reinforced with CB/silica (35/35), and NBR/EPDM rubber blend reinforced with CB/silica (50/20) degradation processes

NBR degradation		
Model	**Variance, S^2**	***F*-test**
F1	0.00903	11141.030
F3/2	0.00718	13645.150
D2	0.00825	12278.244
EPDM degradation		
Model	**Variance, S^2**	***F*-test**
F1	0.02075	2200.808
F3/2	0.02600	2119.809
NBR/EPDM 70 phr SiO_2 degradation		
Model	**Variance, S^2**	***F*-test**
F1	0.01717	10201.561
F3/2	0.01694	12274.933
D2	0.02848	9897.479
D3	0.05598	6901.458
NBR/EPDM 35 phr CB/35 phr SiO_2 degradation		
Model	**Variance, S^2**	***F*-test**
R2	0.04979	3103.400
D2	0.00469	78629.498
D3	0.01593	30976.088
D4	0.00529	77140.411
NBR/EPDM 50 phr CB/20 phr SiO_2 degradation		
Model	**Variance, S^2**	***F*-test**
D1	0.04108	3358.700
D2	0.02807	5793.325
D3	0.01679	11589.348
D4	0.02382	7261.020

NBR/EPDM rubber blend reinforced with 70 phr of silica degradation process, under nitrogen atmosphere.

For the NBR/EPDM rubber blend reinforced with CB/silica (35/35) degradation, we can see from Table 9.20 that the D2 represents the best reaction model, for the mechanistic description of the investigated system.

Finally, for the NBR/EPDM rubber blend reinforced with CB/silica (50/20) degradation process, we have that the three-dimensional diffusion mechanism represents the best kinetic model (Table 9.20). In Table 9.20, the best statistical selected kinetic models are marked with bold letters.

Figure 9.20 shows the comparison between experimental (color lines) and calculated (symbols) differential conversion curves at the different heating rates, for the NBR, EPDM, and NBR/EPDM with 50 phr CB degradation processes, respectively. The calculated rate curves for both investigated systems are evaluated using F3/2 and F1 models and the values of kinetic parameters from the IKP method.

From Fig. 9.20a, we can see that there is a very good agreement between experimentally obtained $d\alpha/dt - T$ curves and numerically calculated rate curves for the NBR and EPDM degradation processes, using the corresponding kinetic triplets. From the obtained results, it follows that the thermal degradation of EPDM can be described by the random scission mechanism, in terms of the first-order kinetics. In this case, each bond of the same type has an equal probability for cleavage, with a rate description according to a first-order kinetic model. On the other hand, in the case of the NBR sample, where there is a more complex nature of degradation mechanism, its kinetics is not possible to be described with zero-, first-, or second-order kinetic equations. Consequently, with the single-step reaction model (Fn-type toward n = 3/2), a complex and multistep degradation reaction system as the NBR can be successfully approximated.

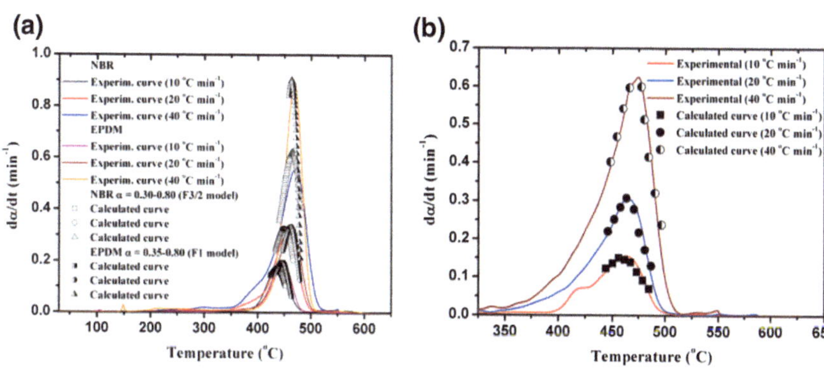

Fig. 9.20 The comparison between experimental (*color lines*) and calculated (*symbols*) differential conversion curves, at the different heating rates (β = 10, 20, and 40 °C min^{-1}) for the **a** NBR (α = 0.30–0.80 (F3/2)), **b** EPDM (α = 0.35–0.80 (F1)), and **c** NBR/EPDM with 50 phr CB (α = 0.45–0.95 (*D*3)) degradation processes

Figure 9.20b shows the comparison between experimental (color lines) and calculated (symbols) differential rate (dα/dt) curves, at the heating rates of 10, 20, and 40 °C min^{-1}, for the NBR/EPDM 50 phr CB degradation process, in the range of $0.45 \le \alpha \le 0.95$. From Fig. 9.20b, we can see that there is a very good agreement between experimentally obtained $d\alpha/dt - T$ curves and numerically calculated differential conversion curves for the investigated degradation process, using the analytical function for the D3 reaction model. Therefore, the degradation mechanism can primarily be presented by a diffusion model with a spherical symmetry (D3), which would be in accordance with the formation of a physical barrier. In this case, for the carbon black-loaded polymer blend, we have a strong self-protective behavior leading to the formation of a protective barrier [102].

9.7 DSC Measurement of Ethylene–Propylene–Diene Rubber Nanoblends

A DuPont 990 thermal analyzer and DuPont differential scanning calorimeter model 910 were used for testing the sample materials. DSC analysis of the samples was performed under a nitrogen atmosphere using a flow of 50 cm^3 per min in temperatures ranging from −110 to 110 °C (heating rate, 20 °C per min). The examined sample weights ranged from 10 to 15 mg. The glass transition temperature (T_g) was determined from the primary point that intersected the tangent discontinuity in the DSC data. The T_g of the polymer depend on the structure and cooperative mobility of its segments. This behavior is reflected in a single T_g for miscible blends. In the case of partially miscible blends, the T_g of the blend components remain separated but are shifted toward each other, as compared with pure components. In completely immiscible polymer blends, the T_g values of all components remain largely unaltered. Therefore, DSC can be considered as a practical method for investigating the miscibility of polymer blends. DSC curves of the silica-reinforced nanocomposites based on EPDM and NBR rubbers and their EPDM/NBR rubber blends are shown in Fig. 9.21.

Two separate *Tg* values of the silica-reinforced nanocomposites based on EPDM/NBR rubber blends can be observed (Fig. 9.22). The first *Tg* (Tg1) results from the EPDM rubber with a temperature range of −60 to −52 °C. The second *Tg* (Tg2) is produced by the NBR rubber with a temperature range of −30 to −23 °C. Therefore, as indicated by the DSC curves of the silica-reinforced EPDM/NBR rubber blends, the EPDM/NBR polymer blends are immiscible.

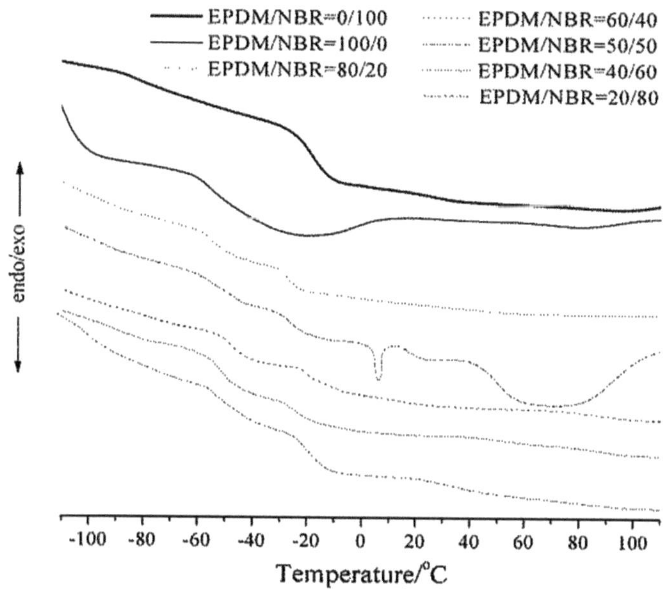

Fig. 9.21 DSC curves of the silica-reinforced nanocomposites based on EPDM and NBR rubbers and EPDM/NBR rubber blends

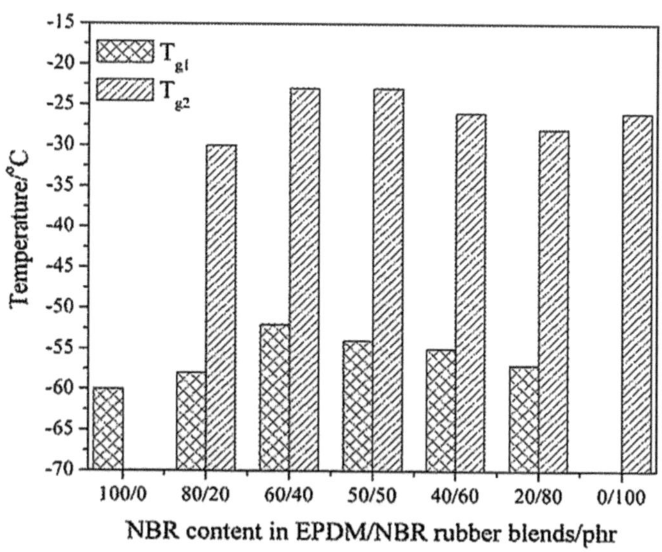

Fig. 9.22 Glass transition temperature variations of the silica-reinforced nanocomposites based on EPDM and NBR rubbers and EPDM/NBR rubber blends, measured by DSC

9.8 Morphological Studies of Ethylene–Propylene–Diene Rubber Nanoblends

The fractured surfaces of the blended materials were imaged using a JEOL JSM-5400 model SEM. The samples were sputter-coated with gold for 3 min under high vacuum with image magnifications of 2000X. From our preliminary experiments, we found that the fractured surface did not change even after storing the specimen for 72 h (without gold coating) prior to the SEM observations. Morphology is a major factor for rubber blends, which determine the extent to which the blends are compatible. It is well known that the phase structure of the blend is influenced by several factors, including the surface characteristics, blend ratio, viscosity of each component, and compounding process. The primary factor that determines the final morphology of the mixes is their composition. NBR, with strong molecular polarity, has higher surface tension than that of EPDM, resulting in their incompatibility.

Figure 9.23 shows the morphology of NBR/EPDM composite with 70 phr CB filler. The dispersion of rubber components in the blends is not uniform, and its heterogeneous nature is indicative. Two phases with irregular shape can be observed. This means that NBR/EPDM blends are completely immiscible and large EPDM domains are dispersed in the NBR matrix. The interphase adhesion between NBR and EPDM is very weak. According to poor compatibility between NBR and EPDM, they have poor adhesion with carbon black.

In the EPDM/NBR = 80/20 rubber blend, NBR, with higher viscosity, is inclined to form the dispersion phase so that it can be dispersed as round particles with maximum surface energy. As the NBR content increases, it becomes more difficult for NBR to be dispersed in the EPDM rubber [106]. Particle surfaces of silica possess hydrophilic silanol (or hydroxyl) groups, which results in a strong filler/filler interaction via hydrogen bonding. Silica forms a strong hydrogen bond

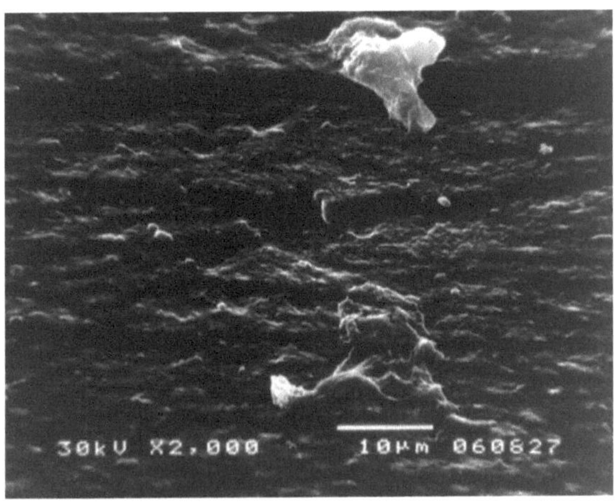

Fig. 9.23 SEM micrograph with magnification X = 2000 of NBR/EPDM composite containing 70 phr CB

with basic materials because of its acidic surface. NBR possesses a nitrile group (ACN), and because the nitrile group is basic, it will bond with the silanol group of silica, resulting in a strong silica/NBR interaction. These types of rubbers were chosen partly because of their differences in relative polarity and solubility, as well as the ease of distinguishing them under an electron microscope due to their electron density contrast. Figure 9.24a–c presents SEM micrographs of EPDM, NBR, and the EPDM/NBR = 20/80 rubber composites.

As shown, the component dispersion nature in the blends was not uniform, which is indicative of their heterogeneous nature. The changes in the domain sizes of the dispersed phase are due to the volume fraction and viscosity differences of the components. These results further demonstrate that the interphase adhesion between NBR and EPDM is very weak.

Aside from their poor compatibility, NBR and EPDM could not make good use of the large surface area-to-volume ratio of nanoparticles to improve their interphase adhesion. Architectural homogeneity and the appearance of strong ridgelines indicate a better dispersion state with efficient interfacial cross-linking. The compact

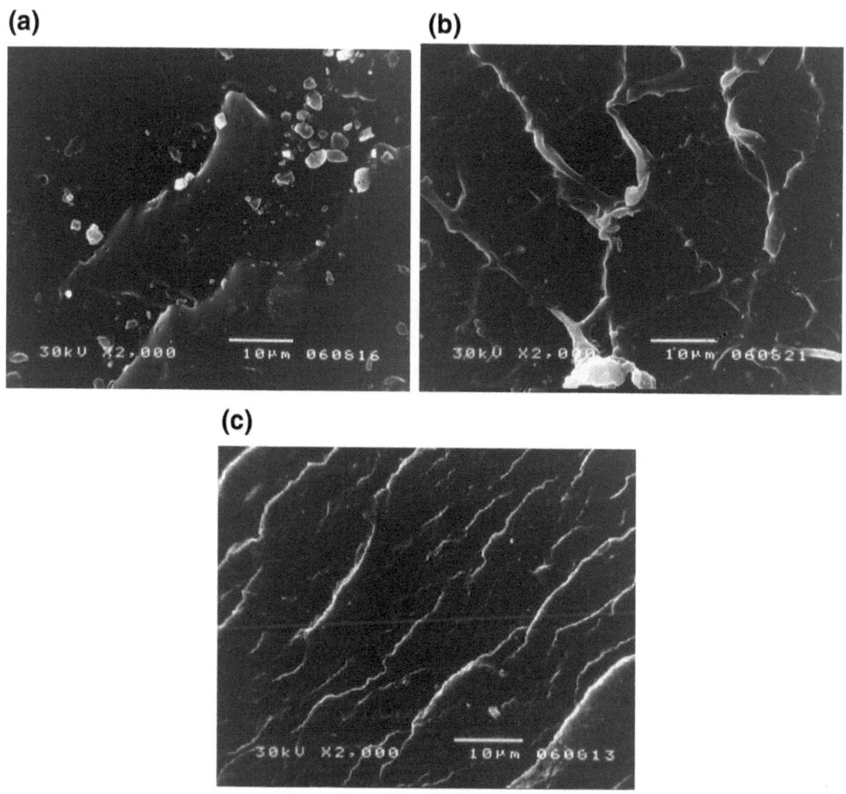

Fig. 9.24 SEM micrographs of **a** EPDM and **b** NBR rubbers without filler and **c** EPDM/NBR rubber blend with 50 phr SIL-1

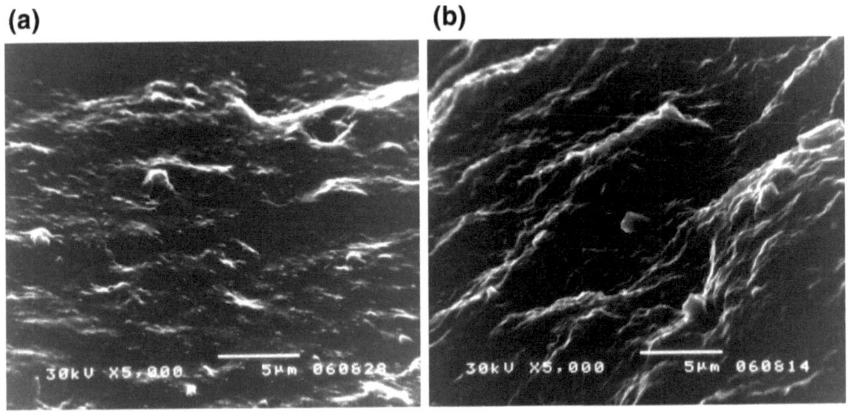

Fig. 9.25 SEM micrographs of NBR/EPDM rubber blend filled with 50 phr CB (**a**) and 50 phr silica (**b**) (magnification X = 5000)

nature of blend systems, as revealed from the study of blend morphology, is manifested in the significant enhancement of the mechanical properties. Vertical steps and stairs are more frequent and more prevalent in cross-linked rubber blends with higher tensile strength values [107]. SEM was employed to study the fractured nanocomposites based on the EPDM/NBR rubber blends and confirmed that certain features of the fracture can be correlated with the mechanical properties.

The SEM micrographs of rubber blends reinforced with 50 phr of CB and precipitated silica are given in Fig. 9.25a and b. The fractured surface of NBR/EPDM silica-reinforced blend is more polished than carbon black-reinforced blend. Also, some amount of fibrosity in the matrix, voids and crazes, as well as microcracks in all samples, have been noticed. The crystallinity contributes toward failure of the network structure. From Fig. 9.25, it is evident that the crack formation is coupled with small holes.

9.9 Some Specific Applications of Ethylene–Propylene–Diene Rubber Nanoblends

Valentini et al [108] investigated graphene-based EPDM nanocomposites for its usage in space or military components subjected to high-frequency, high-magnitude shock pulses. In aerospace missions, pyroshocks occur due to controlled explosions of ordnance devices enabling the functionality of space modules. These shocks result from deployment mechanisms or opening solar sails and can cause failures of electronic devices and structures. Components made of composite materials are of crucial importance for assuring the reliability of aerospace modules, and pyroshock tests for the completion of strict requirements have to be met. Consequently, it is significant to predict the pyroshock responses on composite materials. The pyroshock prediction is still a complex prediction but one of the most challenging issues in composite science and technology. Rubber is an important commercial material widely being used not only for automobile and industrial applications but also for space components. Since rubber for space application requires good reinforcement at the minimum possible filler loading, incorporation of few percentages of graphene nanoplatelets is a viable alternative to obtain low-density high-performance composites. The mechanical properties of such matrix containing equal amount of graphene nanoplatelets and carbon black were the optimum compared to those containing either graphene nanoplatelets or carbon black [108]. Ethylene–propylene–diene terpolymer rubber and graphene nanoplatelet nanocomposites were prepared via a vulcanization process. It was observed how such hybrid nanocomposite has a higher number of cross-links than carbon black nanocomposites. The hybrid nanocomposites showed a higher hardness values demonstrating their higher stiffness. Pyroshock test and analysis are used to extract damping properties of hybrid EPDM nanocomposites. It is observed that increasing the GNP content along with the impact energy causes an increase in the damping properties. The damping ratio of hybrid composites depending on the GNP content and on the impact energy was rationalized in terms of the stick–slip concept in nanocomposites.

The use of EPDM as an insulator for case-bonded solid rocket motors was evaluated more than a decade ago and found feasible [109]. Being a material with a low specific gravity, EPDM has emerged as a novel material for manifold applications including insulation for solid rocket motors. EPDM with dienes such as 1,4 hexadiene (HD) or ethylene norbornene (ENB) has been reported as a suitable primary polymer in successfully developed insulation compositions. A typical formulation for EPDM insulation is given in Table 9.21.

The relationship between ethylene content and crystallinity has been discussed. Crystallinity and curing rates were found to increase with the increase in the ethylene content. Further, ethylene content is also found to have pronounced effect on mechanical properties. The reported literature [109] indicates that higher ethylene content is a better candidate for a primary polymer (among the various grades of EPDM) for rocket motor insulation compositions. Among the commercially available grades, EPDM based on 1,4 hexadiene (NORDEL 1040, 2522, 2722)

Table 9.21 Typical formulation for EPDM insulation

Ingredients	phr
Primary EPDM terpolymer	70–80
Secondary polymer	15–25
Tackifier	5–10
Antioxidant	1–3
Wetting agent	0–1
Cure activator	5–10
Silica filler	40–50
Pigment	0–3
Plasticizer	15–25
Curing agent	10–20

ENB (KELTAN 4506, 1446 A, 2308, NORDEL IP NDR 4520, and NORDEL IP NDR 4640) have reportedly been found suitable for rocket motor insulation. With EPDM as a primary polymer base, polymers such as liquid polybutadiene, liquid EPDM, and adhesion-promoting polymers such as hydroxyterminated polybutadiene (HTPB), neoprene, and hypalon have been found successful as secondary polymers [109].

Thermal protection materials are necessary to protect structural components of launch vehicles during liftoff of launching system [110]. Different reinforcing fillers filled with EPDM rocket motor insulation compound have been widely reported in patents and many research articles [111–115]. Light polymer nanostructured materials are necessary for improved performance for use in rocket insulation applications. The properties needed for such applications are low density, high specific strength, ablation resistance, and associated thermal and aging properties. Ethylene–propylene–diene monomer (EPDM) has been established as a base elastomer in solid propellant rocket insulator compound due to its low density (0.85–0.87 g/cm^3), excellent heat resistance, good low-temperature properties, high resistance to permanent deformation, and processability. EPDM is also used as a binder resin for the insulation compound. Superior aging property of EPDM with high ethylene content is of interest to study the thermal rocket motor insulator compounded with a variety of fillers.

The using potential of polymeric materials and their applications within the fields of ionizing radiation are related to their structural characteristics, defined by changes in the level of degradation. In the case of favorable interactions between the polymer matrix and the reinforcing element, composite materials may have a considerably higher radiation resistance than pure polymers, which considerably broadens the field of application of polymeric materials in the nuclear industry. Polymeric materials become damaged by exposure over time to ionizing radiation. Despite the limited lifetime, polymers have unique engineering material properties and polymers continue to be used in tritium-handling systems. In tritium-handling systems, polymers are employed mainly in joining applications such as valve sealing surfaces (e.g., stem tips, valve packing, and O-rings). The O-rings used in the H1616 shipping package are fabricated in accordance with Sandia specification SS395668 [116]. The service temperature range for the H1616 vessel is −40–76 °C. The normal condition of transport (NCT) temperature 76 °C is based on an ambient temperature of 38 °C with solar heating. The maximum temperature at the flange closest to the O-rings is 67 °C with solar heating and 47 °C in the shade. The maximum temperature judged to be most applicable to the O-rings is 67 °C. Typical seal temperatures are expected to be lower. Current expectations are that the O-rings will maintain a seal at bounding normal temperatures in service 67 °C for at least 12 months. The baseline aging data review suggests that the EPDM O-rings are likely to retain significant mechanical properties and sealing force at bounding service temperatures to provide a service life of at least 2 years [116]. The H1616 is a certified "Type B" package for the transport of radioactive tritium reservoirs by the Department of Defense (DoD), United Kingdom (UK), Atomic Weapons Establishment (AWE), and Authorized Users (Savannah River Site (SRS), Pantex), Fig. 9.26. Aluminum packing material used to position the tritium reservoir is added to the containment vessel (CV) which is constructed from Type 304 stainless steel. The CV is placed inside a 16-gallon stainless steel drum (Fig. 9.26). Two EPDM O-rings are housed between the CV lid and body and are used to seal each containment vessel [116].

Comparison of aging data for different EPDM compounds from multiple sources using varying approaches and parameters to define failure is complex [117]. However, the collective data suggest that the H1616 EPDM O-rings can likely retain sufficient properties to remain leaktight for at least two years at or below 76 °C, the bounding NCT (normal conditions of transport) vessel temperature. Conservative interpretation of accelerated aging data from examined references suggests possible lifetimes of at least 4 years at 76 °C and at least 6 years at 67 °C, which is the peak O-ring temperature with solar heating.

The trend line shown in Fig. 9.27 [117] (which is a lower bound to all failure data) is intended to show reasonable and conservative lifetimes that might be predicted from the available data. Actual lifetimes could vary. At 76 °C, the trend line suggests a lifetime of at least 4.2 years. At 67 °C, the trend line suggests a lifetime of ~5.8 years. At 47 °C (max O-ring temperature in shade), the estimated lifetime is increased to about 19 years. In the absence of solar heating and ambient temperatures greater than 38 °C, the 47 °C O-ring temperature is likely more realistic (but still bounding) for typical exposure.

Review of relevant aging data for EPDM elastomers suggests that the H1616 O-rings should remain functional for at least 2 years at bounding conditions. Direct comparison of data from various references with different compounds, approaches, and failure criteria is difficult. However, a conservative interpretation of the available data suggests that the H1616 O-rings can have lifetimes of nearly 6 years at 67 °C (maximum temperature at the flange closest to the O-rings with solar heating). This temperature is conservative relative to typical O-ring temperatures in service.

Fig. 9.26 H1616 container showing O-ring location

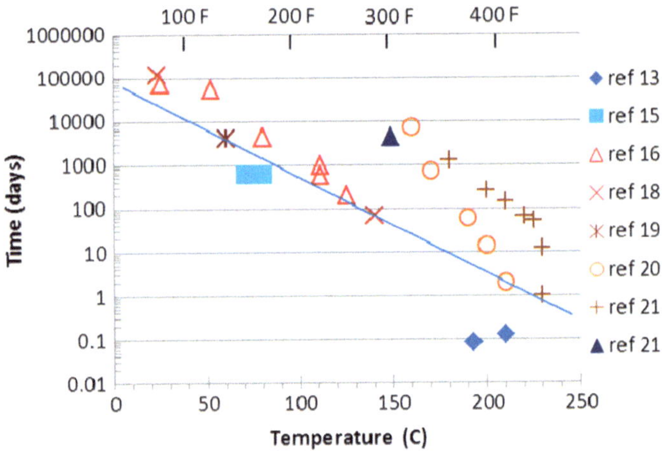

Fig. 9.27 EPDM O-ring lifetime data from literature references [117]

9.10 Conclusions

Rubber industry is an important resource-based industry in the world. Being a material of low modulus, rubber needs reinforcement for enhancing its strength and performance. EPDM elastomers are one of the most versatile, fastest growing, and interesting synthetic rubber polymers. Excellent resistance to heat, oxidation, ozone, and weather aging are expected to provide continued value to demanding automotive, construction, and mechanical goods applications. Current and emerging advanced polymerization and catalyst technologies also provide the ability to design polymers to meet application and processing needs that are important to meeting the ever-increasing demands for product quality, uniformity, and performance. EPDM ranks as the third largest synthetic rubber, contributing about 12 % of the world rubber production an automobile, almost two-thirds of the total rubber content is used in tires. Remaining rubber components are mainly gaskets, sealing materials, hoses, vibration isolators, and plugs. Almost 25–50 % of these non-tire rubber parts are produced with EPDM. Other largest single use of EPDM is in roofing. The wide ranges of particulate fillers are used in the rubber industry to improve and modify its physicomechanical properties. Carbon black is the most important reinforcing filler used so far owing to its superior rubber–filler interactions due to the larger surface area and higher surface activity leading to bound rubber formation. Second to carbon black is the silica, both precipitated and pyrogenic and used as reinforcing filler under the non-black category also has lower particle size and larger surface area but relatively less surface activity for enhancing technical properties of rubber products. The unique advantages of precipitated silica-reinforced rubber compounds include improvement in tear, flex, abrasion, heat resistance, hardness, modulus, resilience, and rolling resistance.

The mechanical properties of silica-reinforced NBR/EPDM rubber blends show higher tensile strength, hardness, and rebound resilience, but lower elongation at break than carbon black (CB)-reinforced rubber blends. Compared to pure NBR and EPDM rubbers, all mechanical properties are improved. At low filler loading, relatively large distances between the filler particles exist. In this case, the rubber chains are attached to one filler particle because of a single attachment or multiple attachments of chain segments of one rubber molecule. With the increase in the concentration of the filler interparticle attachment predominates, rubber molecules connect two or more filler particles. Silica-reinforced nanocomposites exhibited higher changes in all mechanical properties than carbon black-reinforced nanocomposites.

The poor compatibility between NBR and EPDM rubber and poor adhesion with carbon black are detected by SEM. The fractured surface of NBR/EPDM silica-reinforced blend is more polished than carbon black-reinforced blend.

Using different calculation procedures, the kinetic triplets (A, E_a, and $f(\alpha)$) are calculated for the non-isothermal degradation of the acrylonitrile–butadiene (NBR), ethylene–propylene–diene monomer (EPDM) polymers, and NBR/EPDM rubber blends reinforced with CB/silica fillers. For the studied samples in this chapter, using different isoconversional methods, it was found that the NBR, EPDM degradations and NBR/EPDM rubber blend reinforced with 70 phr of silica, NBR/EPDM rubber blend reinforced with CB/silica (35/35) and NBR/EPDM rubber blend reinforced with CB/silica (50/20) degradations represent the complex processes, where they were identified the conversion regions with a constant value of the apparent activation energy (E_a). It was found that NBR and EPDM degradations under nitrogen atmosphere can be described by the one-and-a-half order ($n = 3/2$) and the first-order ($n = 1$) reaction kinetics, respectively. For the investigated systems, a linear dependence was found between $\ln A$ and E_a, known as the compensation effect. Degradation process of the NBR/EPDM rubber blend reinforced with 70 phr of silica under nitrogen atmosphere can be described by the one-and-a–half-order ($n = 3/2$; F3/2) kinetics, while the degradation of the NBR/EPDM rubber blend reinforced with CB/silica (35/35) and NBR/EPDM rubber blend reinforced with CB/silica (50/20) can best be described with two-dimensional (D2) and three-dimensional (D3) diffusion mechanisms. For all investigated systems, the artificial compensation effect (*art*-CE) was found. Good agreement was observed between experimental and calculated conversion ($\alpha - T$) curves (shown at the heating rate of 20 °C min^{-1}), for all investigated degradation processes. It was found that the sample of the polymer blend, which contains the highest CB content (50 phr), shows the greatest self-protective behavior (i.e., in the formation of a protective barrier). Because of this effect, it was concluded that the NBR/EPDM matrix can be strengthened and physical properties such as thermal stability are significantly improved, if there is an increased degree of CB addition.

Acknowledgments This research work was partially supported by the Ministry of Science and Environmental Protection of Serbia, under the projects 45020, 45022, and 172015.

References

1. Samaržija-Jovanović S, Jovanović V, Marković G, Konstantinović S, Marinović-Cincović M (2011) Nanocomposites based on silica-reinforced ethylene–propylene–diene–monomer/acrylonitrile–butadiene rubber blends. Compos B 42:1244–1250
2. Brydson JA (1978) Rubber chemistry. Applied Science Publishers, London
3. Marinović-Cincović M, Janković B, Jovanović V, Samaržija-Jovanović S, Marković G (2013) The kinetic and thermodynamic analyses of non-isothermal degradation process of acrylonitrile–butadiene and ethylene–propylene–diene rubbers. Compos B 45:321–332
4. Deuri SA, De PP, Bhowmick AK, De SK (1988) Studies on the ageing of EPDM based rocket insulator compound by stress relaxation and the effect of propellant binder. Polym Degrad Stab 20(2):135–148
5. Baldwin FP, Ver Strate G (1972) Polyolefin elastomers based on ethylene and propylene rubber. Chem Technol 45:709–881
6. Ver Strate G, Bikales NM, Over- berger CG, Menges, G (Ed), vol 6. Wiley, New York, pp 522–565
7. Cady L, Denton D, Edmondson S, Glass S, Mangold D, Flexpo (1999) International conference June 1999, Houston
8. Kole S, Chaki TK, Bhowmich AK, Tripathy DK (1993) Polym Degrad Stab 41:109–115
9. Smith DBW, Laird JL, Larson SF (1996) In paper presented at the 150 Technical Meeting, Rubber Division ACS1996, Louisville 8–9, Paper No 27
10. Gamlin CD, Dutta NK, Choudhury NR (2003) Mechanism and kinetics of the isothermal thermodegradation of ethylene-propylene-diene (EPDM) elastomers. Polym Degrad Stab 80 (3):525–531
11. Allen, Ralph D (1982) Fundamental of compounding EPDM for cost/performances. Rubber World 186:36–41
12. Jovanović V, Samaržija-Jovanović S (2013) Nanocomposites of elastomers based on different network precursors, monograph, Academic mind, Belgrade, Faculty of Natural Science and Mathematics, University of Priština, Kosovska Mitrovica (in Serbian language)
13. Gwaily SE, Abdel Aziz MM, Madani M (1998) Thermal and electrical studies on irradiated silica–ethylene–propylene diene monomer (SiO_2/EPDM) composites. Polym Testing 17:265–287
14. Jovanović V, Samaržija-Jovanović S, Budinski-Simendić J, Marković G, Marinović-Cincović M (2013) Composites based on carbon black reinforced NBR/EPDM rubber blends. Compos Part B 45:333–340
15. Sau K, Chaki TK, Khastgir D (1998) Carbon fibre filled conductive composites based on nitrile rubber (NBR), ethylene propylene diene rubber (EPDM) and their blend. Polymer 39:6461–6471
16. Grigoryeva OP, Karger-Kocsis J (2000) Melt grafting of maleic anhydride onto an ethylene–propylene–diene terpolymer (EPDM). Eur Polym J 36:1419–1429
17. Papke N, Karger-Kocsis J (2001) Thermoplastic elastomers based on compatibilized poly (ethylene terephthalate) blends: effect of rubber type and dynamic curing. Polymer 42:1109–1120
18. Wu D, Wang X, Jin R (2004) Toughening of poly(2,6-dimethyl-1,4-phenylene oxide)/nylon 6 alloys with functionalized elastomers via reactive compatibilization: morphology, mechanical properties, and rheology. Eur Polym J 40:1223–1232
19. Kang TK, Kim Y, Lee WK, Park HD, Cho WJ, Ha CS (1999) Properties of uncompatibilized and compatibilized poly(butylene terephthalate)–LLDPE blends. J Appl Polym Sci 72:989–997
20. Pukanszky B, Tudos F, Kolarik J, Lednicky F (1990) Ternary composites of polypropylene, elastomer, and filler – analysis of phase-structure formation. Polym Compos 11:98–104
21. Aminabhavi TM, Manjeshwar LS, Cassidy PE (1986) Water permeation through elastomer laminates IV. NBR/EPDM. J Appl Polym Sci 32(2):3719–3723

22. Vara R, Dunn JR (1991) Developments in fuel hoses to meet changing environmental needs. Rubber World 209:24–31
23. Namboodiri CSS, Tripathy DK (1992) Strain-dependent isothermal damping behaviour of filled EPDM rubber: effect of vulcanizing system. Plast Rubber Compos Process Appl 17 (3):171–178
24. Maity SK, Chakraborty KK (1993) Studies on curing characteristics of natural rubber-, nitrile rubber- and silicone rubber-based gum vulcanizates in the presence of boron compounds. J Elastomers Plast 25(4):358–380
25. Samaržija-Jovanović S, Jovanović V, Marković G, Marinović-Cincović M (2009) The effect of different types of carbon blacks on the rheological and thermal properties of acrylonitrile butadiene rubber. J Therm Anal Calorim 98(1):275–283
26. Samaržija-Jovanović S, Jovanović V, Konstantinović S, Marković G, Marinović-Cincović M (2011) Thermal behavior of modified urea-formaldehyde resins. J Therm Anal Calorim 104(3):1159–1166
27. Gatos KG, Karger-Kocsis J (2010) Rubber/clay nanocomposites: preparation, properties and applications in rubber nanocomposites: preparation, properties and applications. In: Thomas S, Stephen R (eds) Wiley (Asia) Pte Ltd
28. Bhuvaneswari CM, Kakade SD, Deuskar VD, Dange AB, Gupta M (2008) Filled ethylene–propylene diene terpolymer elastomer as thermal insulator for casebonded solid rocket motors. Defence Sci J 58:94–102
29. Subramanian V, Ganapathy S (1998) Aging of vulcanizates of formulations for rubber seals. J Appl Polym Sci 70(5):985–994
30. Hassander H, Tornell B (1985) Methods for testing morphology–property relationships in rubber blends. Polym Test 5(1):11–25
31. Jovanović V, Samaržija-Jovanović S, Marković G, Marinović-Cincović M, Budinski-Simendić J (2011) Mechanical and Morphological Properties Rubber blends reinforced with nanofillers. KGK 9:52–56
32. Arrighi V, Gagliardi S, Higgins JS, Triolo A, Zanotti J-M (2002) Quasielastic neutron scattering as a probe of molecular motion in polymer-filler systems. E-MRS spring meeting 2002. Strasbourg (France). N-15
33. Persello J (2002) Designing nanostructured particular fillers for elastomers. Role of nanostructure and polymer filler interactions in rubber reinforcement. E-MRS spring meeting 2002. Strasbourg (France). N-8
34. Rajeev RS, De SK (2002) Crosslinking of rubbers by filler. Rubber Chem Technol 75 (3):475–509
35. Gerspacher M, Advanced CB (2009) Characterizations to better understand polymer–filler interaction. Kaut gummi kunst 5:233–239
36. Zaborski M, Donnet JB (2003) Activity of fillers in elastomer networks of different structure. Macromol Symp 194:87–100
37. Marković G, Jovanović V, Samaržija-Jovanović S, Marinović-Cincović M, Budinski-Simendić J (2013) Thermal stability of gamma irradiated chlorinated isobutylene-isoprene copolymer/chlorosulphonated polyethylene rubber blend/carbon black nanocomposites. J Thermoplast Compos Mater 28(8):1071–1108
38. Frohlich J, Niedermeier W, Luginsland H-D (2005) The effect of filler–filler and filler–elastomer interaction on rubber reinforcement. Compos A 36:449–460
39. Kohls DJ, Beaucage G (2002) Rational design of reinforced rubber. Curr Opin Solid State Mater Sci 6:183–194
40. Abadir EF (2013) Mechanism and kinetics of the non-isothermal degradation of ethylenepropylene diene monomer (EPDM). J Therm Anal Calorim 114(3):1409–1413
41. Gamlin CD, Dutta NK, Choudhury NR (2003) Mechanism and kinetics of the isothermal thermodegradation of ethylene-propylene-diene (EPDM) elastomers. Polym Degrad Stab 80:525–531
42. Rezende Duek EA, De Paoli MA (1990) The photo-oxidation of EPDM rubber: Part III—mechanistic aspects and stabilization. Polym Degrad Stab 30:283–292

43. Guzzo M, De Paoli M (1992) The photo-oxidation of EPDM rubber: Part IV—degradation and stabilization of vulcanizates. Polym Degrad Stab 36:169–172
44. Zaharescu T, Giurginca M, Jipa S (1999) Radiochemical oxidation of ethylene–propylene elastomers in the presence of some phenolic antioxidants. Polym Degrad Stab 63:245–251
45. Zaharescu T, Mihalcea I (1997) Behaviour of ethylene-propylene elastomers in salt solution —III. Gel content data. Polym Degrad Stab 55:265–268
46. Zaharescu T (1996) Irradiation effects on ethylene-propylene elastomers in an aqueous environment. Polym Testing 15:69–73
47. Zaharescu T, Giurginca M, Mihalcea I (1995) Effects of some antioxidants on the stability of ethylene-propylene elastomers. Polym Degrad Stab 48:151–154
48. Zaharescu T, Vilcu R, Podina C (1998) Some kinetic and thermodynamic aspects of thermal degradation of lightly stabilised elastomers. Polym Testing 17:587–596
49. Ginic-Markovic M, Choudhury NR, Dimopoulos M, Matisons JG (2000) Weatherability of coated EPDM rubber compound by controlled UV irradiation. Polym Degrad Stab 69:157–168
50. Zaharescu T, Podina C, Wurm D (2001) Effect of metallic oxides on thermal stability of ethylene–propylene terpolymer. J Appl Polym Sci 82:2155–2158
51. Delor-Jestin F, Lacoste J, Barrois-Oudin N, Cardinet C, Lemaire J (2000) Photo-, thermal and natural ageing of ethylene–propylene–diene monomer (EPDM) rubber used in automotive applications. Influence of carbon black, crosslinking and stabilizing agents. Polym Degrad Stab 67:469–477
52. Samaržija-Jovanović S, Jovanović V, Marković G (2008) Thermal and vulcanization kinetic behaviour of acrylonitrile butadiene rubber reinforced by carbon black. J Therm Anal Calorim 94:797–803
53. Gamlin C, Markovic MG, Dutta NK, Choudhury NR, Matisons JG (2000) Structural effects on the decomposition kinetics of EPDM elastomers by high-resolution TGA and modulated TGA. J Therm Anal Calorim 59:319–336
54. Rajeev RS, De SK, Bhowmick AK, Baby J (2003) Studies on thermal degradation of short melamine fibre reinforced EPDM, maleated EPDM and nitrile rubber composites. Polym Degrad Stab 79:449–463
55. Amraee IA (2009) The effect of heat history on thermal degradation of elastomers containing butadiene units. J Appl Polym Sci 113:3896–3900
56. Kissinger HE (1957) Reaction kinetics in differential thermal analysis. Anal Chem 29:1702–1706
57. Jovanović V, Budinski-Simendić J, Samaržija-Jovanović S, Marković G, Marinović-Cincović M (2009) The influence of carbon black on curing kinetics and thermal aging of acrylonitrile–butadiene rubber. CI&CEQ 15(4):283–289
58. Bakobza L (2009) In: Karger-Kocsis J, Fakirov S (eds) Rubber nanocomposites: new developments, new opportunities—nano- and micro- mechanics of polymer blends and composites, first edition, chapter 4. Carl Hanser Verlag, Munich, Germany, pp 141–162
59. McNeill IC, Mohammed MH (1995) Thermal analysis and degradation mechanisms of blends of LDPE, polyethylacrylate and ethylene-ethylacrylate copolymer with calcium carbonate. Polym Degrad Stab 49:263–273
60. Ikeda Y, Tanaka A, Kohjiya S (1997) Effect of catalyst on in situ silica reinforcement of styrene-butadiene rubber vulcanizate by the sol-gel reaction of tetraethoxysilane. J Mater Chem 7:455–458
61. Ikeda Y, Kohjiya S (2000) Reinforcement of general purpose grade rubbers by silica generated in situ. Rubber Chem Tech 73:534–550
62. Montserrat S, Málek J, Colomer P (1998) Thermal degradation kinetics of epoxy–anhydride resins: I.: influence of a silica filler. Thermochim Acta 313:83–95
63. Venter SAS, Kunita MH, Matos R, Nery RC, Radovanovic E, Muniz EC, Girotto EM, Rubira AF (2005) Thermal and scanning electron microscopy/energy-dispersive spectroscopy analysis of styrene–butadiene rubber–butadiene rubber/silicon dioxide and styrene–

butadiene rubber–butadiene rubber/carbon black–silicon dioxide composites. J Appl Polym Sci 96:2273–2279
64. Thavamani P, Sen AK, Khastgir D, Bhowmick AK (1993) The effect of crosslink density, curing system, filler and resin on the decomposition of hydrogenated nitrile rubber and its blends. Thermochim Acta 219:293–304
65. Matheson MJ, Wampler TP, Simonsick WJ Jr (1994) The effect of carbon-black filling on the pyrolysis behavior of natural and synthetic rubbers. J Anal Appl Pyrol 29:129–136
66. Pandey JK, Reddy KR, Kumar AP, Singh RP (2005) An overview on the degradability of polymer nanocomposites. Polym Degrad Stab 88:234–250
67. Janković B, Marinović-Cincović M, Jovanović V, Samaržija-Jovanović S, Marković Gordana (2012) The comparative kinetic analysis of non-isothermal-degradation process of acrylonitrile-butadiene/ethylene-propylene-diene rubber blends reinforced with carbon black/silica fillers. Part II, Thermochim Acta 543:304–312
68. Senum GI, Yang RT (1977) Rational approximations of the integral of the Arrhenius function. J Therm Anal Calorim 11:445–447
69. Wypych G (2000) Handbook of fillers, 2nd edn. ChemTec Publishing, Toronto, Ontario M1E 1C6, Canada, pp 131–141
70. Vyazovkin S, Wight CA (1999) Model-free and model-fitting approaches to kinetic analysis of isothermal and nonisothermal data. Thermochim Acta 340–341:53–68
71. Vyazovkin S, Wight CA (1998) Isothermal and non-isothermal kinetics of thermally stimulated reactions of solids. Int Rev Phys Chem 17:407–433
72. Flynn JH (1997) The 'temperature integral'—its use and abuse. Thermochim Acta 300:83–92
73. Khawam A, Flanagan DR (2006) Solid-state kinetic models: basics and mathematical fundamentals. J Phys Chem Part B 110:17315–17328
74. Vyazovkin S (2006) Model-free kinetics: staying free of multiplying entities without necessity. J Therm Anal Calorim 83:45–51
75. Pratap A, Lilly Shanker Rao T, Lad KN, Dhurandhar HD (2007) Isoconversional vs. model fitting methods: a case study of crystallization kinetics of a Fe-based metallic glass. J Therm Anal Calorim 89:399–405
76. Akahira T, Sunose T (1971) Joint convention of four electrical institutes. Res Rep Chiba Inst Technol 16:22–31
77. Friedman HL (1964) Kinetics of thermal degradation of char-forming plastics from thermogravimetry—application to a phenolic resin. J Polym Sci Part C Polym Lett 6:183–195
78. Coats AW, Redfern JP (1964) Kinetic parameters from thermogravimetric data. Nature 201:68–69
79. Budrugeac P, Segal E, Pérez-Maqueda LA, Criado JM (2004) The use of the IKP method for evaluating the kinetic parameters and the conversion function of the thermal dehydrochlorination of PVC from non-isothermal data. Polym Degrad Stab 84:311–320
80. Vyazovkin S, Linert W (1995) False isokinetic relationships found in the nonisothermal decomposition of solids. Chem Phys 193:109–118
81. Málek J (1992) The kinetic analysis of non-isothermal data. Thermochim Acta 200:257–269
82. Senum GI, Yang RT (1977) Rational approximations of the integral of the Arrhenius function. J Therm Anal Calorim 11:445–447
83. Amraee IA (2009) The effect of heat history on thermal degradation of elastomers containing butadiene units. J Appl Polym Sci 113:3896–3900
84. Deuri SA, Bhowmick AK, Ghosh R, John B, Sriram T, De SK (1988) Thermal and ablative properties of rocket insulator compound based on EPDM. Polym Degrad Stab 21(1):21–28
85. Janković B, Marinović-Cincović M, Jovanović V, Samaržija-Jovanović S, Marković G (2012) The comparative kinetic and thermodynamic studies of non-isothermal degradation process of acrylonitrile-butadiene and ethylene-propylene-diene rubber compounds. Part I Thermochim Acta 543:295–303

86. Vyazovkin S (2008) In: Brown ME, Gallagher PK (eds) Handbook of thermal analysis and calorimetry, vol. 5: recent advances, techniques and applications. Isoconversional kinetics. Elsevier B.V., Amsterdam, pp 503–538
87. Chrissafis K, Paraskevopoulos KM, Stavrev SY, Docoslis A, Vassiliou A, Bikiaris DN (2007) Characterization and thermal degradation mechanism of isotactic polypropylene/carbon black nanocomposites. Thermochim Acta 465(1–2):6–17
88. Peterson JD, Vyazovkin S, Wight CA (2001) Kinetics of the thermal and thermo-oxidative degradation of polystyrene, polyethylene and poly(propylene). Macromol Chem Phys 202 (6):775–784
89. Brazier DW, Schwartz NV (1978) The effect of heating rate on the thermal degradation of polybutadiene. J Appl Polym Sci 22(1):113–124
90. McCreedy K, Keskkula H (1978) Application of thermogravimetric analysis to the thermal decomposition of polybutadiene. J Appl Polym Sci 22(4):999–1005
91. Blanksby SJ, Ellison GB (2003) Bond dissociation energies of organic molecules. Acc Chem Res 36(4):255–263
92. Zhou D, Mark JE (2004) Preparation and characterization of trans-1,4-polybutadiene nanocomposites containing in situ generated silica. J Macromol Sci Part A Pure Appl Chem 41(11):1221–1232
93. Gamlin C, Dutta N, Choudhury NR, Kehoe D, Matisons J (2001) Influence of eyhylene propylene ratio on the thermal degradation behaviour of EPDM elastomers. Thermochim Acta 367–368:185–193
94. Riedel JA, Laan RV (1990) Ethylene propylene rubbers: the vanderbilt rubber handbook, 13th edition, Norwalk, CT, R.T. Vanderbilt Co., Inc., pp 123–148
95. Vlaev LT, Markovska IG, Lyubchev LA (2004) Kinetics compensation effect at thermal degradation of rice husk. Oxid Commun 27:444–452
96. De Thomas Kurian PP, Khastgir D, Tripathy DK, De SK, Peiffer DG (1995) Reinforcement of EPDM-based ionic thermoplastic elastomer by carbon black. Polymer 36:3875–3884
97. Ismail H, Rusli A, Azura AR, Ahmad Z (2008) The effect of partial replacement of paper sludge by commercial fillers on natural rubber composites. J Rein Plast Comp 27:1877–1891
98. Ramesan MT (2011) Effect of silica on uncompatibilized and compatibilized styrene butadiene rubber and nitrile rubber blends. Int J Polym Mater 60:1130–1146
99. Turmanova SCh, Genieva SD, Dimitrova AS, Vlaev LT (2008) Non-isothermal degradation kinetics of filled with rise husk ash polypropene composites. Express Polym Lett 2:133–146
100. Montserrat S, Málek J, Colomer P (1998) Thermal degradation kinetics of epoxy-anhydride resins: I. Influence of a silica filler. Thermochim Acta 313:83–95
101. Burnham AK, Braun RL (1999) Global kinetic analysis of complex materials. Energy Fuels 13:1–22
102. Siat C, Bourbigot S, Le Bras M (1997) Thermal behaviour of polyamide-6-based intumescent formulations-a kinetic study. Polym Degrad Stab 58:303–313
103. Albano C, Rodríguez B, Karam A, Hernández M, Ichazo M, González J, Covis M (2012) Nitrile rubber–bentonite composites: a thermal degradation study. Polym Bull 68:1935–1950
104. Freund JE, Perles BM (2006) Modern elementary statistics, 12th edn. Pearson Prentice Hall, Upper Saddle River
105. Vyazovkin S, Burnham AK, Criado JM, Pérez-Maqueda LA, Popescu C, Sbirrazzuoli N (2011) ICTAC kinetics committee recommendations for performing kinetic computations on thermal analysis data. Thermochim Acta 520:1–19
106. Zhang LQ, Li T, Lu YL, Tang YW, Qiao JL, Tian M (2006) The morphology and property of ultra-fine full-vulcanized acrylonitrile butadiene rubber particles/EPDM blends. J Appl Polym Sci 100:3673–3679
107. Gent AN, Pulford CRT (1984) Micromechanics of fracture in elastomers. J Mater Sci 19:3612–3619
108. Valentini L, Bolognini A, Alvino A, Bittolo Bon S, Martin-Gallego M, Lopez-Manchado MA (2014) Pyroshock testing on graphene based EPDM nanocomposites. Compos Part B Eng 60:479–484

109. Bhuvaneswari CM, Sureshkumar SM, Kakade SD, Gupta M (2006) Ethylene-propylene diene rubber as a futuristic elastomer for insulation of solid rocket motors. Defense Sci J 56:309–320
110. Sangita S, Guchhait PK, Bandyopadhyay GG, Chaki TK (2013) Development of polyimide–nanosilica filled EPDM based light rocket motor insulator compound: influence of polyimide–nanosilica loading on thermal, ablation, and mechanical properties. Compos Part A 44:8–15
111. Fan JL, Tsai Shr-Hau, Tu Fu-Hua, Tu Yao Tsai. Low density rocket motor insulation, United States patent application, Publication No. US2007/0112091 A1; May 17, 2007
112. Uyarel AY, Pektas I (1996) A thermal analysis investigation of new insulator compositions based on EPDM and phenolic resin. J Therm Anal 46:163–176
113. Deuri AS, Bhowmick AK (1986) Ageing of rocket insulator compound based on EPDM. Polym Degrad Stabil 16:221–239
114. Deuri AS, Bhowmick AK (1987) Degradation of rocket insulator at high temperature. J Therm Anal 32:755–770
115. Harvey AR, Ellertson JW (2003) Rocket motor insulation containing hydrophobic particles, United States patent no. US 6,606,852 B1 August, 2003
116. Stefek TM, Daugherty WL, Skidmore TE (2012) "Status report for aging studies of EPDM o-ring material for the H1616 shipping package", SRNL-STI-2012-00348, August 2012
117. Skidmore TE (2012) Review of aging data on EPDM O-rings in the H1616 Shipping Package (U)", SRNL-STI-2012-00149, March 2012

MIX
Papier aus verantwortungsvollen Quellen
Paper from responsible sources
FSC® C105338

If you have any concerns about our products,
you can contact us on
ProductSafety@springernature.com

In case Publisher is established outside the EU,
the EU authorized representative is:
Springer Nature Customer Service Center GmbH
Europaplatz 3, 69115 Heidelberg, Germany

Printed by Libri Plureos GmbH
in Hamburg, Germany